# Lecture Notes in Computer Science    10675

Commenced Publication in 1973
Founding and Former Series Editors:
Gerhard Goos, Juris Hartmanis, and Jan van Leeuwen

More information about this series at http://www.springer.com/series/7409

Zhe Wang · Anni-Yasmin Turhan
Kewen Wang · Xiaowang Zhang (Eds.)

# Semantic Technology

7th Joint International Conference, JIST 2017
Gold Coast, QLD, Australia, November 10–12, 2017
Proceedings

Springer

*Editors*
Zhe Wang
Griffith University
Brisbane, QLD
Australia

Anni-Yasmin Turhan
Dresden University of Technology
Dresden
Germany

Kewen Wang
Griffith University
Brisbane, QLD
Australia

Xiaowang Zhang
Tianjin University
Tianjin
China

ISSN 0302-9743 ISSN 1611-3349 (electronic)
Lecture Notes in Computer Science
ISBN 978-3-319-70681-8 ISBN 978-3-319-70682-5 (eBook)
https://doi.org/10.1007/978-3-319-70682-5

Library of Congress Control Number: 2017957983

LNCS Sublibrary: SL3 – Information Systems and Applications, incl. Internet/Web, and HCI

Printed on acid-free paper

This Springer imprint is published by Springer Nature
The registered company is Springer International Publishing AG
The registered company address is: Gewerbestrasse 11, 6330 Cham, Switzerland

# Preface

These are the proceedings of the 7th Joint International Semantic Technology Conference (JIST) that took place at the Gold Coast in Australia during November 10–12, 2017. This conference series is dedicated to applications of semantic technologies, theoretical results, new algorithms, and tools to facilitate the adoption of semantic technologies.

In response to the call for papers, this edition of JIST received more than 40 submissions from ten different countries to the research track, in-use track, and the special track on "Semantic Processing for Knowledge Graphs". The submissions were assessed by an international Program Committee (PC) with members from 19 countries. Each paper received at least three reviews. The papers accepted for presentation at the JIST conference are mainly from Asia, Europe and, as one would expect, from the hosting country Australia. The set of accepted papers is an interesting mix of contributions from long-standing semantic technology groups and some newcomer groups.

As is customary for the JIST conference series, there were awards for the best paper and the best student paper. The award winners of this year were selected based on the reviews supplied by the PC members. The paper chosen for the best paper award was "Refinement-Based Closed-World OWL Class Induction with Optimistic Bounded Convex Measures" by David Ratcliffe and Kerry Taylor, both from the Australian National University. The paper chosen for the best student paper award was "Preliminary Results on the Identity Problem in Description Logic Ontologies" by Franz Baader, Adrian Nuradiansyah, and Daniel Borchmann, all from Dresden University of Technology.

We are indebted to many people who made this event possible. We would like to thank the members of the PC and their appointed reviewers for having contributed to JIST by means of their reviews and their lively discussions about the submissions. This committee's competence and effort are expedient to making conferences like this year's JIST a success. In particular we would like to thank the following reviewers who supplied emergency reviews for some of JIST submissions: Sebastian Binnewies, Pouya Ghiasnezhad Omran, Peng Xiao, and Zhiqiang Zhuang all from Griffith University and Stefan Borgwardt, Andreas Ecke, Patrick Koopmann, and Benjamin Zarrieß all from Dresden University of Technology. We would also like to thank Griffith University for supporting JIST and the team of local organizers for setting up such a diverse event.

The continuing success of the Joint International Semantic Technologies conference means that planning can now proceed with confidence for the next event to be held in this conference series in 2018.

September 2017

Anni-Yasmin Turhan
Zhe Wang

Preface

# Organization

## General Chairs

Thepchai Supnithi          National Electronics and Computer Technology Center, Thailand
Kewen Wang          Griffith University, Australia

## Program Chairs

Anni-Yasmin Turhan          Dresden University of Technology, Germany
Zhe Wang          Griffith University, Australia

## In-Use Track Chairs

Hong-Gee Kim          Seoul National University, Korea
Hideaki Takeda          National Institute of Informatics, Japan
Thanaruk Theeramunkong          SIIT, Thailand

## Special Track Chairs

Kouji Kozaki          Osaka University, Japan
Juanzi Li          Tsinghua University, China

## Poster and Demo Chairs

Hanmin Jung          Korea Institute of Science and Technology Information, Korea
Yuefeng Li          Queensland University of Technology, Australia

## Workshop Chairs

Marut Buranarach          NECTEC, Thailand
Jianfeng Du          Guangdong University of Foreign Studies, China

## Tutorial Chairs

Jeff Z. Pan          University of Aberdeen, UK
Yuan-Fang Li          Monash University, Australia

## Industrial Chairs

Michael Lawley          CSIRO, Australia
Alan Wu                 Oracle, USA

## Proceedings Chair

Xiaowang Zhang          Tianjin University, China

## Publicity Chairs

Nopphadol Chalortham    Chiang Mai University, Thailand
Riichiro Mizoguchi      Japan Advanced Institute of Science and Technology,
                        Japan

## Local Organizers

Sebastian Binnewies     Griffith University, Australia
Junhu Wang              Griffith University, Australia
Lian Wen                Griffith University, Australia
Zhiqiang Zhuang         Griffith University, Australia

## Program Committee

Fernando Bobillo             University of Zaragoza, Spain
Chantana Chantrapornchai     Kasetsart University, Thailand
Huajun Chen                  Zhejiang University, China
Gong Cheng                   Nanjing University, China
Stefan Dietze                L3S Research Center, Germany
Dejing Dou                   University of Oregon, USA
Jianfeng Du                  Guangdong University of Foreign Studies, China
Alessandro Faraotti          IBM, Italy
Zhiyong Feng                 Tianjin University, China
Marcelo Finger               Universidade de Sao Paulo, Brazil
Naoki Fukuta                 Shizuoka University, Japan
Volker Haarslev              Concordia University, Canada
Armin Haller                 Australian National University, Australia
Wei Hu                       Nanjing University, China
Eero Hyvönen                 Aalto University, Finland
Ryutaro Ichise               National Institute of Informatics, Japan
Ernesto Jimenez-Ruiz         University of Oxford, UK
Takahiro Kawamura            Japan Science and Technology Agency, Japan
Evgeny Kharlamov             University of Oxford, UK
Martin Kollingbaum           University of Aberdeen, UK

# Contents

## Information Retrieval and Knowledge Discovery

## Knowledge Graphs

## Applications of Semantic Technologies

# Ontology and Data Management

# Building Wikipedia Ontology with More Semi-structured Information Resources

Tokio Kawakami[✉], Takeshi Morita, and Takahira Yamaguchi

Keio University, 3-14-1 Hiyoshi, Kohoku-ku, Yokohama, Kanagawa 223-8522, Japan
kawakami0412@keio.jp, {t_morita,yamaguti}@ae.keio.ac.jp

**Abstract.** Wikipedia has been recently drawing attention as a semi-structured information resource for the automatic building of ontology. This paper describes a method of building general-purpose "lightweight ontology" by semi-automatically extracting the Is-a relation (rdfs:subClassOf), class-instance relation (rdf:type), concepts such as Triple, and a relation between concepts from information that includes category trees, define statements, lists and Wikipedia infoboxes. Also, we evaluate the built ontology by comparing it with other Wikipedia ontologies, such as YAGO and DBpedia.

**Keywords:** Ontologies · Wikipedia · Semi-structured information resource

## 1 Introduction

The usability of building a large-scale ontology has been the focus for various fields, such as information searches, data integration, and question answering. WordNet[1] and DBpedia Ontology[2] have been recognized as large-scale ontologies. However, as they have been manually built, they require a huge development cost and are difficult to maintain and update. Under the circumstances, research on automatic ontology development with a focus on a semi-structured information resource, such as Wikipedia and Folksonomy, has recently been drawing attention. Particularly, as Wikipedia has excellent vocabulary coverage and immediate updating to allow for easy maintenance and updates of the developed ontology, many research studies consider the automatic development of ontology based on Wikipedia. YAGO [Suchanek 08] is a well-known example of automatic ontology development from Wikipedia.

YAGO extracts the class-instance relation by linking the end class of Word-Net to a Wikipedia category to extract the Is-a relation and set the article, using the category as the instance. This method allows all articles using categories to be set as an instance. However, as the information in the main text is not used, in cases where the article does not exist, but the information is contained, or

---

[1] https://wordnet.princeton.edu/.
[2] http://wiki.dbpedia.org/.

© Springer International Publishing AG 2017
Z. Wang et al. (Eds.): JIST 2017, LNCS 10675, pp. 3–18, 2017.
https://doi.org/10.1007/978-3-319-70682-5_1

where information not reflected in the categories exists in the main text, it is impossible to extract such information as the instance, and this is a problem for this method.

In this paper, we perform string processing for the Wikipedia category tree, matching with the infobox template, and scraping content titles to extract the Is-a relation (rdfs:subClassOf) from Wikipedia. We perform list scraping to extract the class-instance relation (rdf:type). We extract information from infoboxes to extract Triple. We conduct morphological analysis of define statements to extract the upper-lower relations. With these methods, we perform large-scale and general-purpose ontology learning and conduct an evaluation. We also evaluate by comparing with other Wikipedia ontologies, such as YAGO and DBpedia, to determine whether the extracted relation is included there.

## 2    Related Work

YAGO2 [Hoffart 10] aimed for further expansion of ontology with the knowledge base expansion of YAGO; not only by the previous linkage between WordNet and Wikipedia category, but also by the extraction of spatiotemporal information from Wikipedia and GeoNames[3]. Also, in the expansion version, YAGO3 [Mahdisoltani 15], Wikipedia (in other languages as well as English) is used to allow multilingual expansion. YAGO focused on nonhierarchical relations and built an advanced ontology, incorporating a spatiotemporal relation, not just based on the hierarchical relation. However, it has not utilized Wikipedia's unique structure, such as information in the main text, define statements, and Wikipedia lists.

DBpedia is an ontology that was built based on frequently used infoboxes in Wikipedia. As it manually describes the class hierarchy and defines ontology property mappings, there is a high development cost.

Flati [Flati 14] extracted relationships from Wikipedia definition sentences by morphological analysis. However, it didn't examine whether Isa relation or class instance relation. In this research, we try to classify extracted upper and lower relation.

Melo [Melo 10] joined Infobox templates and Wikipedia categories to extract relationships, but they extracted only class instance relationship. In this research, Infobox template is used as one method of extracting Isa relation.

Kuhn [Kuhn 16] complemented DBpedia by extracting relationships from Wikipedia list articles and comparing the relationships of instances in articles. In this research, we improved the accuracy by comparison with the extracted Isa relation, and it does not depend on DBpedia.

Gupta [Gupta 16] extracted Isa relation from Wikipedia category by using heuristics. Although this approach is similar to "Backward string matching", they do not use "Backward string matching part removal" proposed by this research.

---

[3] http://www.geonames.org/.

Ponzetto et al. [Ponzetto 07] tried to extract both the Is-a relation and not-is-a relation by using string matching and vocabulary syntax patterns for categories in the English Wikipedia.

Wu and Weld [Wu 08] built an ontology by combining the Wikipedia infobox template and WordNet class hierarchy. The properties owned by the combined infobox template are inherited by the Is-a relation. Although this is an advanced ontology that has built the property domain from the infobox template, the property type is not considered.

# 3    English Wikipedia Ontology Building Method

The following paragraphs describe the details of method proposed by this research.

## 3.1    Is-a Relation Extraction Method

**String matching to category hierarchy**
"Backward string matching" and "Backward string matching part removal" are used here to extract the Is-a relation from the category hierarchy of Wikipedia. Backward string matching is a method used to extract a child category name, which is set as "any string + parent category name" after comparing the parent category name and child category name as the category hierarchy is configured. For example, the category hierarchy of "Directors" - "Women directors" exists as shown in Fig. 1, and is extracted as the Is-a relation. In cases where the child category name is "any string + preposition + parent category name" as with the parent category name: "Japan" and child category name: "People from Japan" a clearly incorrect Is-a relation is often extracted. Therefore, such cases are not covered here. The backward string matching method was previously proposed by [Ponzetto 07].

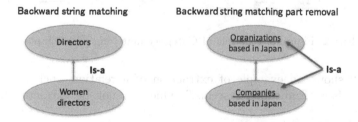

**Fig. 1.** Example of Backward string matching and Backward string matching part removal

Backward string matching part removal is a method that extracts an Is-a relation by finding a matching modifier following a noun in the category hierarchy, and removing the matching part. The reason it is limited to a modifier following the noun is that previous experiments have shown that the noun's meaning is often limited when followed by a modifier. Thus, it is highly likely that a correct Is-a relation is extracted. For example, a noun before the phrase "based in" is often a noun for an organization, as in Fig. 1. Thus, the correct Is-a relation, "Companies" Is-a "Organizations" is obtained.

**Matching the infobox template name with the category name**
This method extracts the Is-a relation by matching the template name with the category name, focusing on the relation between the abstract infobox template and category, which has many concrete concepts in some domains. The extraction procedures are shown below:

1. Perform simple string matching between the category name and template name.
2. Remove all categories, except those of which the articles having a matching target template belong, from the sub-categories that exist under the matching target categories.
3. Extract the category hierarchy from step 2 as it is the Is-a relation.
4. Eliminate the Is-a relation in such case where the child class takes the form of "any string + preposition + parent class".

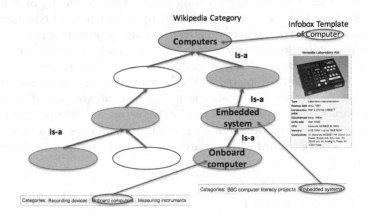

**Fig. 2.** Figure 3: Example of Category matching with template

Figure 2 shows an example of extraction of a relation, such as "Onboard computers" Is-a "Embedded systems," which cannot be extracted by string matching.

**Scraping of Content titles**
The articles that include a word meaning "Classification" or "Genre" in the title, often correctly describe the classification hierarchy. By focusing on this, the hierarchical relation within the articles containing "Classification," "Taxonomy," and "Genre" is extracted by scraping an Is-a relation. For example, from the lower hierarchical structure existing in the article titled "Bear," an Is-a relation, such as "Family Ursidae" - "Subfamily Ailuropodinae" can be extracted, as in Fig. 3.

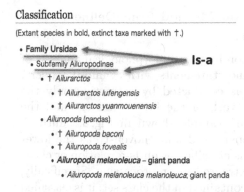

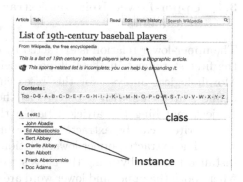

**Fig. 3.** Main text under the title word "Classification" of the article for "Bear"

**Fig. 4.** Example of lists

## 3.2   Class-Instance Relation Extraction Method

Lists are related articles that are selected according to a certain standard. The English Wikipedia has lists under the title of "List of ." as shown in Fig. 4. We focus on these lists and use scraping to extract the class-instance relation by setting the list name as class and matters listed within the list as instance.

## 3.3   Triple Extraction Method

A three-piece set of "article-item-value" that the infobox owns is extracted as "instance-property-property value". In the case that the triple set is extracted directly from the dump data, the meaning of these properties may seem to be difficult to understand at a glance, or the properties may not be integrated. Accordingly, properties are converted into HTML by Java Wikipedia API (Bliki engine) [4] here, and the portions equivalent to properties are integrated and extracted. Also, the property type is examined to see whether the property object is instance or literal, to classify it as owl:ObjectProperty or owl:DatatypeProperty. Specifically, in such case that the property object is the subject of a different Triple, or in the case that it is contained in a set of instances already extracted, it is classified as owl:ObjectProperty. On the other hand, in the case where the property object takes the form of "number + unit", it is classified as owl:DatatypeProperty. Taking this irregularity into account, in the case where 75% of the Triple set using the property is instance, it is determined to be owl:ObjectProperty, and in the case that 75% is literal, it is determined to be owl:DatatypeProperty. A property object that cannot yet be classified as either one, is determined as Unknown.

---

[4] http://code.google.com/p/gwtwiki.

### 3.4  Upper-Lower Relation Extraction Method from Define Statements

The upper-lower relation is extracted by conducting a morphological analysis of the first statement in the Wikipedia define statements with Stanford Parser[5]. Morphological analysis for this research is conducted by extracting only the define statements with a be-verb as the verb of the first statement and the subject matching the article title. In the example shown in Fig. 5, "author" and "writer" can be extracted as the upper word and "novelist" as the lower word. The extracted upper-lower relation is classified as an Is-a relation or class-instance relation by using the class set extracted in preprocessing. Specifically, where both the upper and lower word are contained in the class set, it is classified as an Is-a relation. On the other hand, where only the upper word is contained in the class, it is classified as a class-instance relation. Finally, where these do not apply, it is classified as Unknown.

> A **novelist** is an author or writer of novels, though often novelists also write in other genres of both fiction and non-fiction. Some novelists are professional novelists, thus make a living writing novels and other

**Fig. 5.** Define statements of the article for "Novelist"

## 4  Evaluation of Built Ontology

In order to construct ontology using the proposed method, we use the English Wikipedia dump data (January 20, 2017 version).

### 4.1  Result of Is-a Relation Extraction

**Result and review of extraction by string matching for category hierarchy**

We extracted 281,394 Is-a relation pieces by performing backward string matching, and 23,516 Is-a relation pieces by performing backward string matching part removal, from the category hierarchy. After eliminating overlapping parts, a total of 302,425 relation pieces were extracted. By extracting 1,000 samples from the extracted relations, a 95% confidence interval was estimated according to the following equations. $N$ is the population, $n$ is the number of samples, $\hat{p}$ is the estimate of the true correct answer rate, which is the number of samples of the correct answer divided by the total number of samples.

$$[\hat{p} - 1.96\sqrt{(1 - \frac{n}{N})\frac{\hat{p}(1 - \hat{p})}{n - 1}}, \hat{p} + 1.96\sqrt{(1 - \frac{n}{N})\frac{\hat{p}(1 - \hat{p})}{n - 1}}] \tag{1}$$

---

[5] http://nlp.stanford.edu/software/lex-parser.shtml.

Thus, the 95% confidence interval for backward string matching was 96.2 ± 1.18%, and the 95% confidence interval for backward string matching part removal was 80.2 ± 2.42%. The total result was 93.4 ± 1.54%. Table 1 shows the relation extracted for the upper three lines by the backward string matching method and for the lower three lines by the backward string matching part removal method. Table 1 indicates that a relation independent from the string can be extracted by the backward string matching part removal method.

**Table 1.** Correct example of Is-a relations (rdfs:subClassOf) extracted by string matching

| Parent class | Child class |
| --- | --- |
| Propaganda films | American World War II propaganda films |
| Surgery | Ear surgery |
| Music | Wedding music |
| Food companies | Fast-food chains |
| Rulers | Kings |
| Non-fiction writers | Science writers |

Next, Table 2 shows the erroneously extracted Is-a relation. Errors on the first and second lines in Table 2 show an example of class-instance relation extracted by mistake. As Wikipedia has categorized famous instances, class-instance relations are extracted by string matching in some cases. This can be eliminated by using the result extracted by the class-instance relation extraction method described in Sect. 3.2 above. An error on the third line in Table 2 shows an example of extracting an incorrect Is-a relation where the hierarchy has an abstract parent category existing in the upper hierarchy of Wikipedia Category. As the upper Categories of English Wikipedia consist of 21 major Categories such as "Arts," "Culture," "Health," and "Politics," these are the core classifications of Wikipedia hierarchy. This error can be eliminated by a limitation according to the depth of hierarchy from the root.

**Table 2.** Example of erroneous Is-a relations (rdfs:subClassOf) extracted by string matching

| Parent class | Child class |
| --- | --- |
| Asia | Laos |
| Landshut | EV Landshut |
| Politics | Sector |

**Result and review of extraction by matching of infobox template name with category name**

According to the method described in Sect. 3.1 (2), 6,315 pieces of Is-a relation were extracted. The result of extracting 1,000 samples from the extracted relations with a 95% confidence interval was $80.1 \pm 2.27\%$. The correct example is as shown in Table 3.

**Table 3.** Correct example of Is-a relations (rdfs:subClassOf) extracted by matching of infobox template name with category name

| Parent class | Child class |
| --- | --- |
| Wearable computers | Smartwatches |
| Embedded systems | Real-time computing |
| Computer hardware | Computing output devices |

Table 3 shows that many relations independent from the string were extracted. On the other hand, many class-instance relations were extracted by mistake. This is because many of the extracted relations had a category related to geography, such as "Mountains" "Islands" and "Seas" as the root, which includes many instances in the sub category. String matching was used here for matching the infobox template with the category name; however, even when the string does not match, semantic matching can be identified in some cases. For example, the "instrument" template is used for articles on musical instruments; however, there is no category for "instrument" in the category hierarchy. Instead, there is a category called "Musical instrument." To perform this extraction, it is necessary to limit it to phrases by morphological analysis in the preprocessing, and execute string matching or other appropriate processing.

**Result and review of extraction by scraping Content titles**

According to the method described in the Sect. 3.1 - "Scraping of Content titles", 83,003 pieces of Is-a relation were extracted. The result of extracting 500 samples from the extracted relations with a 95% confidence interval was $65.8 \pm 2.92\%$. The correct example of extraction is as shown in Table 4.

**Table 4.** Correct example of Is-a relations (rdfs:subClassOf) extracted by scraping Content titles

| Parent class | Child class |
| --- | --- |
| Field artillery | Mountain gun |
| Family Delphinidae | Genus Deophinus |
| Idiophone | Slit drum |

Table 4 shows that relations not extracted by string matching have been extracted here. Also, many biological relations that were not extracted by infobox template matching with the category name, as described above, have been extracted. On the other hand, many relations have been extracted by mistake. Unlike the other Is-a relation extraction methods, this method uses only a part of the main text. Certain articles might be written using word combinations that are uncommon and unexpected. Accordingly, an inappropriate classification hierarchy would be extracted as it is an Is-a relation and thus, will reduce the accuracy.

## 4.2 Result of Class-Instance Relation Extraction

According to the method described in Sect. 3.2, we have executed extraction of the class-instance relation for lists extracted from the Wikipedia dump data. The number of instances obtained was 1,767,124 pieces, number of classes was 33,806 pieces, and number of class-instance relations was 2,705,573 pieces. 1,000 samples were extracted from the above class-instance relations to estimate the accuracy rate interval. This result was $89.2 \pm 1.92\%$. Table 5 shows correctly extracted relations, and Table 6 shows erroneously extracted relations. Table 5 shows that a wide variety of instances, such as person and programming language have been extracted. On the other hand, Table 6 shows that an instance of the class called "Composition for piano and orchestra" includes a person. Figure 6 shows a list extracting erroneous class-instance relations. As shown in the Fig. 6, this article is a list of compositions; however, it includes a player as the item, and thus, an erroneous class-instance has been extracted. When lists have items with such an hierarchical relation, appropriate processing, such as extraction of the lowest items alone, will be necessary.

**Table 5.** Correct example of relations extracted from lists

| Class | Instance |
|---|---|
| Rice varieties | African rice |
| Rivers of Massachusetts | Green Harbor River |
| IMAX films | Ultimate Wave Tahit 3D |
| Educational software | eCollege |
| Wildlife artists | John Abbot |
| Armoured fighting vehicles by country | MB-3 Tamoyo |

**Table 6.** Example of erroneous relations extracted from lists

| Class | Instance |
|---|---|
| Compositions for piano and orchestra | Nikolai Kapustin |
| Bulldog mascots | Carthage Independent School District |

List of compositions for piano and orchestra

From Wikipedia, the free encyclopedia

- Nikolai Kapustin
  - Concertino for piano and orchestra, Op. 1 (1957)
  - Concerto for piano and orchestra No. 1, Op. 2 (1961)
  - Concerto for piano and orchestra No. 2, Op. 14 (1974)

**Fig. 6.** List extracting erroneous class-instance relation

| Teams managed | |
|---|---|
| **Years** | **Team** |
| 2009 | Al-Majd[1] |
| 2011–2012 | Al-Oruba[2] |

**Fig. 7.** Example of infobox with a failure in Triple extraction

### 4.3   Result of Triple Extraction

From the Wikipedia dump data, we have extracted 8,311,427 infobox, 12,039 infobox templates, and 22,767,071 pieces of infobox Triple. The number of property types in the infobox Triple was 12,088 pieces.

According to the property type estimation, approximately 60% of the properties were classified as owl:ObjectProperty or owl:DatatypeProperty. The reason 40% of the properties were classified as "Unknown" was that the literal classification was determined by whether it falls under the classification of "number + unit". In the future, we need to take into consideration that property values such as "small" can be literal. 1,000 samples were extracted from all Triples to estimate the accuracy rate interval of $93.1 \pm 1.57\%$. An example of a common error is shown in Fig. 7. For this infobox, "Years" was extracted as the property and "Team" was extracted as the property value. Because the parts extracted as the property for Triple here can be specified by each infobox template, this error can be handled by changing the extraction method according to the infobox template.

### 4.4   Result of Upper-Lower Relation Extraction from Define Statements

Because of the extraction described in Sect. 4.1, we have obtained 3,114,222 upper-lower relations from define statements. Also, because of the classification as Is-a relation or class-instance relation, 2,036,428 pieces have been classified as the class-instance relation and 2,898 pieces as the Is-a relation. Accordingly, we have classified approximately 65% of upper-lower relations. On the other hand, the reason approximately 30% of upper-lower relations have been classified as unknown is that the number of classes extracted thus far is insufficient. This can be improved by further developing the Is-a relation extraction method.

Moreover, the determination of accuracy for the class-instance relations and Is-a relations after classification was $93.7 \pm 1.51\%$ and $80.0 \pm 2.01\%$, respectively. Table 7 shows the correctly extracted Is-a relations, and Table 8 shows the correctly extracted class-instance relations. These relations had not been extracted by the methods described earlier.

Table 9 shows the erroneously extracted class-instance relations, and Table 10 shows the erroneously extracted Is-a relations. As for the class-instance relations, because the items that should be included in the class are not contained, Is-a

**Table 7.** Correct example of Is-a relations extracted from define statements

| Parent class | Child class |
|---|---|
| Road vehicle | Bus |
| String instrument | Electric guitar |

**Table 8.** Correct example of class-instance relations extracted from define statements

| Class | Instance |
|---|---|
| Protein | F-box protein 16 |
| Golf tournament | Tall City Open |

**Table 9.** Inappropriate class-instance relations extracted from define statements

| Class | Instance |
|---|---|
| Bishop | Anglican bishop |
| Field | Gender study |

**Table 10.** Inappropriate Is-a relations extracted from define statements

| Parent class | Child class |
|---|---|
| Capital | London |
| Programming language | C++ |

relations were extracted by mistake in many cases. On the contrary, in regards to the Is-a relation, because the items that should not be included in the class are contained, class-instance relations have been extracted in many cases. To increase the accuracy, it is necessary to improve the accuracy in Is-a relation extraction.

## 5  Evaluation by Comparison with Existing Ontology

### 5.1  Comparison with YAGO

YAGO3 is currently multilingual. This research covers the relations extracted only from the English Wikipedia to compare with the ontology built by the same resources (Table 11).

**Table 11.** Comparison of extraction method between this research and YAGO

| | This research | YAGO |
|---|---|---|
| Method to extract Is-a relations and class-instance relations | Category string matching | Article classification according to Wikipedia categories by using WordNet |
| | Matching of infobox template with category | |
| | Is-a relation extraction from content titles | |
| | Class-instance relation extraction from lists | |
| | Upper-lower relation extraction | |

As for the extraction method, YAGO has built a hierarchy by linking the WordNet class and Wikipedia categories to extract an article under the corresponding category as the instance. As this method may identify all articles

**Table 12.** Comparison of the number of relations extracted and accuracy between this research and YAGO

|  | This research | | YAGO | |
|---|---|---|---|---|
|  | Number extracted | Accuracy | Number extracted | Accuracy |
| Is-a relation | 349,982 | 92.5 ± 1.53% | 367,040 | 93.4% |
| Class-instance relation | 4,228,849 | 92.4 ± 1.64% | 8,414,398 | 97.7% |

having a category as the instance, it can extract many instances. On the other hand, this research uses not only the category hierarchy, but also the information in the main text, such as content titles, define statements, and items in the lists. Although the accuracy decreases, this allows the extraction of instances and classes without article, and as such, is the difference from YAGO.

Table 12 provides a comparison of the number of relations extracted and accuracy. This table shows the result after integrating the relations extracted by the processing as described in Sect. 3, and after eliminating inappropriate ones, such as class-instance relations erroneously included in Is-a relations. As for YAGO, the target is limited to the relations extracted from Wikipedia. Although the result of both the number extracted and accuracy of IS-a relations was closer to that of YAGO, the number of class-instance relations extracted was far inferior to YAGO. This is because the extraction of class-instance relations for this research was limited to those from the articles that are described on the lists or that have define statements. On the other hand, in the case where relations are described only on define statements or lists, this research method can extract the relations not extracted by YAGO. Table 13 shows the result of relations extracted. In this table, the 1,000 relations that have been correctly extracted by this research have been examined to determine whether relations with semantic similarity exist in YAGO and DBpedia Ontology.

**Table 13.** Comparison of relations extracted by between this research and YAGO (out of 1000 pieces)

|  | Number of relations that do not exist in YAGO | Number of relations that do not exist in DBpedia | Number of relations that do not exist either in YAGO or in DBpedia |
|---|---|---|---|
| Is-a relations | 589 | 998 | 589 |
| Class-instance relations | 422 | 826 | 408 |

Table 13 shows that many of the relations that are not extracted by YAGO are actually extracted by our proposed method. Examples shown in the Tables 1, 3, 4, 5, 7 and 8 are relations that do not exist in YAGO. The following items are simple descriptions of the relations that are not extracted by YAGO.

1. Is-a relation including a class that does not exist in WordNet
2. Class-instance relation including an instance that is described in lists, however, the article itself does not exist
3. Class-instance relation that can be extracted from an article whose define statements includes information which is not reflected in a category

As for the first Is-a relation above, as the parent class of YAGO is the class of WordNet, YAGO cannot extract any Is-a relation whose parent class does not exist in WordNet. As shown in the example of "Non-fiction writers"-"Science writers" and "Food companies"-"Fast-food chains" in Table 1, the Is-a relation having a lower class as the parent class does not exist in YAGO, in many cases. In another example, as the class of "voice actor" does not exist in WordNet, Is-a relations regarding "voice actor" are not extracted. As our proposed method does not depend on WordNet, it can extract these relations.

As for class-instance relations that YAGO cannot extract, many of these relations include instances (in red) that are described in a list, but the article itself does not exist, as shown in Fig. 8.

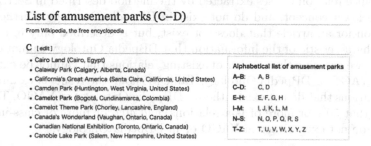

**Fig. 8.** Example of list including class-instance relations that YAGO cannot extract. (Color figure online)

Moreover, where information not reflected in the category is described in define statements, new class-instance relations not seen in YAGO have been extracted. As exemplified in Fig. 9, the class-instance relation of "Method"-"Broma process" can be extracted from the article regarding "Broma process" by the method described in Sect. 4.4. However, this relation is not reflected in the article category; YAGO cannot extract the relation.

## 5.2   Comparison with DBpedia Ontology

DBpedia Ontology is also multilingual, and extraction from Wikipedia of various languages has been tried. In this research, only relations extracted from English Wikipedia are used for comparison. Table 13 shows the result of examination on 1,000 relations that have been correctly extracted by this research to determine whether the relations with semantic similarity exist in DBpedia Ontology.

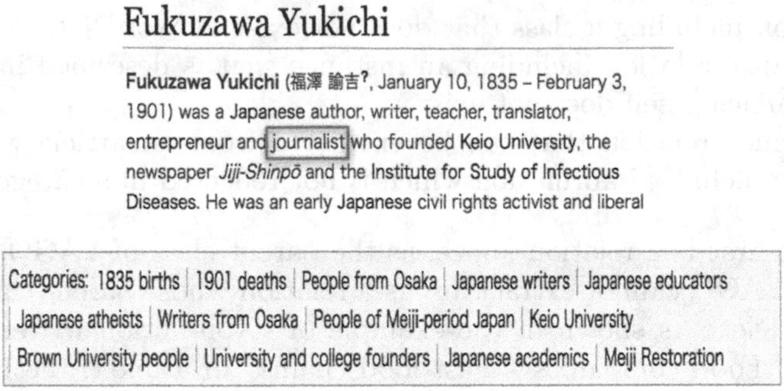

## Fukuzawa Yukichi

**Fukuzawa Yukichi** (福澤 諭吉?, January 10, 1835 – February 3, 1901) was a Japanese author, writer, teacher, translator, entrepreneur and journalist who founded Keio University, the newspaper *Jiji-Shinpō* and the Institute for Study of Infectious Diseases. He was an early Japanese civil rights activist and liberal

Categories: 1835 births | 1901 deaths | People from Osaka | Japanese writers | Japanese educators | Japanese atheists | Writers from Osaka | People of Meiji-period Japan | Keio University | Brown University people | University and college founders | Japanese academics | Meiji Restoration

**Fig. 9.** Category column and Define statements of the article on "Fukuzawa Yukichi"

DBpedia Ontology manually describes the class hierarchy. Therefore, the DBpedia Ontology class has many upper relations. On the other hand, the ontology class built by our proposed method has many lower concepts; corresponding Is-a relations are rarely seen in DBpedia Ontology. Also, many of the class-instance relation classes extracted by the method described in Sect. 3.2 fall under the lower concept, and do not exist in DBpedia Ontology. Furthermore, information for an article that does not exist, but is described only in a list, is another characteristic of the information that DBpedia Ontology cannot obtain.

Lastly, we examine the number of existing relations that cannot be extracted either by YAGO or DBpedia Ontology. Table 13 shows the result of examination on the relations that do not exist either in DBpedia Ontology or YAGO. Table 13 indicates that 50% or more of Is-a relations, and 40% or more of class-instance relations, do not exist either in YAGO or in DBpedia Ontology.

## 6   Conclusion

This paper has proposed a method for building large-scale ontology from Wikipedia, and conducted an evaluation of the method. Also, by comparing the proposed method with overseas Wikipedia Ontology, or YAGO and DBpedia Ontology, this paper confirmed the usability by showing that the proposed method can extract relations that cannot be extracted by YAGO or DBpedia Ontology. The method used for this research can be applied to Japanese Wikipedia too. We have applied the method described by [Tamagawa 10] to Japanese Wikipedia, and have built Japanese Wikipedia Ontology, which is a large-scale and general-purpose ontology. Also, [Tamagawa 12] has added property domain extraction, property value range extraction, synonym extraction, and other extractions for special properties, such as symmetric relation property and transitive relation property. Furthermore, research has been conducted to apply Japanese Wikipedia Ontology to question answering, such as [Asano 16].

For future research, we will attempt the extraction of special properties that have been successfully extracted by Japanese Wikipedia Ontology, such as property domain, property value region, symmetric relation property, and transitive relation property. [Morita 14] has integrated the Japanese Wikipedia Ontology and Japanese WordNet by using ontology alignment to supplement upper classes in the Japanese Wikipedia Ontology. We will advance our research by setting the ultimate target to building this type of cluster schema hierarchy in the English Wikipedia.

# References

[Suchanek 08]  Suchanek, F.M., Kasneci, G., Weikum, G.: YAGO: a large ontology from Wikipedia and WordNet. J. Web Semant. **6**(3), 203–217 (2008). Elsevier

[Hoffart 10]  Hoffart, J., Suchanek, F., Berberich, K., Weikum, G.: YAGO2: a spatially and temporally enhanced knowledge base from Wikipedia, Research Report MPI-I-2010-5007. Max-Planck-Institut fur Informatik (2010)

[Mahdisoltani 15]  Mahdisoltani, F., Biega, J., Suchanek, F.M.: A knowledge base from multilingual Wikipedias. In: CIDR (2015)

[Flati 14]  Flati, T., Vannella, D., Pasini, T., Navigli, R.: Two Is Bigger (and Better) Than One: the Wikipedia Bitaxonomy Project, ACL (2014)

[Melo 10]  de Melo, G., Weikum, G.: MENTA: inducing multilingual taxonomies from Wikipedia. In: Proceedings of the 19th ACM International Conference on Information and Knowledge Management, pp. 1099–1108 (2010)

[Kuhn 16]  Kuhn, P., Mischkewitz, S., Ring, N., Windheuser, F.: Type inference on Wikipedia list pages, vol. P-259. LNI, pp. 2101–2111. GI (2016)

[Gupta 16]  Gupta, A., Piccinno, F., Kozhevnikov, M., Pasca, M., Pighin, D.: Revisiting taxonomy induction over Wikipedia. In: The 26th International Conference on Computational Linguistics (2016)

[Ponzetto 07]  Ponzetto, S.P., Strube, M.: Deriving a large scale taxonomy from Wikipedia. In: Proceedings of National Conference on Artificial Intelligence, pp. 1440–1447 (2007)

[Wu 08]  Wu, F., Weld, D.S.: Automatically refining the Wikipedia infobox ontology. In: Proceedings of the 17th International Conference on World Wide Web, pp. 635–644. ACM (2008)

[Tamagawa 12]  Tamagawa, S., Morita, T., Yamaguchi, T.: Extracting property semantics from Japanese Wikipedia. In: Huang, R., Ghorbani, A.A., Pasi, G., Yamaguchi, T., Yen, N.Y., Jin, B. (eds.) AMT 2012. LNCS, vol. 7669, pp. 357–368. Springer, Heidelberg (2012). https://doi.org/10.1007/978-3-642-35236-2_36

[Tamagawa 10]  Tamagawa, S., Sakurai, S., Tejima, T., Morita, T., Izumi, N.: Learning a Large Scale of Ontology from Japanese Wikipedia, WI/IAT (2010)

[Asano 16]  Asano, H., Morita, T., Yamaguchi, T.: Development and evaluation of an operational service robot using Wikipedia-based and Domain Ontologies. In: Web Intelligence (2016)

[Morita 14]  Morita, T., Sekimoto, Y., Tamagawa, S., Yamaguchi, T.: Building up a class hierarchy with properties by refining and integrating Japanese Wikipedia Ontology and Japanese WordNet. Web Intell. Agent Syst. Int. J. **12**(2), 211–233 (2014). IOS Press

# Data Structuring for Launching Web Services Triggered by Media Content

Makoto Urakawa[✉] and Hiroshi Fujisawa

NHK (Japan Broadcasting Corporation), 1-10-11, Kinuta, Setagaya-ku, Tokyo, Japan
{urakawa.m-gi,fujisawa.h-ja}@nhk.or.jp

**Abstract.** There are some efforts to inspire the viewers to buy something or visit somewhere when they are viewing such objects or places on the TV programs or web sites. It is required to link some specific services, which can be run on smartphones or tablets, to content on TV or a web site. However, simple combination of content and possible services still requires some operations taken by the viewers such as typing search words in the applications or the pages of the services. Such required actions may decline their motivations to use the services. Therefore, the authors propose the data model that allows to seek matching of an entity within content with various services, and to generate service launch information dynamically changed by combination of an entity and a service. The data model is based on combination of ontology class structure and reasoning. In this paper, effectiveness of the data model is shown by the prototype applications which call various web services to inspire viewers for subsequent actions in accordance with media content.

**Keywords:** Ontology · Resource description framework · Knowledge base · Linked data · Semantic web rule language · Inference · Ifttt

## 1 Introduction

Broadcasters are working toward using the Internet to connect with viewers action from broadcasts—one example is displaying related keywords during broadcasts to urge viewers to search for information using the Internet. However, connecting TV programs with user action remains difficult, as television viewers must use their smartphones to type in search keywords. In terms of the web, although users who browse websites using their smartphones can use their web browser's share button to share the browsed information to social media networks (or they can send a link over an email), actions are limited to transmitting information. They are not able to purchase or do some actions directly when viewing web sites. Approaches are therefore in place to connect content with user action for TV programs and websites. However, some issues remain.

An increasing number of services based on the concept of If-This-Then-That (IFTTT)—such as ifttt[1] and zapier[2]—function as frameworks to link a wide range of

---

[1] https://ifttt.com/discover.
[2] https://zapier.com/.

© Springer International Publishing AG 2017
Z. Wang et al. (Eds.): JIST 2017, LNCS 10675, pp. 19–34, 2017.
https://doi.org/10.1007/978-3-319-70682-5_2

web services and Internet of Things (IoT) devices. For example, ifttt recipes have been released that make it possible to send an email or perform some other action, such as a New York Times article serving as a trigger.[3] However, most of the available actions are related to just transmitting information. 224,590 recipes published at ifttt as of September 2015 is ranked higher by Twitter, e-mail, SMS, Facebook, Evernote as an action [1]. Most functions and most actions performed via IFTTT make it possible to transmit information (such as sending an email). It is likely because it is easy to specify services and applications for transmitting information. To connect content with user action that goes beyond transmitting information, it is necessary to gain an understanding of the types of services available and generate queries for launching these services. However, gaining an understanding of the services being developed every day is a difficult task.

In this paper, a data model based on semantic technology is proposed, which use classes that define entities (such as "Itsukushima Shrine") within broadcast program/web site and services (such as Google Maps) including native applications/web applications/web sites, to perform matching these and generate information to launch matched services inherited from property information of entities within content.

In addition to just starting a map application (as the "go" application) when (for example) "Itsukushima Shrine" is introduced at a TV program, the proposed data model can make it possible to seamlessly obtain and configure information required to search for "Itsukushima Shrine" in the map application from the TV program. Other user actions connecting content are also possible. One example is a user who is viewing a cooking recipe that includes "asparagus" on a website. The software can check the user's and place an order on an EC site if there is no asparagus in his/her refrigerator. The idea proposed in this paper will make it possible to create new services that go beyond merely transmitting information.

This paper consists of the following items. Section 2 presents some important issues. Section 3 introduces related research. and Sect. 4 presents a data model for resolving the issues presented in Sect. 2. Section 5 discusses expansion of services with prototype applications. Section 6 provides a summary of this paper. Note that smartphone applications and web services are both referred to as "web services" in this study.

## 2   Issues

The objective of this research is to propose the way of starting web services, which are operated by different industries or companies, with information of entities in media content (such as television programs or websites). This paper premises that content holders (such as broadcasters or website operators) are generally not able to comprehend what kinds of web services can be linked to their content, and service providers are not able to understand what kinds of content they can use for their services as well. In order to launch web services with information of matched entity, 3 data processing is needed. They are extracting entity information from content, matching entity with relevant web services, and generating service launch information inherited from entity.

---

[3] https://ifttt.com/nytimes.

It is possible for users to identify entities within content by selecting items on a device with an interactive screen. However, even if they find the entity, it would be difficult for them to search for a web service that could be launched. Even if entities and services could be matched automatically or if users could select a service that could be launched, the user would need to enter some parameters (such as search keywords) for launching the web service. Such a process does not enable a smooth connection with the user's action.

From the perspective of content holders, it is possible to extract entities from content based on its subtitle information (for television programs) or text information (for websites). However, configuring links to web services is difficult, as a content holder will need to know each provider's web services. Even if a content holder did have information on specific web services, it is not realistic for the content holder to configure information for launching web services, owing to the large amount of work involved in the process.

For example, the NHK World website about Nagasaki[4] provides information on "chanpon noodles" (a type of food) and "Dejima" (a sightseeing area). In this case, it would be necessary to insert links to search results for "chanpon noodles" on an EC site such as Amazon[5] to support the actions of users who might want to purchase this food item. It would likewise be necessary to configure links to search results for "chanpon noodles" on recipe sites[6] to support the actions of users who might want to try cooking this dish themselves. Another link would need to be configured for the "Dejima" entity for launching a service to enable users to find information about visiting the location.

As described in this section, users and content holders can identify entities, however the difficulty is in configuring information to match and launch services provided by other providers.

## 3   Related Work

IFTTT is said to be able to automate actions such as task automation by integrating with Internet service after securing usability [2]. On top of that, there are several researches to introduce Semantic Web technology into IFTTT, which links various Web services and IoT devices in consideration of situation or context. For example, based on relationships and dependencies that can be defined by ontology, a research that cooperates objects such as sensors in space [3] and A tool that allows the user to specify behaviors in the space has been carried out [4]. Evented web ontology (EWE) [5] has been developed so that vocabulary and rules can be shared for task automation such as IFTTT. In addition, research is also being conducted to achieve compatibility of processing performance and expression power on inference [6]. We also see the importance of the

---

[4] https://www3.nhk.or.jp/nhkworld/en/culture/mydestination/20170720.html.

[5] https://www.amazon.co.jp/s/ref=nb_sb_noss?__mk_ja_JP=%E3%82%AB%E3%82%BF%E3%82%AB%E3%83%8A&url=search-alias%3Daps&field-keywords=%E3%81%A1%E3%82%83%E3%82%93%E3%81%BD%E3%82%93&rh=i%3Aaps%2Ck%3A%E3%81%A1%E3%82%83%E3%82%93%E3%81%BD%E3%82%93.

[6] https://www3.nhk.or.jp/nhkworld/en/food/search/?qt=chanpon.

efficiency to describe the linking of services [7]. In this way, by incorporating elements of the semantic Web into IFTTT, more effective service cooperation is possible, and examples of application in concrete scenes such as office space are also shown. However, since the main objective of these researches are to connect various services, these are not considered of inheriting the information of a connected source service. As a research aimed at improving user service, there is research [8] to optimize the existing IFTTT service to the user's situation by utilizing ontology and reasoning. This research is a proposal on optimization considering user's location information and so on, and service collaboration by inheritance of contents information targeted by this paper is not done. In addition, there is research [9] linking information and channels to be offered on a rule basis, to enable content holders to efficiently provide information on-line. In this research, proposal is being made to extend the rule according to the type of content information. However, it is limited to static filtering of rules, and dynamic cooperation is not considered. In terms of adopting OWL into an automatic process, OWL-S is proposed as a framework focusing on describing service and process profile, not dynamic matching services and generating launch information [10].

Therefore, in this paper, we propose a method to match different web services and dynamically generate service launch information that is originated from entity information matched with them.

## 4    Data Model

It has been proposed that data collaboration can be realized by classes instead of entities [11, 12]. This paper focuses on the class structures to which entities belong (rather than the entities themselves) as well as these researches, reflecting the procedure used by humans to launch the web services based on given information. For example, when one wants to buy carrots, he/she would select services where sell foods because carrots are one of food. Likewise, one would select a map application for "Dejima" because it is a "location." Although "Itsukushima Shrine" could be defined as a "shrine," a "shrine" is a subordinate concept of a "location"; therefore, one would select a map application in this case. It should thus be possible to match content entities with web services by abstracting information as entity classes (rather than entities themselves), and then making use of their superordinate/subordinate concepts. In other words, dynamic matching of entity and service can be solved by implementing semantic technology such as ontology. The launch information varies by the type of web service. For example, users would search by the entity name (such as "chanpon noodles") on an EC site, on the other hand, they could use an address or latitude/longitude information (rather than the entity name) on a map website. This dynamic launching services requires data processing of defining what information is required for launching a specific web service, inheriting the required information from the entity, and then automatically generating queries. Especially inheritance of entity information solves dynamic generation of service launching information.

It is assumed that multiple service providers (such as content holders, EC site operators, and map application providers) can be involved to connect user action with TV program or website content. Data are therefore separated into three layers (entity layer, class layer, and matched data layer), and an additional data processing layer is placed between the class layer and matching layer. This allows multiple providers to work together. The following section describes the proposed data model after analysis of actual web services and smartphone applications.

### 4.1 Analysis of Actual Service Categories and Genres

We think that it is possible to use categories and genres already provided by each service as information for class defining in data model. Therefore, we did a survey of 60 services and applications, such as purchasing, map, social network services, e-mail and dial application, identified about 3,000 categories (as of July 2017). For example, Amazon (the major purchasing service) had 20 categories and 255 subcategories[7], one of major Japanese EC sites Rakuten had 35 categories and 463 subcategories[8], and one of major supermarkets Tokyu Store had 19 categories and 106 subcategories[9]. Comparing food-related categories for these services showed that Amazon's AmazonFresh category contained subcategories such as vegetables, fruit, and processed meat/meat processed products, whereas Rakuten's Food category contained subcategories such as vegetables/mushrooms, fruit, and meat/meat processed products. Tokyu Store had a much more segmented category composition, for instance, its vegetables category containing subcategories such as leafy vegetables, beans/stew, and vegetables/root crops, and its meat category containing subcategories such as processed meat and ham/sausage/grilled pork. Category information expressing services depend on service providers. In such a case, a large amount of rule description is necessary to combine category information. Therefore, dividing individual classes for each service provider and common external classes to connect individual classes, which leads the effectiveness of description for linking services and the availability of sharing knowledge for understanding other services.

We also investigated the information necessary for starting the above-mentioned 60 services. There are services (such as Google Maps) that offer two launch methods (keyword [http://maps.google.co.jp/maps?q=%keyword%] and latitude/longitude [http://maps.google.co.jp/maps?q=%latitude%,%longitude%]), most services are launched using one keyword. About 80% of 60 services and applications that were surveyed are able to be launched by inputting entity name within URL, which comes from two notable examples being Amazon (https://www.amazon.co.jp/s?field-keywords=%keyword%) and Rakuten (http://search.rakuten.co.jp/search/mall/%keyword%/).

### 4.2 Overview About Data Model

Figure 1 illustrates the data model for cooperation between content holders (of broadcast and website content) and service providers (of map application and EC website) based on the analysis of actual services. The data model consists of 4 data layers, entity layer,

---

[7] https://www.amazon.co.jp/gp/site-directory/ref=nav_shopall_btn.
[8] https://www.rakuten.co.jp/category/?l-id=top_normal_gmenu_d_list.
[9] http://shop.tokyu-bell.jp/tokyu-store/app/common/index.

class layer, data process layer, matched data layer, to connect content and services flexibly by making it possible for entity, class and inference to be exchanged respectively.

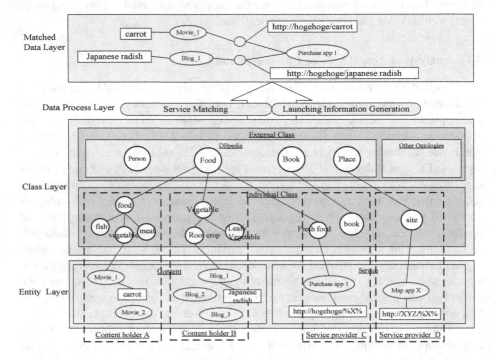

**Fig. 1.** Layered data model

Knowledge about offerings of content holders and service providers is stored at the entity layer. Entities are related to individual classes which can be defined by each service provider at the class layer. Individual classes also can be belonged to external classes. Knowledge about connection of classes are described at the class layer. Inference based on the class structure can be written at the data process layer, and knowledge about combination between content and services can be accessed via the matched data layer.

Figure 1 represents an example where a content holder A distributes a cooking video of "carrots," while a content holder B maintains a website on "Japanese radishes." A web service provider C offers a purchasing service, while a service provider D offers a map service. Each provider defines its own entity data and individual class structure. The scope defined by each company is its own entity data and its own class structure, and it is separated from the external class based on the idea from the actual service analysis mentioned in Sect. 4.1. The data model enables to match independently defined class structures to ones in an external ontology. In the example presented in Fig. 1, individual classes of content holder A, B, and service provider C, are related to the "food" class in the DBpedia ontology as an upper concept. Other ontologies such as GoodRe-lations[10] ontology can also be adopted as the external one.

The data processing layer is responsible for inferring connections between an entity and a service, and for generating service launch information based on the established connections. Section 4.3 will describe how referencing to external classes can optimize inference expressions. The matched data layer stores results generated through inference.

In Fig. 1, a cooking video about "carrots" relates to a purchasing application, and launching information is generated. Similarly, a purchasing application is linked with a website on "Japanese radishes," and service launch information is stored within the data model. Content holder A and B, which distribute the video and maintain the website, can put the links to launch these services within their internal services by retrieving a result from the matched data layer. This makes it possible to induce user action from content. It is also possible to add action-linked functionality to the existing website browsing functionality, by displaying matched layer results in the web browser's share button.

The next sections detail the data model that enables this architecture, and provide information on inference-based data processing.

## 4.3   Structuring Data

In this research, we associate entities within content with services in order for users to act on services when viewing content. That means content is a trigger to launch services. For this reason, we define separate structures for content and services. With the aim of dynamic and efficient data processing by using inference language such as SWRL(Semantic Web Rule Language), we adopt a data structure using an ontology that describes the conceptual structure of information, and then describe media content and web service information using Resource Description Framework (RDF) and Web Ontology Language (OWL).

### 4.3.1   Data Structure for Content

Figure 2 shows an ontology diagram for media content data and Fig. 3 shows an example of instances for broadcast content.

As shown in Fig. 2, media content, which is an instance of cn:Program or cn:webPage, is connected structurally with entities through the cn:hasEntity property relationship. Classes (for which it is assumed that content holders will independently create entities, such as cn:SightseeingSpot and cn:Book) are defined to serve as ranges of values for cn:hasEntity, and these classes are allocated in the individual class layer of the data model. The external classes in the class layer of the data model are associated with individual classes by rdfs:subClassOf relation.

Class definitions for entities are an important part of this data model. In Figs. 2 and 3, entities are associated with TV program scenes for launching services that are relevant

---

[10] https://www.w3.org/wiki/GoodRelations.

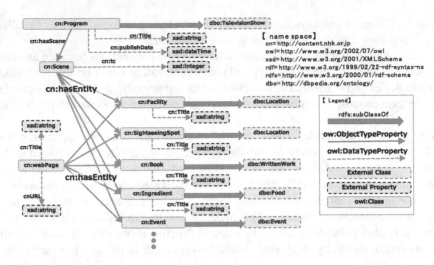

**Fig. 2.** Content ontology

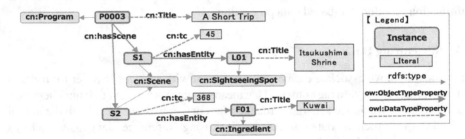

**Fig. 3.** Example of content data instances

to each scene. The "cn" namespace (http://content.nhk.or.jp) shown in Figs. 2 and 3 was created temporarily for this research.

The Protégé ontology editor was used to format data according to the RDF, based on the ontology in Fig. 2. Figure 3 shows an example of these instances. The example in Fig. 3 shows how instances are configured for a television program called "A Short Trip." The "Itsukushima Shrine" instance in the cn:SightseeingSpot class is set for S1, which is a scene that occurs at 45 s of the program. Next, the "Kuwai" (an edible variant of a three-leaf arrowhead) instance in the cn:Ingredient class is set for the 368 s mark. Entities and relevant classes are described using the RDF as content entities, and entities are stored in the entity layer as content knowledge, and the classes are stored in the class layer as category knowledge.

### 4.3.2  Data Structure for Service Description

Web service information is also structured for linking content with web services. More specifically, this describes actions and the targets of actions—such as a map application if the user wants to "go" to a "location," or an online supermarket application if the user wants to "purchase" a "food." This description of action target matches content with web services based on category information (such as "food"), making it possible to launch web services linked with content.

However, users still need to enter entity-related information manually on services if content entities are merely associated with services—content will not be seamlessly connected with user action. To obtain parameters required for launching web services, a framework is needed to obtain information from entities and to set this information as parameters. Descriptions on the service must therefore separately structure information for detecting services and information for launching services.

Figure 4 shows an ontology diagram for web services and Fig. 5 shows an example of web service instances. The "app" namespace (http://app.nhk.or.jp) shown in Figs. 4 and 5 was created temporarily for this research.

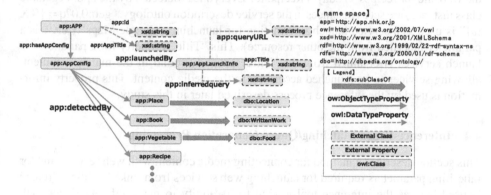

**Fig. 4.** Service description ontology

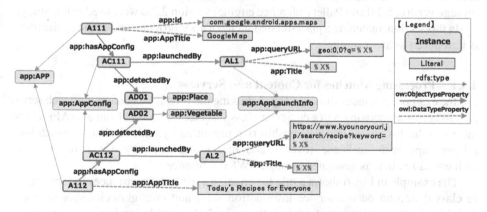

**Fig. 5.** Example of service description instances

The service description ontology makes it possible to configure the class information in order to be matched with content, by using the app:detectedBy property. This class information represents service category offered by service providers. Furthermore, app:queryURL can be retained in instances associated with the app:launchedBy property for launching web services. Data are structured so that the results inferred from app:queryURL are retained in the app:inferredquery property, allowing them to be changed dynamically according to media content. Note that provisional range (rdfs:range) classes are defined for app:detectedBy (in the service description ontology) and for cn:hasEntity (in the content data ontology), under the assumption that these range classes will be created independently by content holders and service providers. In addition, to make the descriptions of inference rule more efficient, classes to which the content entities belong and class information used to detect application information use rdfs:subClassOf to reference external classes. For this reason, DBpedia ontology classes are referenced in this figure. As shown by the data model, it is possible to reference other external ontologies.

In Fig. 5, the "Google Maps" service (detected by the app:Place class instance) and the web site for recipes "Today's Recipe for Everyone" (detected by the app:Vegetable class instance) are set as instances in the service description ontology. "geo:0,0?q = {%X %}" is used as a service launch information for launching "Google Maps" (%X% is a parameter inherited from the other resource). This "Title" property for a parameter to launch services is obtained from the "Title" property found in the matched content, allowing services to be launched according to the media content. This property information is used by the inference process described later in this study.

## 4.4 Inference-Based Matching/Query Generation Processing

This section describes a method for connecting media content with web service, and for inheriting parameters required for launching web services from linked entities. Protégé is used here as the inference tool, owing to its ability to make inferences via RDF schemas and OWL and to generate data using Semantic Web Rule Language (SWRL). Protégé version 5.2.0 and Pallet (inference engine) version 2.2.0 were used in this study.

In this section, matching a place-related scene of a TV program and a map application is introduced as an example.

### 4.4.1 Processing Matches for Content and Services

For the example instances shown in Fig. 3, linking the broadcast content with web services indicates associating S1 (a cn:Scene class instance) with A111 (an app:APP class instance). In this study, this relationship is represented by cn:hasRelation, which has relationships with cn:Scene (over rdfs:domain) and app:APP (over rdfs:range). cn:hasRelation here is generated through SWRL inference.

The example in Fig. 6 shows a rule for creating a triple from instances for app:Place (a class definition on the service information side) and cn:SightseeingSpot (a class definition on the media content side) via the cn:hasRelation relationship. In other words, various rules are required for combining many kinds of classes, and it is necessary to understand all the class definitions in advance. Classes from an external ontology (such as DBpedia) linked via rdfs:SubClassOf can be used for changing the inference rule from Fig. 6, as shown in Fig. 7. Figure 7 shows that they can be standardized in class definitions belonging to the external ontology. Without referring to an external ontology, many inference rules are required to be generated by all the combination of classes, such as app:Place and cn:SightseeingSpot, app:Location and cn:SightseeingSpot, app:Area and cn:SightseeingSpot, app:Place and cn:Address and so on.

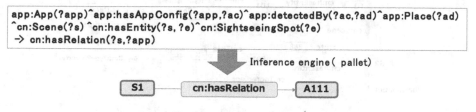

Fig. 6. SWRL rule and generated triple

```
app:App(?app)^app:hasAppConfig(?app,?ac)^app:detectedBy(?ac,?ad)^dbo:Location(?ad)
^cn:Scene(?s) ^cn:hasEntity(?s, ?e)^ dbo:Location(?e)
-> cn:hasRelation(?s,?app)
```

Fig. 7. Updated SWRL rule

### 4.4.2 Processing Query Generation for Launching Services

.Information used for launching web services linked with media content is also generated by inference. More specifically, service launch information is generated as the app:inferredQuery property in the app:LaunchInfo class instance. For example, "geo:0,0?q = Itsukushima Shrine" is generated for launching Google Maps during a scene introducing Itsukushima Shrine in the TV program. Figure 8 shows the SWRL inference rule, the generated triple, and the results screen on Protégé. SWRL contains built-in character string handling (as a subset). As shown in Fig. 8, it is possible to use swrlb:replace to replace parameter variables with media content.

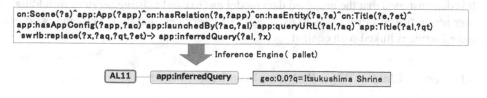

Fig. 8. SWRL rule and generated triple

Figure 9 shows an example of matching content and services described above, as well as the instances that are generated when inference processing is performed for service launch information. As shown in the figure, the "Itsukushima Shrine" entity is linked with "Google Maps" (used to search for directions), rather than with "Today's Recipe for Everyone" (used to search for recipes). The figure also shows that the information for launching Google Maps is generated dynamically as a app:inferredQuery property.

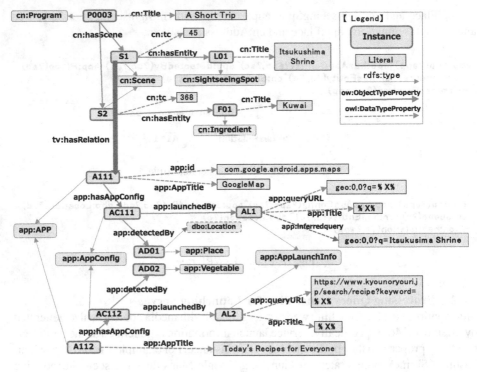

**Fig. 9.** Generated instances

# 5   Discussion

It is demonstrated that the proposed data model can expand existing services by adding user action-linked information to existing IFTTT services, and by developing action-linked services linked with content.

## 5.1   Service Linking When Watching Broadcast Content

To indicate the feasibility of the data model, we prototyped TV-linked application where they can start services triggered by entity shown on TV. This prototype used Fuseki as a data store on the matching layer, making it possible to obtain the data of content and matched web services using SPARQL queries. Figure 10 shows the system architecture. Hybridcast [13] and HbbTV [14] have been launched for the purpose of connecting TV and smartphones/tablets. In this study, to simulate a broadcast, a video playback client was used to play a video of a TV program stored on a web server. The TV-linked application obtains the playback position for identifying the scene being broadcast.

Figure 11 presents a screen from the TV-linked application. On this screen, various kinds of we services are shown by each related action during a scene in which asparagus is introduced. Since an asparagus is an instance of the dbo: Food class, web services and applications related to food are displayed as candidates. Note that action data (such as

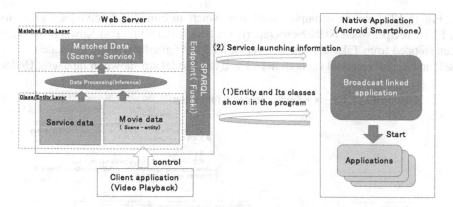

**Fig. 10.**  System architecture of the prototype application

"purchase" or "search for recipes") has been converted to the RDF so that applications to "purchase" "food" or "search for recipes" can be displayed in a group. When the user presses the application's icon on the screen, "Today's Recipe for Everyone" is launched with "asparagus" set as the parameter[11], to allow the user to search for recipes without inputting the keyword "asparagus".

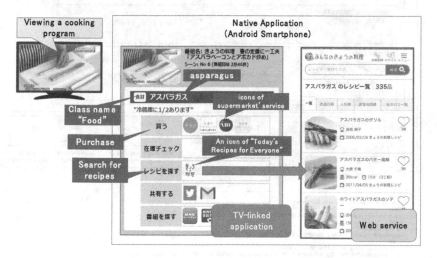

**Fig. 11.**  Prototype application screen

## 5.2  Added Service Linkage to Existing IFTTT Service

This section presents an example of expanding NHK News RSS distribution, where the existing IFTTT service is used for creating a recipe forwarded over an e-mail.

---

[11]  https://www.kyounoryouri.jp/search/recipe?keyword=%E3%82%A2%E3%82%B9%E3%83%91%E3%83%A9%E3%82%AC%E3%82%B9.

Figure 12 shows an example situation in which an e-mail is received according to an FTTT recipe, when NHK News reports via RSS distribution that Kowsi pears have been shipped from Takasaki, Gunma. As the RSS is originally written in Japanese, RSS itself and e-mail received via both IFTTT and this model are shown in Japanese. Under

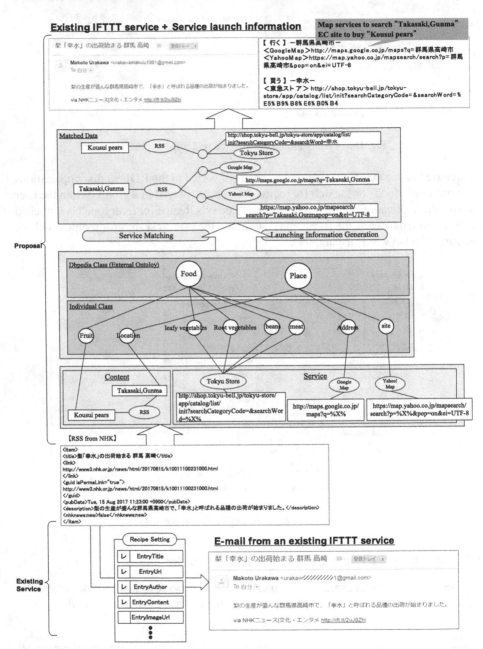

**Fig. 12.** Expansion of IFTTT

the existing IFTTT service where users can select the information to be transmitted by an e-mail, content distributed over RSS can only be transmitted over an e-mail with information configured in the recipe. On the one hand, the data model used in this study matches a map application (such as Google Maps) with "Takasaki, Gunma" in the "Place" class, and then generates launch information as a link. It also appends launch information for an online supermarket (such as Tokyu Store) for "Kosui pear" in the "Food" class. As shown above, the proposed data model is capable of going beyond mere information transmission to expand action-linked functionality connected to user action.

## 6  Summary

In recent years, there have been tremendous efforts to link broadcasts with the Internet and functions to enable actions (such as sharing) when browsing website content. However, users need to enter information manually on their devices such as smartphones or tablets. Although services (such as IFTTT) that use conditional expressions to connect various services have also matured, these services are limited to merely transmitting information. It is needed to match entities within content with services and to generate service launching information based on a related entity, for inducing viewers' action like buying or visiting beyond just information transmitting. Therefore, this study proposed a data model utilizing class structure and inference based on the semantic technology after analyzing 60 major web services and applications. Using this data model enables content holders to link other services even with keeping their own description for services they offer. On the other hand, from content viewers' perspective, they can start relevant web services on their devise like a smartphone without inputting a keyword related to a TV program or web site. That helps viewers to connect their behavior seamlessly.

The use case presented in this study also showed the possibility of expanding IFTTT services and expanding TV-linked user actions by implementing the data model. In future, a data model and architecture capable of deploying services closer to users—such as a framework that gathers user information to automatically select web services with which the user has contracted or that the user frequently uses—will be investigated. In addition, nonmonotonic reasoning can be used for more precise matching. For improvement of media services, we plan to implement the data model based on semantic technology into the existing Hybridcast services.

## References

1. Ur, B., Pak Yong Ho, M., Brawner, S., Lee, J., Mennicken, S., Picard, N., Schulze, D. and Littman, M.L.: Trigger-action programming in the wild: an analysis of 200,000 ifttt recipes. In: Proceedings of the 2016 CHI Conference on Human Factors in Computing Systems, pp. 3227–3231. ACM (2016)
2. Coronado, M., Iglesias, C.A.: Task automation services: automation for the masses. IEEE Internet Comput. **20**(1), 52–58 (2016). https://doi.org/10.1109/MIC.2015.73

3. Gu, T., Wang, X.H., Pung, H.K., Zhang, D.Q.: An ontology-based context model in intelligent environments. In: Proceedings of Communication Networks and Distributed Systems Modeling and Simulation Conference, vol. 2004, 270–275 (2004)
4. Rodríguez, N.D., Lilius, J., Cuéllar, M.P., Calvo-Flores, M.D.: Extending semantic web tools for improving smart spaces interoperability and usability. In: Omatu, S., Neves, J., Rodriguez, J.M.C., Paz Santana, J.F., Gonzalez, S.R. (eds.) Distributed Computing and Artificial Intelligence. AISC, vol. 217, pp. 45–52. Springer, Cham (2013). https://doi.org/10.1007/978-3-319-00551-5_6
5. Coronado, M., Iglesias, C.A., Serrano, E.: EWE ontology specification (2015)
6. De Meester, B., Arndt, D., Bonte, P., Bhatti, J., Dereuddre, W., Verborgh, R., Ongenae, F., De Turck, F., Mannens, E., Van de Walle, R.: Event-driven rule-based reasoning using EYE. In: SSN-TC/OrdRing@ ISWC, pp. 75–86 (2015)
7. Muñoz, S., Llamas, A.F., Coronado, M., Iglesias, C.A.: Smart office automation based on semantic event-driven rules. In: Intelligent Environments Workshops (2016)
8. Ranasinghe, Y.S., Walpola, M.J.: Integrating Context-Awareness with reminder tools. In: 2016 Sixteenth International Conference on Advances in ICT for Emerging Regions (ICTer). IEEE (2016). https://doi.org/10.1109/ICTER.2016.7829921
9. Akbar, Z., García, J.M., Toma, I., Fensel, D.: On using semantically-aware rules for efficient online communication. In: Bikakis, A., Fodor, P., Roman, D. (eds.) RuleML 2014. LNCS, vol. 8620, pp. 37–51. Springer, Cham (2014). https://doi.org/10.1007/978-3-319-09870-8_3
10. OWL-S: Semantic Markup for Web Services. https://www.w3.org/Submission/2004/SUBM-OWL-S-20041122/
11. Kushniretska, I.: Semi-structured data dynamic integration mashup system. In: 2016 XIth International Scientific and Technical Conference Computer Sciences and Information Technologies (CSIT). IEEE (2016)
12. Yamaguchi, A., Kozaki, K., Lenz, K., Yamamoto, Y., Masuya, H., Kobayashi, N.: Data Acquisition by traversing class-class relationships over the linked open data. In: International Semantic Web Conference (Posters & Demos) (2016)
13. Ohmata, H., Takechi, M., Mitsuya, S.: Hybridcast: a new media experience by integration of broadcasting and broadband. In: 2013 Proceedings of ITU Kaleidoscope: Building Sustainable Communities (K-2013). IEEE (2013)
14. Malhotra, R.: Hybrid broadcast broadband TV: the way forward for connected TVs. IEEE Consum. Electron. Mag. 2(3), 10–16 (2013). https://doi.org/10.1109/MCE.2013.2251760

# Refined JST Thesaurus Extended with Data from Other Open Life Science Data Sources

Tatsuya Kushida[1]([✉]) [iD], Yuka Tateisi[1], Takeshi Masuda[2],
Katsutaro Watanabe[3], Katsuji Matsumura[3], Takahiro Kawamura[3],
Kouji Kozaki[3], and Toshihisa Takagi[1,4]

[1] National Bioscience Database Center, Japan Science and Technology Agency,
Tokyo, Japan
{kushida, tateisi}@biosciencedbc.jp,
[2] The Institute of Scientific and Industrial Research, Osaka University,
Suita, Japan
{masuda, kozaki}@ei.sanken.osaka-u.ac.jp
[3] Department of Information Planning, Japan Science and Technology Agency,
Tokyo, Japan
{katsutaro.watanabe, matsumur,
takahiro.kawamura}@jst.go.jp,
[4] Department Biological Sciences, Graduate School of Science,
The University of Tokyo, Tokyo, Japan
tt@bs.s.u-tokyo.ac.jp

**Abstract.** We are developing a refined Japan Science and Technology (JST) thesaurus with thirty-five relations to enable description of rigorous relationships among concepts. In this study, we prepared an environment for performing SPARQL queries and evaluated the JST thesaurus in the life sciences by comparing query results with the originals. Based on the results of the investigation, we constructed a fibrinolysis network from the thesaurus as a collection of concepts connected with fibrinolysis within three steps, and we discovered that fibrinolysis was associated with fifty-four concepts, including sixteen diseases and twelve physiological phenomena. Subsequently, using the sub-classified relations, we divided the sixteen diseases into two diseases that developed after fibrinolysis progressed, seven diseases that shared common molecules in the development mechanism with fibrinolysis, and other associated conditions. Furthermore, we mapped concepts between the JST thesaurus, ChEBI, and Gene Ontology by matching the labels and synonyms. As a result, we could integrate the fibrinolysis network with thirty-seven chemicals, including four antifibrinolytic agents and twenty-seven human gene products that can regulate fibrinolysis. Thus, we were able to handle the information relating to a series of molecules, molecular-level biological phenomena, and diseases by integrating the refined JST thesaurus with information regarding chemicals and gene products from other resources.

**Keywords:** Chemicals · Diseases · Gene products · Refined thesaurus · SPARQL

© Springer International Publishing AG 2017
Z. Wang et al. (Eds.): JIST 2017, LNCS 10675, pp. 35–48, 2017.
https://doi.org/10.1007/978-3-319-70682-5_3

# 1  Introduction

The Japan Science and Technology (JST) thesaurus is an extensive collection of sci-entific and technological concepts. It contains more than two hundred thousand con-cepts and four types of relations among concepts, including broader term, narrower term, synonym, and related term (RT) [1]. For the life sciences, it includes approxi-mately ninety thousand concepts, which are related at different levels, and kinds of concepts, such as diseases and molecular-level phenomena.

The JST thesaurus is used primarily for the purpose of indexing literature [2], and it can perform relaxation search using broader and narrower relations and exploration search using RTs. However, the types of relations are insufficient, with the result that it is impossible to describe the detailed and rigorous relations among concepts; in addition, search results using RT are known to include unintentional information, because RTs are widely used for presenting ambiguous relationships among concepts.

Thus far, we have developed a refined JST thesaurus in which the RTs have been sub-classified into thirty-one relations, such as "has function," "has role," and "has part," and we have decided by majority decision of a panel of life-sciences experts on more than two thousand relationships among biological concepts using the sub-classified relations [3]. As a result, we could identify the following rigorous bio-logical relations in the thesaurus [1].

- Diseases and the preceding biological phenomena

 - Example 1: "fibrinolytic purpura" and "fibrinolysis"
 - Example 2: "thromboembolism" and "platelet aggregation"

- Disease states and the succeeding ones

 - Example 3: "interstitial pneumonia" and "pulmonary fibrosis"

- Diseases and gene products regulating them

 - Example 4: "fibrinolytic purpura" and "PLAT"
 - Example 5: "thromboembolism" and "CLEC2"

In this study, we further improved the refined JST thesaurus in order to be able to describe more rigorous relationships among biological concepts. Moreover, we attempted to integrate the JST thesaurus with information regarding molecules such as chemical substances and gene products from other data sources, and we evaluated the superiority and the effectiveness of the use of the thesaurus in life science research.

The remainder of this paper is structured as follows. In Sect. 2, we describe other ontologies and thesauri for the life sciences. In Sect. 3, we describe our updating of the refined JST thesaurus. In Sect. 4, we present the methods and results of the evaluation of the refined JST thesaurus. In Sect. 5, we summarize our conclusions, present the problems to be solved, and suggest future plans.

## 2 Related Work

We realize that a strength of the JST thesaurus is its handling of direct relationships among concepts of different levels and categories [1]. To highlight the features of the structuring concepts in the JST thesaurus, we now attempt to compare its description of the relationship between platelet aggregation and thromboembolism with that of MeSH [4]. In MeSH, platelet aggregation and thromboembolism are classified as a physiological phenomenon and a disease, respectively, using the relation "broaderDescriptor" corresponding to "broader term" of the JST thesaurus. They have the same structure as in the JST thesaurus. However, in MeSH, platelet aggregation is not directly related to thromboembolism. Moreover, in SNOMED CT [5], one of the largest medical and clinical nomenclatures, the corresponding concepts, Platelet aggregation and Thromboembolic disorder, are structured as in MeSH. There are almost no other ontologies and thesauri that handle direct relationships among the concepts of different levels and categories in the life sciences.

Gene Ontology [6], which consists of three ontologies covering biological processes, molecular functions, and cellular components, is one of the most well-known life science ontologies. Its concepts contain information regarding related gene products. Each relationship between a concept and its related gene product is assigned evidence codes, such as Inferred from Experiment (EXP), Inferred from Sequence, and Structural Similarity (ISS). Gene Ontology is known to life scientists as a reliable data source, and hence is used for interpreting data from high-throughput experiments, including next-generation sequencing and microarrays.

ChEBI [7] is a hierarchy of chemical substances based on structures, roles, and applications. In ChEBI, a specific chemical class based on the information of a role and an application is defined as a collection of chemical substances that have the role and the application in OWL format. NDFRT [8] is an ontology for drugs that contains not only information regarding established drugs but also their physiological effects, interactions, diseases that the drug might treat. Protein Ontology [9], FMA [10], MEDDRA [11], and NCIT [12] are ontologies, terminologies or thesauri regarding proteins and complexes, anatomy, drugs, and medical care, respectively.

The development and maintenance of ontologies and thesauri has typically been conducted with the collaboration of life science experts and ontologists. Mortensen *et al.* pointed out the performance superiority of crowdsourcing for validating the relations among concepts in SNOMED CT and Gene Ontology and the cost-saving effects of combining crowdsourcing with medical experts' curation [13, 14]. LEGO is an ongoing project in which modeling semantic relations among biological processes, molecular functions, cellular components, and the related gene products is conducted using expert crowdsourcing [15]. The objective and approach are similar to that of refining RTs in the JST thesaurus, such as the arranging of biological relations by experts [3].

Existing ontologies and thesauri in the life sciences are collected, organized, and provided by The OBO Foundry [16] and BioPortal [17]. The OBO Foundry publishes information regarding a family of interoperable ontologies that are both logically well-formed and scientifically accurate and that are provided by ontology developers who are committed to collaboration and adherence to shared principles [16]. It clearly displays information regarding the license of each ontology to promote the utilization of ontologies.

BioPortal, a web repository for biomedical ontologies and terminologies, provides an ontology view extraction web service that extracts branches (subsets) of ontologies and assists the ontology developer in constructing a new ontology by reusing existing ones. Ontobee [18] and Linked Open Vocabularies (LOV) [19] are used for finding the vocabularies of classes and properties for semantic web applications.

## 3   Update of Refined JST Thesaurus

First, we reviewed thirty-one sub-classified relations to enable description of more detailed and rigorous relationships [3]. Consequently, we concluded that it was appropriate and reasonable to add four new relations (Table 1). Adding the four relations enabled us to define the relationships between a material such as a gene product, a chemical substance, and the related biological process at the level of a molecular mechanism. Thus far, we have decided on 4,864 relationships out of 41,544 RTs in the life sciences using the thirty-five relations selected by majority decision of a panel of life-sciences experts. Furthermore, we have established correspondences for 15,946 biological concepts in the JST thesaurus to MeSH terms by matching their labels and synonyms. We used a relation "skos[1]:exactMatch" for the relationships.

**Table 1.**  List of new added relations

| Relations | Definitions |
|---|---|
| sio:SIO_000230 (has input) | Relation between a material and a process related to it as the input to the molecular mechanism |
| sio:SIO_000231 (is input in) | Reverse relation of "has input" |
| sio:SIO:000132 (has participant) | Relation between a material or process and a process related to it in the molecular mechanism |
| sio:SIO:000062 (is participant in) | Reverse relation of "has participant" |

sio: <http://semanticscience.org/resource/>

---

[1]  skos: <http://www.w3.org/2004/02/skos/core>.

# 4    Evaluation of the Refined JST Thesaurus

## 4.1    Method

We converted the refined JST thesaurus to the Turtle RDF format by using the improved graphical ontology editor Hozo [20] and stored it in a triplestore in order to be able to perform a SPARQL query. The tentative public SPARQL endpoint is prepared with a CC BY-NC license (http://lod.hozo.jp/repositories/JstNbcdOnt). To evaluate the superiority and effectiveness of the refined JST thesaurus, in which thirty-five relations were sub-classified, we constructed a "Fibrinolysis network" and "Bone Metabolic Turnover (BMT) network" from the JST thesaurus. Fibrinolysis and BMT are related to various cardiovascular diseases and osteoporosis respectively, and these patients are continuously increasing with aging of the population. In addition, increasing of the medical care expenses for citizens has emerged as a social problem, and it is important to collect and arrange the relevant information. These two networks were constructed as collections of concepts connected from fibrinolysis and BMT within three steps in the refined JST thesaurus by performing respective SPARQL queries. Then, we compared both of the networks created from the refined JST thesaurus with those created from the original JST thesaurus in which RTs were not sub-classified.

Furthermore, we attempted to establish correspondences between the components (concepts) in the fibrinolysis and BMT networks and classes of other existing ontologies and thesauri by matching the labels and synonyms using the biological ontology recommendation tool Recommender of BioPortal [21]. The correspondences between them were checked by life science experts. When the experts confirmed them, we created the RDF using skos:exactMatch using the same procedure as we used in mapping MeSH terms (Sect. 3) and stored them in RDF triplestore. The networks were visualized using Cytoscape [22].

## 4.2    Results

Each concept of the JST thesaurus has one or more "subject terms." The subject terms correspond to the scientific and technological categories and contain forty-seven life science terms, including diseases, pathologies, and symptoms. For example, the fibrinolysis network consists of fifty-four concepts categorized into sixteen diseases, pathologies, and symptoms, fourteen hematologies (as a kind of biological phenomenon), seven proteins and peptides, five medicines and forms of medical care, three drugs, three tissues and organs, and the like (Fig. 1 and Table 2).

The BMT network consisted of 132 concepts, which were categorized into twenty-eight diseases, pathologies, and symptoms, twenty tissues and organs, seventeen proteins and peptides, fifteen cytologies, fourteen development and growth patterns, seven medicines and forms of medical care, seven drugs, six genetic patterns (including genes), five hormones, four immunologies, and the like (Fig. 2 and Table 3). Although the subject terms of concepts in the fibrinolysis and BMT networks would provide useful information for understanding the outlines and features of the concepts

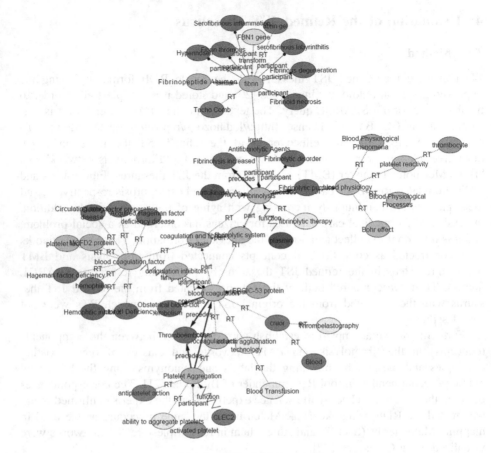

**Fig. 1.** Fibrinolysis network

**Table 2.** Examples of subjects of concepts in the Fibrinolysis network

| Subject | Number of concepts |
|---|---|
| Diseases, pathology, and symptoms | 16 |
| Hematology | 12 |
| Protein and peptide | 7 |
| Medicine and medical cares | 5 |
| Drugs | 3 |
| Tissue and organ | 3 |
| Cytology | 2 |
| Enzyme | 2 |
| Genetics | 2 |
| Immunology | 1 |
| Organic compounds | 1 |
| Pharmacology | 1 |

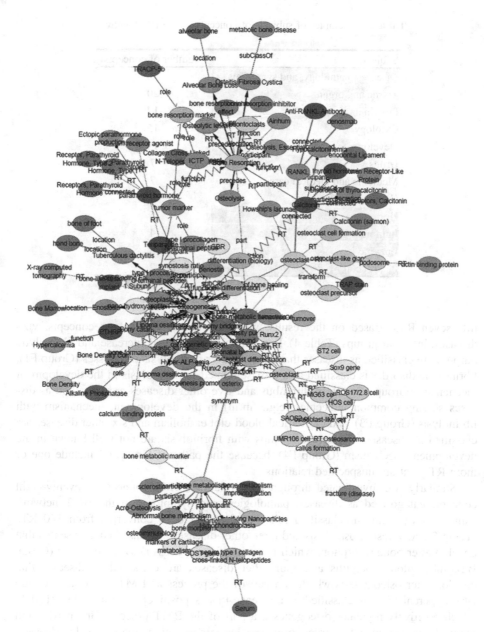

**Fig. 2.** BMT network

for thesaurus users, it is necessary to utilize the information of the sub-classified relations among concepts in order to understand the networks more deeply.

Then, we investigated through what relations fibrinolysis connects to the sixteen concepts categorized into diseases, pathologies, and symptoms in the fibrinolysis network using twenty-eight sub-classified relations that had been converted from

**Table 3.** Examples of subjects of concepts in the BMT network

| Subject | Number of concepts |
| --- | --- |
| Diseases, pathology, and symptoms | 28 |
| Tissue and organ | 20 |
| Protein and peptide | 17 |
| Cytology | 15 |
| Development and growth | 14 |
| Drugs | 7 |
| Medicine and medical cares | 7 |
| Genetics | 6 |
| Hormone | 5 |
| Immunology | 4 |
| Enzyme | 3 |
| Pharmacology | 2 |

fifty-seven RTs. Based on the results, we could divide the sixteen concepts, viz., diseases, into four groups (Table 4). For example, fibrinolysis increase and fibrinolytic purpura are classified as diseases that develop after fibrinolysis progresses (Group F1), fibrinolytic disorder is classified as a disease related to fibrinolysis in the development mechanism (Group F2), fibrin thrombus and five other diseases are classified as diseases sharing common molecules (e.g., fibrin) in the development mechanism with fibrinolysis (Group F3), and Obstetrical blood-clot embolism and six other diseases are classified as diseases whose relationships with fibrinolysis are not well known in the development mechanism (Group F4), because the paths of Group F4 include one or more RTs that are unspecified relations.

Similarly, we investigated through what relations BMT connects to twenty-eight concepts categorized as diseases, pathologies, and symptoms in the BMT network using seventy-eight sub-classified relations that had been converted from 140 RTs (Table 5). As a result, osteolysis and three other diseases are classified as diseases that develop after bone resorption, which is a part of the process as BMT progress (Group B2), tuberculous dactylitis and four other diseases are classified as diseases that develop after osteogenesis which is a part of the process as BMT progresses (Group B6), hypercalcemia is classified as a disease that is positively regulated by PTHRP, which positively regulates osteogenesis, a part of the BMT process (Group B7), and hypocalcitoninemia and disorders of thyrocalcitonin secretion are classified as diseases that have calcitonin as a participant in the development, positively regulating osteogenesis which is a part of the BMT process (Group B8).

Table 6 shows the number of JST thesaurus concepts corresponding to classes of other ontologies. The number of concepts in the JST thesaurus having links to MeSH is greater than that of the others, because in the case of MeSH we had mapped all of the concepts to MeSH terms, while for the other ontologies we had mapped only concepts in the fibrinolysis and BMT networks to other ontologies' classes.

**Table 4.** Classification of diseases connected to fibrinolysis within three steps (relations)

| Group | Concept | Relation | Concept | Relation | Concept | Relation | Concept |
|---|---|---|---|---|---|---|---|
| F1 | | *precedes* | Fibrinolysis increased | | | | |
| | | | Fibrinolytic purpura | | | | |
| F2 | | *is participant in* | Fibrinolytic disorder | | | | |
| F3 | | *has input* | fibrin | *is participant in* | Fibrin thrombus | | |
| | | | | | Fibrinous degeneration | | |
| | | | | | Hyperinosemia | | |
| | | | | | Fibrinoid Necrosis | | |
| | | | | | Serofibrinous inflammation | | |
| | fibrinolysis | | | | serofibrinous labyrinthitis | | |
| F4 | | *RT* | blood coagulation | *precedes* | Obstetrical blood-clot embolism | | |
| | | | coagulation and fibrinolytic system | *RT* | blood coagulation factor | *RT* | Factor XI Deficiency |
| | | | | *succeeds* | platelet aggregation | *precedes* | Thromboembolism |
| | | | blood coagulation | *is function of* | blood coagulation factor | *RT* | Hageman factor deficiency |
| | | | | | | | hemophilia |
| | | | | | | | Acquired Hageman factor deficiency disease |
| | | | | | | | Circulating anticoagulant disease |

In ChEBI, a ChEBI class "antifibrinolytic drug" (CHEBI:48675), which corresponds to the JST thesaurus concept "Antifibrinolytic Agents" has been defined as a concept having the role of antifibrinolytic drug, and we have confirmed that it has four concrete chemical substances (e.g., anagrelide (CHEBI:142290)) as a subclass.

**Table 5.** Classification of diseases connected to BMT within three steps (relations)

| Group | Concept | Relation | Concept | Relation | Concept | Relation | Concept |
|---|---|---|---|---|---|---|---|
| B1 | bone metabolic turnover (BMT) | has part | Bone Resorption | *is participant in* | Osteolysis, Essential osteolysis | | |
| B2 | | | | *precedes* | Osteolytic lesion | | |
| | | | | | Alveolar Bone Loss | | |
| | | | | | Osteitis Fibrosa Cystica | | |
| B3 | | | | *has participant* | Calcitonin | *is participant in* | Disorders of thyrocalcitonin secretion |
| B4 | | | | *is function of* | parathyroid hormone | | Ectopic parathormone production |
| B5 | | | osteogenesis | *is participant in* | Osteoplastic sarcoma | | |
| B6 | | | | *precedes* | Lipoma ossificans | | |
| | | | | | Bony cataract | | |
| | | | | | Osteoplastica | | |
| | | | | | Enostosis | | |
| | | | | | tuberculous dactylitis | | |
| B7 | | | | | PTHRP | *has function* | Hypercalcemia |
| B8 | | | | *is function of* | Calcitonin | | Disorders of thyrocalcitonin secretion |
| | | | | | | *is participant in* | Hypocalcitoninemia |
| B9 | | | | | parathyroid hormone | | Ectopic parathormone production |
| B10 | | *synonym* | bone metabolism | *is participant in* | Abnormal bone metabolism | | |
| | | | | | Acro-Osteolysis | | |

Moreover, Antifibrinolytic Agents relates to fibrinolysis with the relation "has function," indicating that the Antifibrinolytic Agents regulate fibrinolysis. Therefore, we can infer that the four concrete chemical substances that comprise the subclass of antifibrinolytic drugs would also regulate fibrinolysis in the fibrinolysis network.

**Table 6.** Number of JST thesaurus concepts having links to classes of other ontologies in the JST thesaurus and in the Fibrinolysis and BMT networks

| Link to ontologies | JST thesaurus (89,661)[a] | Fibrinolysis network (54)[a] | BMT network (139)[a] |
|---|---|---|---|
| MeSH | 15946 | 21 | 56 |
| ChEBI | 6 | 2 | 4 |
| Gene Ontology | 12 | 3 | 9 |
| MEDDRA | 18 | 7 | 11 |
| SNOMEDCT | 56 | 20 | 36 |
| NCIT | 63 | 10 | 53 |
| NDFRT | 26 | 6 | 20 |
| ICD10 | 3 | 1 | 2 |
| Protein ontology | 10 | 1 | 9 |
| ICD9 | 2 | 0 | 2 |
| OMIM | 20 | 4 | 16 |
| FMA | 7 | 1 | 6 |

[a] The numbers in parentheses donates the numbers of concepts.

Likewise, the JST thesaurus concept "Bone Density Conservation Agents" corresponds to the ChEBI class "bone density conservation agent" (CHEBI:50646), and it relates to "osteogenesis" with the relation "has function" in the BMT network. We have also confirmed that thirty-seven chemical substances (e.g., calcitriol (CHEBI:17823)) were defined as comprising a subclass of bone density conservation agent in ChEBI. Therefore, we can infer that the thirty-seven chemical substances would regulate osteogenesis.

Conversely, a JST thesaurus concept "fibrinolysis" relates to a Gene Ontology class "fibrinolysis" with skos:exactMatch. In Gene Ontology each class (e.g.,., fibrinolysis (GO:0042730)) has information regarding gene products (e.g., PLAUR (UniProtKB: Q03405)) [23] associated with the class (Table 7). Therefore, for example, we infer that twenty-seven human gene products associated with fibrinolysis in Gene Ontology would regulate fibrinolysis in the fibrinolysis network.

Thus, we can handle information relating to a series of molecules, molecular-level phenomena, and diseases in both of the networks extended by the information regarding chemical substances and gene products.

- Example 1: In the fibrinolysis network, a gene product "PLAUR" relates to a molecular-level phenomenon "fibrinolysis," and after fibrinolysis progresses, a disease "fibrinolytic purpura" develops.

- Example 2: In the BMT network, a drug "calcitriol" relates to the molecular mechanism of a phenomenon "osteogenesis," and after osteogenesis (abnormally) progresses, a disease "tuberculous dactylitis" develops.

These examples include the networks created from the refined JST thesaurus, and they show outcomes of that thesaurus.

ChEBI provides information regarding InChI (InChIKey) [24], an identifier of chemical substances based on structural information. In contrast, UniChem [25] provides mapping information regarding chemical substances registered in approximately thirty public databases, including ChEBI, PubChem [26], ChEMBL [27], and KEGG COMPOUND [28] by using the InChIKey. At present, the mapping information can be used as linked open data (LOD) through NikkajiRDF [29], which provides the SPARQL endpoint. If we use it, we can handle much more information regarding chemical substances.

**Table 7.** List of JST thesaurus concepts corresponding to Gene Ontology classes in the Fibrinolysis and BMT networks, with information regarding the number of associated gene products

| Network | JST_label | JST_id | GO_label | GO_id | No. of gene products |
|---|---|---|---|---|---|
| Fibrinolysis | Fibrinolysis | 200906057747871335 | fibrinolysis | GO:0042730 | 27 |
| Fibrinolysis | blood coagulation | 200906095157928489 | blood coagulation | GO:0007596 | 338 |
| Fibrinolysis | platelet aggregation | 200906034198027470 | platelet aggregation | GO:0070527 | 57 |
| BMT | Bone Resorption | 200906090424766432 | bone resorption | GO:0045453 | 54 |
| BMT | osteogenesis | 200906079594402385 | ossification | GO:0001503 | 372 |
| BMT | Podosome | 200906038956466176 | podosome | GO:0002102 | 32 |
| BMT | osteoblast differentiation | 201106046111356887 | osteoblast differentiation | GO:0001649 | 205 |

## 5   Conclusions

We developed a refined JST thesaurus in which RTs are sub-classified into thirty-five relations in order to enable description of more rigorous and detailed relationships among biological concepts. Furthermore, we attempted to integrate the refined JST thesaurus with information regarding chemical substances and gene products from other data sources. As a result, we can handle information regarding relationships among the concepts of various levels and categories by using thirty-five sub-classified relations having meaning. There are almost no other ontologies and thesauri that collect and organize information regarding the concepts of different levels and categories and have direct relationships among them in the life sciences. This is a strength of our refined JST thesaurus.

However, the refined JST thesaurus does not comprehensively include information regarding each category in the life sciences. In other words, the thesaurus was originally designed to collect information in the life sciences broadly and shallowly. For example, the fibrinolysis network constructed from the JST thesaurus includes only nine concepts categorized into genes and gene products (e.g., the FBN1 gene). On the other hand, we can handle information regarding 422 gene products (e.g., PLAUR) using the fibrinolysis network by incorporating information from Gene Ontology into the thesaurus (Table 7). To take advantage of the thesaurus and to compensate for this deficiency, we are planning to extend the refined JST thesaurus by incorporating information from other databases and LOD in the life sciences. We expect that the refined JST thesaurus will then be able to contribute to the discovery of new knowledge.

We think that it would be a benefit and a contribution for the semantic web research community to convert a thesaurus into an ontology, and to provide a knowledge base for reasoning, such as DL reasoning. This study is a challenge with regard to sub-classifying relations with shallow semantics into relations with deeper semantics, and it requires consideration of the method. Conversely, Kless *et al.* pointed out that merely specifying the relation of the thesaurus is inappropriate for ontology construction [30]. Thus it is necessary to carefully design the structure of the relationship between concepts to convert the thesaurus into a more rigorous and solid ontology. Moreover, we need to consider removing the existing relations in order to avoid duplication and conflict. To establish a general methodology for extending/refining an existing thesaurus or ontology is our future work.

In the process of developing thesauri and ontologies, selecting concepts and determining the relationships among them are performed on the basis of experts' knowledge and experience, and these tasks are laborious and time consuming. In the future, we plan to consider introducing machine learning and cloud-sourcing methods, and we expect that combining them with experts' curation will efficiently improve the process.

**Acknowledgment.** This work was supported by an operating grant from the Japan Science and Technology Agency and JSPS KAKENHI Grant Number JP17H01789.

# References

1. Kushida, T., Masuda, T., Tateisi, Y., Watanabe, K., Matsumura, K., Kawamura, T., Kozaki, K., Takagi, T.: Refining JST thesaurus and discussing the effectiveness in life science research. In: Proceedings of the IESD 2016 (2016)
2. J-GLOBAL. http://jglobal.jst.go.jp/en/. Accessed 10 Aug 2017
3. Kushida, T., Kozaki, K., Tateisi, Y., Watanabe, K., Masuda, T., Matsumura, K., Kawamura, T., Takagi, T.: Efficient construction of a new ontology for life sciences by sub-classifying related terms in the Japan Science and Technology Agency thesaurus. In: Proceedings of ICBO 2017 (2017, in press)
4. Bodenreider, O., Nelson, S.J., Hole, W.T., Chang, H.F.: Beyond synonymy: exploiting the UMLS semantics in mapping vocabularies. In: Proceedings of AMIA Symposium, pp. 815–819 (1998)

5. SNOMED CT. http://www.ihtsdo.org/snomed-ct. Accessed 10 Aug 2017
6. Gene Ontology Consortium: Creating the gene ontology resource: design and implementation. Genome Res. **11**(8), 1425–1433 (2001)
7. Hastings, J., de Matos, P., Dekker, A., Ennis, M., Harsha, B., Kale, N., Muthukrishnan, V., Owen, G., Turner, S., Williams, M., Steinbeck, C.: The ChEBI reference database and ontology for biologically relevant chemistry: enhancements for 2013. Nucleic Acids Res. **41**, D456–D463 (2013)
8. NDFRT (in BioPortal). http://bioportal.bioontology.org/ontologies/NDFRT. Accessed 10 Aug 2017
9. Protein Ontology. http://proconsortium.org/pro/. Accessed 10 Aug 2017
10. FMA. http://si.washington.edu/projects/fma. Accessed 10 Aug 2017
11. MEDDRA. https://www.meddra.org. Accessed 10 Aug 2017
12. NCIT (in BioPortal). http://bioportal.bioontology.org/ontologies/NCIT. Accessed 10 Aug 2017
13. Mortensen, J.M., Minty, E.P., Januszyk, M., Sweeney, T.E., Rector, A.L., Noy, N.F., Musen, M.A.: Using the wisdom of the crowds to find critical errors in biomedical ontologies: a study of SNOMED CT. J. Am. Med. Inform. Assoc. **22**, 640–648 (2015)
14. Mortensen, J.M., Telis, N., Hughey, J.J., Fan-Minogue, H., Van Auken, K., Dumontier, M., Musen, M.A.: Is the crowd better as an assistant or a replacement in ontology engineering? An exploration through the lens of the Gene Ontology. J. Biomed. Inform. **60**, 199–209 (2016)
15. LEGO. http://geneontology.org/page/connecting-annotations-lego-models. Accessed 10 Aug 2017
16. OBO. http://www.obofoundry.org. Accessed 10 Aug 2017
17. BioPortal. http://bioportal.bioontology.org/. Accessed 10 Aug 2017
18. Ontobee. http://www.ontobee.org/. Accessed 10 Aug 2017
19. Linked Open Vocabularies. http://lov.okfn.org/dataset/lov/. Accessed 10 Aug 2017
20. Hozo: Ontology Editor. http://www.hozo.jp/. Accessed 10 Aug 2017
21. Martinez-Romero, M., Jonquet, C., O'Connor, M.J., Graybeal, J., Pazos, A., Musen, M.A.: NCBO ontology recommender 2.0: an enhanced approach for biomedical ontology recommendation. J. Biomed. Semant. **8**, 21 (2017)
22. Shannon, P., Markiel, A., Ozier, O., Baliga, N.S., Wang, J.T., Ramage, D., Amin, N., Schwikowski, B., Ideker, T.: Cytoscape: a software environment for integrated models of biomolecular interaction networks. Genome Res. **13**(11), 2498–2504 (2003)
23. UniProt Knowledgebase. http://purl.uniprot.org/. Accessed 10 Aug 2017
24. Heller, S., McNaught, A., Stein, S., Tchekhovskoi, D., Pletnev, I.: InChI-the worldwide chemical structure identifier standard. J. Cheminform. **5**, 7 (2013)
25. Chambers, J., Davies, M., Gaulton, A., Hersey, A., Velankar, S., Petryszak, R., Hastings, J., Bellis, L., McGlinchey, S., Overington, J.P.: UniChem: a unified chemical structure cross-referencing and identifier tracking system. J. Cheminform. **5**, 3 (2013)
26. PubChem. https://pubchem.ncbi.nlm.nih.gov. Accessed 10 Aug 2017
27. ChEMBL. https://www.ebi.ac.uk/chembl/. Accessed 10 Aug 2017
28. KEGG COMPOUND. http://www.genome.jp/kegg/compound/. Accessed 10 Aug 2017
29. NikkajiRDF. https://integbio.jp/rdf/?view=detail&id=nikkaji. Accessed 10 Aug 2017
30. Kless, D., Jansen, L., Milton, S.: A content-focused method for re-engineering thesauri into semantically adequate ontologies using OWL. Semant. Web **7**(5), 543–576 (2016)

# Refinement-Based OWL Class Induction
# with Convex Measures

David Ratcliffe[1,2(✉)] and Kerry Taylor[1]

[1] College of Engineering and Computer Science, Australian National University,
Canberra, ACT 2601, Australia
{david.ratcliffe,kerry.taylor}@anu.edu.au
[2] CSIRO Data61, GPO Box 1700, Canberra, ACT 2601, Australia

**Abstract.** Beam-search may be used to iteratively explore and evaluate *refinements* of candidate hypotheses expressed in logical formalisms such as description logic. In this paper, we analyse heuristics for beam search methods over OWL classes and present a novel search algorithm, OWL-MINER which leverages the properties of convex measure functions to deliver an improved memory-bounded beam search. We present performance results on the mutagenesis benchmark problem and demonstrate superior performance relative to another state-of-the-art implementation, and present 10-fold cross-validated accuracy results which are comparable with those from a variety of other methods. Our improvements to the space and time-based efficiency of refinement-based learning algorithms are significant for expanding the size of learning problems that can be feasibly addressed by refinement learning and the quality of solutions that can be found with limited resources.

## 1 Introduction

Automated methods for the analysis of data and knowledge represented directly in the Resource Description Framework (RDF) and Web Ontology Language (OWL) are of current interest [1]. Concept induction is a technique for discovering descriptions of data, such as OWL class expressions generalising and describing RDF training data. These class expressions capture patterns in the data which can be used for classification of unseen data.

The semantics of OWL2-DL is underpinned by Description Logics (DLs), a family of expressive and decidable fragments of first-order logic [2]. In this paper we employ the syntax and semantics of the DL that underlies OWL2-DL and RDF, namely $\mathcal{SROIQ}(\mathcal{D})$ [3].

Recently, methods of concept induction which are well studied in the field of Inductive Logic Programming have been applied to DLs for supervised classification [4]. A *concept* in DL corresponds to a *class* in OWL, so concept induction in

---

The original version of this chapter was revised: The authors corrected errors (mainly in Section 5) regarding the results reported for OWL-Miner and DL-Learner. An erratum to this chapter can be found at https://doi.org/10.1007/978-3-319-70682-5_24

© Springer International Publishing AG 2017
Z. Wang et al. (Eds.): JIST 2017, LNCS 10675, pp. 49–65, 2017.
https://doi.org/10.1007/978-3-319-70682-5_4

DL equivalently learns OWL2-DL class expressions that are inductive hypotheses to describe RDF training data. For examples of concepts expressed in DL syntax, refer to Sect. 5.

Refinement structures the search space of DL concepts and progressively generalises or specialises candidate concepts to cover example data as guided by quality criteria such as accuracy. However, the current state-of-the-art in this area is limited in that such methods were not primarily designed to scale over large knowledge bases or do not support class languages as expressive as OWL2-DL. Our work addresses these limitations by increasing the efficiency of these learning methods whilst permitting a concept language up to the expressivity of OWL2-DL classes (that is, $\mathcal{SROIQ}(\mathcal{D})$ concepts).

We introduce a general beam search algorithm to search the space of concepts generated by refinement. We focus on improving the efficiency of a basic beam search algorithm by introducing a family of convex utility heuristics for ranking concepts in the beam frontier. Supplemented with these heuristics, we propose a time and memory-bounded top-$k$ concept search algorithm.

We have implemented our methods as the learning system called OWL-MINER and show that our methods outperform a state-of-the-art system for DL learning in terms of search efficiency.

## 1.1 Preliminaries

Our work in this paper is independent of the specific DL language chosen for concepts, which we denote as $\mathcal{L}$. We use a refinement operator, in particular a *downward* refinement operator, to traverse concepts of a language $\mathcal{L}$ in the space of concepts ordered by subsumption. Typically we expect a refinement operator to be complete for $\mathcal{L}$, that is, the transitive closure of applications of the operator from a starting point ($\top$) should include all the concepts in $\mathcal{L}$.

**Definition 1 (Downward Refinement Operator).** *Given a quasi-ordered space $(\mathcal{L}, \sqsubseteq)$, a **downward refinement operator** is a mapping from $\mathcal{L}$ to $2^{\mathcal{L}}$ where, $\forall C \in \mathcal{L}$, $\rho^{\downarrow}(C) : \{D \,|\, D \in \mathcal{L} \wedge D \sqsubseteq C\}$. Each $D \in \rho^{\downarrow}(C)$ are called **specialisations** of $C$.*

*Supervised* learning problems take a set of examples $\mathcal{E}$ for which each member $e_{\omega} \in \mathcal{E}$ has been attributed with some label $\omega \in \Omega$ where $|\Omega| \geq 2$. In this way, the set of examples can be partitioned into sets containing examples with a common label $\omega$, denoted $\mathcal{E}^{\omega}$ where $\mathcal{E} = \bigcup_{\forall \omega \in \Omega} \mathcal{E}^{\omega}$. We aim to construct *hypotheses* $h$ which describe certain proportions of each of the labelled examples of each set $\mathcal{E}^{\omega}$. We say that a hypothesis *covers* an example, denoted by the boolean function $covers(h, e_{\omega})$, if $h$ describes example $e_{\omega}$ where $e_{\omega} \in \mathcal{E}^{\omega}$. With respect to the set of all examples $\mathcal{E}$, we denote the *cover* of hypothesis $h$ as the set $cover(h, \mathcal{E}) = \{e \in \mathcal{E} \mid covers(h, e)\}$. This is a loose definition of *covers* that is sufficient for the study here although specific learning algorithms may compute *cover* in different ways, usually using a background theory or *knowledge base* to support the descriptive power of the hypothesis.

We focus on the *binary classification* problem with two labels $|\Omega| = \{+, -\}$, where $\mathcal{E}^+$ are the *positive* examples and $\mathcal{E}^-$ are the *negative* examples. Hypotheses are sought which cover all *positive* examples $\forall e \in \mathcal{E}^+ : covers(h, e)$ and no *negative* examples $\forall e \in \mathcal{E}^- : \neg covers(h, e)$.

We use a *measure function* to assess hypothesis performance and common examples for binary classification are *accuracy* and *relative frequency* as follows.

**Definition 2 (Accuracy).** *Given a labelled set of examples $\mathcal{E}$ partitioned into positive examples $\mathcal{E}^+$ and negative examples $\mathcal{E}^-$, a hypothesis $h$ and its cover $C$ where $C = cover(h, \mathcal{E})$, the **accuracy** function is defined as:*

$$acc(h, \mathcal{E}) = \frac{TP + TN}{TP + FP + FN + TN}$$

*where $TP = \mathcal{E}^+ \cap C$ (true positives); $FP = \mathcal{E}^- \cap C$ (false positives); $TN = \mathcal{E}^- \setminus C$ (true negatives); and $FN = \mathcal{E}^+ \setminus C$ (false negatives).*

**Definition 3 (Relative Frequency).** *Given a hypothesis $C$ and a set of examples $\mathcal{E}$, **relative frequency** is defined as $relFreq(C, \mathcal{E}) = \frac{|cover(C, \mathcal{E})|}{|\mathcal{E}|}$*

## 2 Beam Search for Concepts

The space of concepts in some language $\mathcal{L}$ is typically unbounded and naive generate-and-test methods for finding good hypotheses are impractical. In refinement based search, a downward refinement operator $\rho^\downarrow$ enumerates concepts in a general-to-specific order so that more general concepts are found earlier (to avoid overfitting) and so that many unsatisfiable concepts are never generated. For example, if a concept is unsatisfiable, then each of its refinements are also unsatisfiable and need not be generated.

A general purpose downward refinement learning algorithm searches the space of concepts that are expressible in the hypothesis language $\mathcal{L}$ starting from the top concept $\top$. A *frontier* of unexplored concepts is maintained and a concept is iteratively selected from the frontier for evaluation and further refinement, whereby its specialised concepts generated by one application of the refinement operator are added back into the frontier. Depending on the approach for selecting concepts from the frontier, this generalised architecture behaves as breadth-first, depth-first, or beam-search algorithm over $\mathcal{L}$.

While the use of $\rho^\downarrow$ may trivially exclude many low value concepts, the space of concepts which $\rho^\downarrow$ structures may remain impractically large for knowledge bases containing a large number of concepts and roles. Therefore, we are required to implement some simple pruning strategies. Firstly, we bound the total number of hypotheses under consideration at any stage of the algorithm to limit memory usage. Secondly, once we have evaluated a candidate hypothesis and found it to be acceptable as a solution for the learning problem, we avoid generating any of its refinements as they will be redundant with respect to the accepted hypothesis. We evaluate the acceptability of a solution by a threshold test on a *measure* function, such as accuracy over the training data. Thirdly, we employ a *utility* function to select the best concepts for refinement and to prune unwanted refinements of a concept from the search. A basic *beam search* algorithm by downward refinement is shown as Algorithm 1.

---

**Algorithm 1.** A basic *beam search* with refinement operator $\rho^{\downarrow}$ to search for hypotheses in $\mathcal{L}$ using examples $\mathcal{E}$. $\mathbf{u}$ is a utility function evaluating hypotheses and $\mathcal{Q}$ is a quality function identifying solutions. $b_{max}$ is the beam width.

| | |
|---|---|
| 1: $B := \{\top\}$ | ▷ The hypothesis frontier beam of search candidates |
| 2: $S := \emptyset$ | ▷ The set of solutions |
| 3: **while** $\lvert B \rvert > 0$ **do** | ▷ While the frontier beam is non-empty |
| 4:     $E = \emptyset$ | ▷ Initialise the expansion set |
| 5:     **for all** $C \in B$, $D \in \rho(C)$ **do** | |
| 6:         **if** $\mathcal{Q}(D, \mathcal{E}) = true$ **then** | ▷ Hypothesis $D$ is a sufficient candidate |
| 7:             $S := S \cup \{D\}$ | ▷ Capture solution $D$ |
| 8:         **else** | |
| 9:             $E := E \cup \{D\}$ | ▷ Include $D$ in the expansion set |
| 10:         **end if** | |
| 11:     **end for** | |
| 12:     $B := \emptyset$ | ▷ Reinitialise the beam |
| 13:     **while** $\lvert E \rvert > 0$ and $\lvert B \rvert < b_{max}$ **do** | |
| 14:         $D \in \arg\max_{D \in E} \mathbf{u}(D)$ | ▷ Arbitrary best refinement |
| 15:         $E := E \setminus \{D\}$ | |
| 16:         $B := B \cup \{D\}$ | ▷ Include $D$ in the next beam |
| 17:     **end while** | |
| 18: **end while** | |

---

Algorithm 1 traverses the space of concepts by maintaining a beam set $B$ of maximum cardinality $b_{max}$ to maintain the *search frontier* containing the set of all candidate concepts to consider for further refinement. The algorithm places refinements of the candidates in $B$ into a temporary *expansion set* $E$, before repopulating the beam set $B$ with at most $b_{max}$ of the best candidates from $E$ as measured with the utility function $\mathbf{u}$. The basic beam search is conditioned on a boolean *quality* function $\mathcal{Q}(D, \mathcal{E})$ which evaluates candidate hypotheses by some measure function together with a threshold value to determine whether an hypothesis is *good enough* to be considered as a (non-unique) solution to the learning problem. A quality function over accuracy (Definition 2), a typical case, may be defined as $\mathcal{Q}(h, \mathcal{E}) : acc(h, \mathcal{E}) > 0.95$ which holds when hypothesis $h$ has an accuracy over 95%.

## 3   Utility and Quality Functions in Beam Search

Heuristic utility and quality functions over candidate hypotheses are critical to the performance of a beam search such as Algorithm 1. Heuristic functions are often defined in terms of accuracy, or structural complexity measured simply by the number of terms in the expression could be used. Such functions can be combined into one heuristic utility function, for example:

$$\mathbf{u}(h, \mathcal{E}) = acc(h, \mathcal{E}) - \beta \cdot length(h)$$

where $length : \mathcal{L} \mapsto \mathbb{N}$ maps concept expressions to their length as the number of terms in the expression $h \in \mathcal{L}$, and where $\beta \in [0, 1]$ is a fixed, experimentally chosen parameter which captures the user-defined importance of concept length relative to accuracy.

A utility function $\mathbf{u}$ induces a *ranking* over concepts where, for two concepts $C, D$ and a set of examples $\mathcal{E}$, that $\mathbf{u}(C, \mathcal{E}) < \mathbf{u}(D, \mathcal{E})$ implies that $D$ is preferred, or is *stronger* than, concept $C$. In a search algorithm which selects candidate expressions such as $C$ or $D$ to refine towards concepts which are solutions, the utility function can be used to select the most promising concepts to refine first over others. For example, in a beam search with a beam set $B$ of cardinality $n$ for which all members have been refined into a set $E = \{h \mid \forall b \in B : h \in \rho^{\downarrow}(b)\}$, repopulation of the beam $B$ may be performed by selecting the $n$-best candidates of $E$ according to their utility $\mathbf{u}$.

In addition to a utility function, a search algorithm will impose a minimum threshold $\tau$ on a measure function to define a boolean *quality* function $\mathcal{Q}$ which can indicate when a hypothesis $h$ may be considered a solution to a learning problem, such as $\mathcal{Q}(h, \mathcal{E}) = acc(h, \mathcal{E}) > 0.95$ to describe solutions as having at least 95% accuracy. A search algorithm which assesses the performance of hypotheses based on a measure function $f$ may also be able to leverage certain properties of $f$ to determine if *any* refinements of any particular candidate hypothesis $h$ could satisfy the quality function. Such properties of the measure functions are those of *anti-monotonicity* or *convexity*, as we will describe in the next two sections. These properties permit us to define lower bounds on the coverage of certain sets of labelled examples for any hypothesis $h$ which indicate if any of the refinements $\rho^{\downarrow}(h)$ could ever be considered a solution. If they cannot, the space of concepts defined by the set $\rho^{\downarrow}(h)$ may be effectively pruned from the search space, thus improving efficiency.

**Anti-monotonic Quality Criteria.** A boolean quality function $\mathcal{Q}$ is defined as being *anti-monotonic* if, for any two concepts $C, D$ where $C \sqsubseteq D$, when $\mathcal{Q}(C)$ succeeds, $\mathcal{Q}(D)$ necessarily succeeds. If $\mathcal{Q}$ is defined in terms of a threshold $\tau$ on a measure function $f(C, \mathcal{E}) \geq \tau$ which fails for all refinements $E \in \rho^{\downarrow}(C)$ and all subsequent refinements, we conclude that the condition $f(h, \mathcal{E}) \geq \tau$ is anti-monotonic.

The *relFreq* function which computes *relative frequency* of Definition 3 is an example of an anti-monotonic measure function which we can use to apply to a search. Basically, *relFreq* can be used to ensure that any candidate concept $C$ must strictly cover a minimum proportion of examples from all examples $\mathcal{E}$ as:

$$relFreq(C, \mathcal{E}) \geq \tau_{min}$$

where $\tau_{min} \in [0,1]$ and represents a minimum threshold on relative frequency. For example, $\tau_{min} = 0.1$ requires that any candidate $C$ must cover at least 10% of all examples. Clearly, this is anti-monotonic as any concept $D$ where $C \sqsubseteq D$ will also have $relFreq(D, \mathcal{E}) \geq \tau_{min}$ as $D$ cannot cover any fewer examples than $C$. This condition can be used for pruning the search space, as if we find that $relFreq(C, \mathcal{E}) < \tau_{min}$, then for all concepts $E \in \rho^{\downarrow}(C)$ and all subsequent refinements to any concept $E'$, we will find $relFreq(E', \mathcal{E}) < \tau_{min}$.

Measures such as the accuracy function that are *not* anti-monotonic cannot be used to prune away parts of the search space based on simply thresholding on their values. Fortunately, we may still define similar conditions under which refinements of any concept may never be considered of sufficient quality if the measure function has the property of *convexity*, in which case we may impose certain minimum bounds on the cover of hypotheses such that we ensure that their refinements may still contain a candidate of sufficient quality.

**Convex Quality Criteria.** Generally, if a quality function $\mathcal{Q}$ is defined over a measure function $f$ which can be shown to be *convex*, we may conclude that $f$ is anti-monotonic which will assist us in understanding when to prune parts of a search space away to improve efficiency. To define the convexity of a function such as $f$, we first define the notion of a *convex set*, as follows.

**Definition 4 (Convex Set).** *A set $S$ in some vector space such as $\mathbb{R}^d$ is* **convex** *if, for any two vectors $s_1, s_2 \in S$ and any $t \in [0,1]$, then the vectors described by $(1-t)s_1 + ts_2$ must also belong to $S$. Intuitively, this means that all vectors which lie on the straight line between any two vectors $s_1$ and $s_2$ appear in $S$.*

**Definition 5 (Convex, Concave Function).** *A function $f : S \mapsto \mathbb{R}$ is* **convex** *iff $S \subseteq \mathbb{R}^d$ is a convex set and $\forall s_1, s_2 \in S, t \in [0,1] : f(ts_1 + (1-t)s_2) \leq tf(s_1) + (1-t)f(s_2)$. If $f$ is convex, then we say that $-f$ is* **concave***, and vice-versa [5].*

In order to analyse such functions $f$ in terms of the covers of the concepts, we consider the *stamp point* function $sp : \mathcal{L} \times S \mapsto \mathbb{Z}^d$ which maps concepts $C \in \mathcal{L}$ and the *population* set of examples $\mathcal{E} \in S$ where each example $e \in \mathcal{E}$ is labelled with one of $d$ distinct labels as $e_{\omega_i}$ where $\omega_i \in \Omega$ and $d = |\Omega|$ and $1 \leq i \leq d$ as:

$$sp(C, \mathcal{E}) = \langle x_{\omega_1}, \ldots, x_{\omega_d} \rangle$$

where $x_{\omega_i} = |cover(C, \mathcal{E}) \cap \mathcal{E}^{\omega_i}|$. The function $sp$ maps the *example cover* of some concept expression $C$ to a so-called *stamp point* $\langle x_{\omega_1}, \ldots, x_{\omega_d} \rangle$ where each component $x_{\omega_i}$ represents the number of examples labelled $\omega_i$ for each label $1 \leq i \leq d$ in $d$-dimensional *coverage space* [6,7]. Measure functions $f$ can then be redefined in terms of functions in coverage space with $\sigma_f$ where:

$$\sigma_f(sp(C, \mathcal{E})) = f(C, \mathcal{E})$$

For example, consider the accuracy function as per Definition 2 where $\Omega = \{+, -\}$ which is mapped into coverage space as follows:

$$sp(C, \mathcal{E}) = \langle x, y \rangle \text{ where } x = |cover(C, \mathcal{E}) \cap \mathcal{E}^+|, \text{ and } y = |cover(C, \mathcal{E}) \cap \mathcal{E}^-|$$

where $x$ describes the number of examples labelled $+$ and $y$ describes the number of examples labelled $-$ in the cover of $C$. In terms of two-dimensional coverage space, the accuracy function $acc(C, \mathcal{E})$ where $P = TP + FN = |\mathcal{E}^+|$ and $N = FP + TN = |\mathcal{E}^-|$ becomes:

$$\sigma_{acc}(\langle x, y \rangle) = \frac{x + (N - y)}{x + y + (P - x) + (N - y)} = \frac{x + N - y}{P + N}$$

Consider the case where any concept $C$ is considered a solution to a learning problem by imposing a minimum threshold $\tau_{min}$ over accuracy as $\mathcal{Q}(C, \mathcal{E}) = acc(C, \mathcal{E}) \geq \tau_{min}$. By rearranging for $y$, the definition of accuracy in coverage space becomes $y \leq x - \tau_{min}(N + P) + N$. By this equation, concept $C$ meets the criteria to be considered a solution given the numbers $x, y$ of covered examples labelled $+, -$ when this equation is satisfied. In Fig. 1, we plot this as a function in coverage space for various values of minimum accuracy $\tau_{min}$, which each represent *isometric lines* passing through all points $\langle x, y \rangle$ for a fixed value of $\tau_{min}$. Here, the isometric curves are linear, which is the condition under which a function is both convex and concave.

From Fig. 1 we see the isometric lines for the accuracy function plotted for both $\tau_{min} = 0.75$ and $\tau_{min} = 0.95$, where those candidate concepts with covers $\langle x, y \rangle$ which lie in the area between each respective isometric line and the $x$-axis

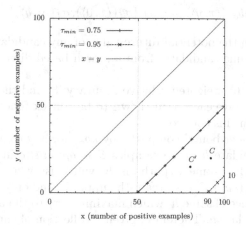

**Fig. 1.** A *coverage space* plot representing the number of positive examples labelled $+$ on the $x$-axis and negative examples labelled $-$ on the $y$-axis. Two *isometric lines* for accuracy threshold values $\tau_{min} \in \{0.75, 0.95\}$ for a problem with 100 positive and 100 negative examples are shown, along with the diagonal line at $x = y$. Also plotted are the *stamp points* of two hypotheses, $C$ at $\langle 92, 20 \rangle$ and $C'$ at $\langle 80, 15 \rangle$.

representing solutions which meet or exceed the minimum accuracy $\tau_{min}$. The point at which isometric lines in coverage space cross an axis such as where $y = 0$ provides us with *lower bounds* on the other variables, such as $x$. In this example, the isometric line for $\tau_{min} = 0.75$ crosses the $x$-axis at $x = 50$ when $y = 0$, therefore no candidate with a stamp point $\langle x, y \rangle$ can ever be a solution where $\sigma_{acc}(\langle x, y \rangle) \geq 0.75$ when $x < 50$. This analysis provides us with conditions with which we may use to prune candidates and all of their specialisations from a downward-refinement search, as concept covers $\langle x, y \rangle$ may only ever be reduced.

Given any candidate $h$ with stamp point $\langle x, y \rangle$ in coverage space, we can also define *upper bounds* on the potential future value of any downward refinement of $h$ by inspecting the value of the measure function $\sigma_f$ by setting one of the variables of the stamp point to 0. For example, consider the two candidates $C$ at stamp point $\langle 92, 20 \rangle$ and $C'$ at point $\langle 80, 15 \rangle$ from Fig. 1. The upper bounds on the potential future value of accuracy for any number of refinements of $C$ or $C'$ is found by assuming the refinements have covers which contain no negative examples, where $C : \sigma_{acc}(\langle 92, 0 \rangle) = 0.96$, and $C' : \sigma_{acc}(\langle 80, 0 \rangle) = 0.9$.

With respect to the minimum threshold $\tau_{min} = 0.95$, we find downward refinements of $C$ *might* reach a candidate with accuracy 0.96, so should be considered for future refinement. However, as $C'$ has an upper bound of 0.9, we may safely prune it from the search as $C'$ and all of its downward refinements $\rho^\downarrow(C')$ can never satisfy $\sigma_{acc} \geq 0.95$. The idea of using an upper bound on the future potential of convex measure functions $\sigma_f$ was first formalised as follows [8].

**Definition 6 (Upper Bounds on Convex Measure Functions).** *The upper bound on values of a convex measure function $\sigma_f$ with respect to a candidate hypothesis $h$ with stamp point $\langle x, y \rangle$ is given by the function $ub_{\sigma_f}$ as:*

$$ub_{\sigma_f}(\langle x, y \rangle) = max \{ \sigma_f(\langle x, 0 \rangle), \sigma_f(\langle 0, y \rangle) \}$$

Upper bounds on the potential future value of any candidate computed with $ub_{\sigma_f}$ permit us to prune candidates from a search based directly on a minimum threshold $\tau_{min}$ on $\sigma_f$.

The $\chi^2$ statistic [6], weighted relative accuracy [8] and the Matthews correlation coefficient [9] have each been shown to be convex functions and can also be effectively used in this context.

Note that the upper bound computed by $ub_{\sigma_f}$ for any convex measure function $f$ given a candidate $h$ over examples $\mathcal{E}$ is *optimistic* in the sense that it is not guaranteed that some hypothesis $h'$ will exist where $h' \in \rho^\downarrow(h)$ and $\sigma_f(sp(h', \mathcal{E})) = ub_{\sigma_f}(sp(h, \mathcal{E}))$. Instead, the upper bound only indicates the possibility of the existence of some $h'$ which maximises $\sigma_f$ to the upper bound value for the parent candidate $h$. Therefore, given a collection of candidates such as a beam set which is maintained in a beam search, we may order candidates based on their upper bounds in order to refine those which may lead to the strongest solutions. In this way, the upper bound of any candidate may be incorporated as a heuristic into a utility function $\mathbf{u}_{OM}$ to indicate the strength of potential solutions in the set of refinements of any candidate.

**Definition 7 (Utility Function $u_{OM}$).** *Given a concept d refined from c from the set of all concepts $\mathcal{L}$, a set of labelled examples $\mathcal{E}$, a convex measure function $\sigma_f$, the upper bound function $ub_{\sigma_f}$, the stamp point function sp, the length function to compute the number of symbols in any concept, and real-valued parameters $\alpha, \beta, \gamma \in [0,1]$, the utility function $u_{OM}(d, \mathcal{E})$ is defined as:*

$$u_{OM}(d, \mathcal{E}) = |\sigma_f(sp(d, \mathcal{E}))| + \gamma \cdot ub_{\sigma_f}(sp(d, \mathcal{E}))$$
$$+ \alpha \cdot (|\sigma_f(sp(d, \mathcal{E}))| - |\sigma_f(sp(c, \mathcal{E}))|) - \beta \cdot length(d)$$

The utility function $\mathbf{u}_{OM}$ of Definition 7 incorporates the following:

- The current performance of $d$ according to the convex measure function $\sigma_f$ in the range $[-1, 1]$ where $|\sigma_f|$ reflect weak solutions at 0 and the strongest solutions at 1;
- The optimistic upper bound on $\sigma_f$ for $d$;
- Any stepwise gain in the performance of $d$ relative to its parent concept, $c$, according to $\sigma_f$;
- The complexity of the expression $d$ as the number of symbols it contains.

Note that $\mathbf{u}_{OM}$ ranges over the set of reals, where larger values correspond to stronger candidates. This utility function captures gain in the measure function and penalises long concepts, but also incorporates the optimistic upper bound $ub_{\sigma_f}$ on $\sigma_f$ to capture the potential strength of candidates amongst the set of refinements. The user-defined parameter $\gamma$ controls the importance of the optimistic upper bound of future potential of a candidate $d$. By incorporating this upper bound as a heuristic in the utility function, we boost the utility of currently low-performing candidates with high future potential, without lowering the utility of currently high-performing candidates also with high future potential. Experimentally, we have typically used the settings $\alpha = 0.5, \beta = 0.02, \gamma = 0.2$.

In a learning algorithm, once any solution $C$ has been found which exceeds a minimum threshold $\tau_{min}$ for some measure function $\sigma_f$, we may opt to terminate the search, or continue to look for other concepts $C'$ which exceed the performance of the last found solution $C$. So-called *anytime* algorithms work in this way, where the last best solution to the problem is maintained and is potentially improved given more computation time. If a solution is found with value $\sigma_f = \tau$ where $\tau \geq \tau_{min}$ and computation is allowed to continue to search for better solutions, the minimum bound $\tau_{min}$ may be reset to $\tau$ which has the effect of permitting the algorithm to prune more concepts from the search space. For example, consider a search algorithm which finds a solution with measure value $\tau_{best}$. Then, any subsequent candidate $D$ under consideration which has an optimistic upper bound $t$ where $t \leq \tau_{best}$ may be pruned, as none of the specialisations in $\rho^{\downarrow}(D)$ can have a measure value which exceeds $\tau_{best}$. Similarly, if a learning algorithm is designed to locate the top-$k$ concepts for some fixed $k$, the threshold $\tau_{min}$ may simply be set to that of the weakest of the current set of $j$ solutions where $1 \leq j \leq k$.

## 4    A Top-$k$ Beam Search for Supervised Concept Learning

Now we improve the basic beam search to permit more control over the beam set $B$ and expansion set $E$ which supports *both* memory-bounded beam and *best-first* search. We introduce a maximum bound on the expansion set $E$ as $E_{max}$, and maintain it by only permitting the best $E_{max}$ candidates at any one time. When the algorithm has refined all candidates in the beam $B$ populating the expansion set $E$, we permit control over how any *remaining* candidates in $E$ are treated. In a beam search such as Algorithm 1, any remaining candidates in the expansion set $E$ are discarded once the beam $B$ is repopulated, as per line 4 of Algorithm 1. However, if we permit remaining candidates to reside in $E$, subsequent refinement of candidates of the beam $B$ into $E$ can be chosen on a best-first basis. This behaviour allows candidate concepts of varying lengths in the beam at any one time, and are selected based on the strength of their utility, which is a *best-first* search approach. We will permit the search to maintain up to $k \geq 1$ of the best solutions found at any point, and limit the time the algorithm may spend searching for solutions with a maximum computation time $t_{max}$ to support an *anytime* strategy. Candidates with the weakest utility $\mathbf{u}_{OM}$ may be pruned from $E$.

Algorithm 2 is our general-purpose anytime search algorithm for supervised classification, and takes the following:

- $\mathcal{E}$: A set of binary labelled examples as $\mathcal{E} = \bigcup_{\omega \in \Omega} \mathcal{E}^\omega$ for $\Omega = \{+, -\}$;
- $\rho^{\downarrow}$: Downward refinement operator defined against a concept language $\mathcal{L}$;
- $k$: The maximum number of concepts to find as solutions where $k \geq 1$;
- $\mathbf{u}_{OM}$: Our real-valued heuristic utility function;
- $sp$: The stamp point function mapping a concept $C$ to a stamp point $\langle x, y \rangle$ in coverage space relative to $\mathcal{E}$ and $\Omega$;
- $\sigma_f$: A convex measure function defined over coverage space;
- $ub$: The upper bound function defined over $\sigma_f$;
- $\tau_{min}$: A minimum threshold on $\sigma_f$ which signifies when candidates with stamp points $\langle x, y \rangle$ are solutions when $\sigma_f(\langle x, y \rangle) \geq \tau_{min}$;
- $t_{max}$: The maximum time for which the algorithm may execute;
- $B_{max}$: The maximum width of an *open/beam set* $B$ where $|B| \leq B_{max}$ of candidates to refine, so as to bound memory consumption;
- $E_{max}$: The maximum width of an *expanded/successor set* $E$ where $|E| \leq E_{max}$, also to bound memory consumption;
- *beam*: A boolean variable which, when *beam* = *true*, indicates that the search should use a *beam search* strategy similar to that of Algorithm 1 which rejects expanded candidates every time the beam is repopulated. Otherwise when *beam* = *false*, we revert to a *best-first search* strategy which constantly maintains at most $E_{max}$ best candidates.

The function REPOPULATEBEAM as used on line 28 of Algorithm 2 regenerates the beam set $B$ by selecting candidates from $E$. We simply select the best $B_{max}$ candidates from $E$ to repopulate $B$, where best is defined by the utility

---

**Algorithm 2.** Time- and Memory-Bounded Top-$k$ Concept Search Algorithm

1: **function** TOP-$k$-SEARCH($\mathcal{E}, k, \sigma_f, \mathbf{u}_{OM}, sp, ub, \tau_{min}, t_{max}, B_{max}, E_{max}, beam$)
2:     $S := \emptyset$                                                    ▷ Solution set
3:     $B := \{\top\}$                          ▷ Open/beam set starting with the top concept $\top$
4:     $E := \emptyset$                                      ▷ Expanded/successor set
5:     $\tau := \tau_{min}$                    ▷ Initialise user-defined minimum quality on $\sigma_f$
6:     $t := t_0$                                   ▷ Start computation timer at $t_0$
7:     **while** $(|B| > 0) \wedge (t < t_{max})$ **do**       ▷ Beam not empty and time not exceeded
8:         $h := h \in B$                               ▷ Arbitrary candidate hypothesis
9:         $B := B \setminus \{h\}$
10:        $C := \rho^{\downarrow}(h)$                  ▷ Generate all single-step specialisations of $h$
11:        $E := E \cup \{c \in C \mid \sigma_f(sp(c, \mathcal{E})) < \tau \wedge ub(sp(c, \mathcal{E})) \geq \tau\}$
12:        $S := S \cup \{c \in C \mid \sigma_f(sp(c, \mathcal{E})) \geq \tau\}$       ▷ Add any solutions to $S$
13:        $S := S \setminus \{s \in S \mid \exists s' \in S : s \neq s' \wedge s \sqsubseteq s' \wedge sp(s, \mathcal{E}) = sp(s', \mathcal{E})\}$
14:        **while** $|S| > k$ **do**
15:            $s \in \arg\min_{s \in S} \sigma_f(sp(s, \mathcal{E}))$       ▷ Remove arbitrary weakest solution
16:            $S := S \setminus \{s\}$
17:        **end while**
18:        **if** $|S| = k$ **then**
19:            $s \in \arg\min_{s \in S} \sigma_f(sp(s, \mathcal{E}))$
20:            $\tau := \sigma_f(sp(s, \mathcal{E}))$                  ▷ Update minimum quality threshold
21:        **end if**
22:        $E := \{e \in E \mid ub(sp(e, \mathcal{E})) \geq \tau\}$       ▷ Filter on minimum quality threshold
23:        **while** $|E| > E_{max}$ **do**
24:            $e \in \arg\min_{e \in F} ub(sp(e, \mathcal{E}))$             ▷ Arbitrary weakest candidate
25:            $E := E \setminus \{e\}$
26:        **end while**
27:        **if** $(|B| = 0) \wedge (|E| > 0)$ **then**
28:            $B := \text{REPOPULATEBEAM}(E, B_{max}, \mathbf{u}_{OM}, \mathcal{E})$
29:        **end if**
30:        **if** beam **then**
31:            $E := \emptyset$                  ▷ Reject remaining candidates (beam search strategy)
32:        **end if**
33:        $t := t + n$                  ▷ Increment timer with $n$ time units for this loop
34:    **end while**
35:    **return** $S$                                    ▷ Return up to $k$ top solutions
36: **end function**

---

function $\mathbf{u}_{OM}$ while other methods, such as stochastic beam search, can be used to increase diversity in the search to avoid getting trapped in local maxima.

Learning Algorithm 2 can be easily parallelised although our performance results presented here depend on a single-threaded implementation. We refer the reader to [10] for an overview of the refinement operator and method of coverage checking, noting that we also re-use the convex measure function $\sigma_f$ to define minimum bounds on concept performance to achieve fast-failure of coverage checking.

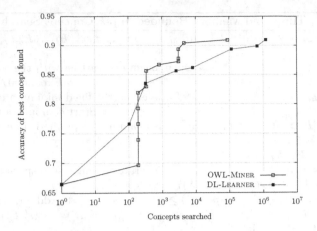

**Fig. 2.** The performance of OWL-MINER and DL-LEARNER over the mutagenesis dataset, plotting the number of concepts searched by each system versus the accuracy of the best performing candidate.

## 5  Experimental Evaluation

Mutagenesis is a well-known benchmark problem in machine learning [11]. A variety of techniques have been applied to the mutagenesis dataset to construct classification models, from ILP to kernel-based methods [12], however the application of DL learning to this problem has not been described before. The mutagenesis dataset contains examples of chemical compounds and their characteristics, such as the atomic structure including functional groups, and various real-valued measures such as a water/octanol partition coefficient, *log P*. The so-called 'regression friendly' dataset contains 188 example compounds, 125 of which are labelled positive for mutagenicity, and the remaining 63 are labelled negative. The OWL ontology for this problem[1] contains 88 classes, 11 properties and 14,145 individuals.

We applied both the OWL-MINER and DL-LEARNER v1.2 systems to this dataset, running each experiment in isolation on the same machine with 16 Gb RAM and one CPU thread for a maximum runtime of 15 min. The same hypothesis concept language was used in both systems. Algorithm 2 was set to locate one best solution ($k = 1$) with beam ($B_{max}$) and expansion set ($E_{max}$) sizes of 10,000 each. During experimentation, we began testing with a minimum threshold on accuracy of 99%, but only found solutions at around 90% accuracy, which we set as the minimum accuracy for the experiments described below. In order to compare the performance of OWL-MINER with DL-LEARNER, we observed the accuracy of newly discovered best-performing candidates and the number of concepts which had been tested up to that point of discovery, as shown in Fig. 2.

From Fig. 2, we see that OWL-MINER locates a concept with around 89% accuracy after searching through around 3,000 concepts, and eventually locates

[1] https://github.com/AKSW/DL-Learner/tree/develop/examples/mutagenesis.

a concept with around 91% accuracy after searching through around 90,000 concepts. DL-LEARNER locates a concept of 86% accuracy after searching through around 8,000 concepts, and eventually locates a concept with around 91% accuracy. DL-LEARNER searched through over 1.2 million concepts to achieve that 91% accuracy, more than 13 times the number of concepts OWL-MINER took to reach a similar result. OWL-MINER located its best concept at 90.96% accuracy after less than 30 s, as follows:

$$Compound \sqcap \geqslant^4 hasStructure.(\neg Methyl \sqcap \neg HeteroAromatic5Ring \sqcap$$
$$\neg HeteroAromatic6Ring \sqcap \neg Benzene) \sqcap \geqslant^4 hasAtom.(Hydrogen_3) \sqcap$$
$$\exists lumo.[\geq -3.768 \wedge \leq -1.102]$$

This concept can be read as: *mutagenic compounds are those with at least four structures which are not methyl, benzene or hetero-aromatic 5 or 6 rings, and which have at least four hydrogen-3 atoms, and which have a lumo value of between $-3.768$ and $-1.102$.* Similarly, the best concept produced by DL-LEARNER also with an accuracy of 90.96% was:

$$Compound \sqcap \exists hasAtom.(\exists charge.[\leq -0.368]) \sqcap$$
$$\geqslant^4 hasStructure.(\neg Benzene \sqcap \neg Methyl) \sqcap \exists logp.[\geq 1.91]$$

This concept can be read as: *mutagenic compounds are those with at least one atom with a charge of less than or equal to $-0.368$, and which has at least four structures which are not methyl or benzene, and which has a logp value of at least 1.91.* Both of these concepts are expressive and easily comprehensible, highlighting the applicability of DL learning as a suitable method for this problem.

To test if the best concepts generated by OWL-MINER were over-fitting the data, we computed the 10-fold cross validation accuracy and $F_1$ scores for various minimum accuracy thresholds $\tau_{min}$ as shown in Table 1.

A sample of previously reported best accuracies of a variety of methods which have been applied to the mutagenesis problem can be found in Table 2 from Lodhi and Muggleton [12], to which we have added results for OWL-MINER and also ALCHEMY, a higher-order logic learning system [13].

From Table 2 we see that the 10-fold cross validation accuracy for the mutagenesis problem ranges from around 85% to 95% accuracy, so the best accuracy produced in our experiments by OWL-MINER are comparable. We attribute the strong performance of OWL-MINER over DL-LEARNER to the efficiency

**Table 1.** 10-fold cross validation accuracy and $F_1$ scores for various minimum thresholds over accuracy for the mutagenesis dataset with the OWL-MINER system.

| $\tau_{min}$ | $Acc. \pm \sigma$ (%) | $F_1 \pm \sigma$ (%) | Runtime $\pm \sigma$ (s) | Length $\pm \sigma$ |
|---|---|---|---|---|
| 0.88 | $86.50 \pm 0.09$ | $89.34 \pm 0.08$ | $0.79 \pm 0.81$ | $11.4 \pm 0.97$ |
| 0.89 | $88.25 \pm 0.06$ | $91.20 \pm 0.04$ | $13.19 \pm 34.92$ | $12.5 \pm 2.01$ |
| 0.90 | $\mathbf{90.50 \pm 0.09}$ | $\mathbf{92.25 \pm 0.07}$ | $77.09 \pm 104.21$ | $15.7 \pm 1.25$ |

**Table 2.** Various accuracy results for the mutagenesis problem taken from [12], we refer the reader to this paper for citations. We include Higher-Order Learning (HOL) [13] and DL-Learning (DLL) here. We were unable to produce any 10-fold cross validation results for DL-LEARNER. The best reported result as highlighted in bold was achieved by an Aleph-based system [12].

| Type | System | Eval. Method | Accuracy (%) |
|---|---|---|---|
| Inductive Logic Programming (ILP) | P-Progol | 10-fold | $88.0 \pm 2.0$ |
| | FOIL | 10-fold | 86.7 |
| | STILL | One train/test: 90/10 | $93.6 \pm 4.0$ |
| Propositionalisation-based ILP | MFLOG | 10-fold | 95.7 |
| | RSD | 10-fold | 92.6 |
| | SINUS | 10-fold | 84.5 |
| | RELAGGS | 10-fold | 88.0 |
| Ensemble Methods in ILP | Aleph+RS | 10-fold | $\textbf{95.8} \pm \textbf{3.3}$ |
| | Boosted FOIL | 10-fold | 88.3 |
| Kernels | MIK | N/A | 93.0 |
| | RK | 10-fold | 85.4 |
| | $GK^3$ | leave-one-out | 96.1 |
| Naive Bayes + ILP | nFOIL | 10-fold | $78.3 \pm 12.0$ |
| | Aleph+NB | 10-fold | $72.8 \pm 11.7$ |
| Others | Neural Networks | 10-fold | $89.0 \pm 2.0$ |
| | Linear Regression | 10-fold | $89.0 \pm 2.0$ |
| | CART | 10-fold | $88.0 \pm 2.0$ |
| Higher-Order Learning | ALCHEMY | 10-fold | 89.39 |
| DL-Learning | OWL-MINER | 10-fold | $90.50 \pm 0.09$ |

achieved by restricting the set of concepts available to the refinement operator which are suitable for testing as search candidates, together with the method of early pruning based on upper bound estimation, and the simultaneous learning of lower and upper bounds on datatype property restrictions with numerical ranges. These fundamentally different approaches are the main advantages the OWL-MINER system has over the DL-LEARNER system which enable it to locate high performing concepts efficiently.

## 6   Related Work

Motivated by earlier techniques in ILP such as [14], *refinement operators* for traversing the hypothesis search space have been researched for a number of DLs including $\mathcal{ALER}$ [15], $\mathcal{EL}$ [16] and $\mathcal{SROIQ}(D)$ [4]. Systems which implement induction with refinement operators include DL-LEARNER [17] and DL-FOIL [18], the latter of which implements a covering approach for $\mathcal{ALC}$ based on the well-known FOIL algorithm [19]. Other systems include YINYANG [20] which combines top-down and bottom-up refinement search for $\mathcal{ALC}$, and FRONT [21] uses top-down refinement for $\mathcal{EL}$ for discovering frequent patterns.

The DL-LEARNER system [17] is specifically notable as many of our methods and implementation OWL-MINER were designed around improvements to DL-LEARNER. Our novel utility function $u_{OM}$ incorporates aspects of the OCEL and CELOE heuristics used in DL-LEARNER which capture gain in the measure function and penalise long concepts [4], but ours also incorporates the optimistic upper bound $ub_{\sigma_f}$ on convex $\sigma_f$ to capture the potential strength of candidates.

Our major contribution in this paper is the development of Algorithm 2, an anytime best-first search algorithm, which develops the basic beam search Algorithm 1 with ideas from the CG algorithm [8]. Our algorithm reflects the CG algorithm in that it dynamically increases the minimum bound $\tau$ on a convex measure function $\sigma_f$ if any solutions are found such that candidates which have an upper bound $ub_{\sigma_f} < \tau$ may be pruned from the search space, as they can never be stronger than any of the current solutions. Additionally, any solution $s$ which is subsumed by some pre-existing solution $s'$ where $s \sqsubseteq s'$ which shares the same stamp point $\langle x, y \rangle$ will be excluded as $s$ is unlikely to provide any additional information about the cover. The main difference between our Algorithm 2 and the CG algorithm is that the former permits bounds on the size of the set of candidates currently under consideration, as the set of candidates which have upper bounds on $\sigma_f$ at any one time may be impractically large to store in memory. This is why the expansion set $E$ is also limited with maximum size $E_{max}$, and when $|E| > E_{max}$, candidates with the worst upper bounds are pruned, as per line 24.

In limited space we have evaluated our work on the long-standing relational mutagenesis problem and present comparative results among 18 other systems using a diverse range of learning methods [12].

# 7    Conclusion

Driven by analysis of heuristics for beam search methods over OWL concepts, we derive a novel search algorithm which leverages the properties of convex measure functions to deliver a memory-bounded best-first beam search. We present performance results that demonstrate how our heuristics enable our system OWL-MINER to discover hypotheses more efficiently than the state-of-the-art system DL-LEARNER, while producing comparable 10-fold cross-validated accuracy when compared with a diverse range of methods. Our improvements to the space and time-based efficiency of refinement-based learning algorithms are significant for expanding the size of learning problems that can be feasibly addressed by refinement learning and also the quality of solutions that can be found with limited resources. In future work we will expand the scope of OWL-MINER to address learning problems beyond binary classification as discussed here, and present in detail our methods for fast evaluation of hypothesis cover.

# References

1. d'Amato, C., Fanizzi, N., Esposito, F.: Inductive learning for the Semantic Web: what does it buy? Semant. Web **1**(1, 2), 53–59 (2010)
2. Baader, F., Calvanese, D., McGuinness, D.L., Nardi, D., Patel-Schneider, P.F. (eds.): The Description Logic Handbook: Theory, Implementation, and Applications. Cambridge University Press, New York (2003)
3. Horrocks, I., Kutz, O., Sattler, U.: The even more irresistible SROIQ. In: Proceedings of the 10th International Conference on Principles of Knowledge Representation and Reasoning KR 2006, pp. 1–36 (2006)
4. Lehmann, J., Hitzler, P.: Concept learning in description logics using refinement operators. Mach. Learn. **78**(1–2), 203–250 (2009)
5. Rudin, W.: Principles of Mathematical Analysis. International Series in Pure and Applied Mathematics. McGraw-Hill, New York (1976). Includes index. Bibliography: pp. 335–336
6. Morishita, S., Sese, J.: Transversing itemset lattices with statistical metric pruning. In: Proceedings of the Nineteenth ACM SIGMOD-SIGACT-SIGART Symposium on Principles of Database Systems, PODS 2000, pp. 226–236. ACM, New York (2000)
7. Fürnkranz, J., Flach, P.A.: ROC 'n' rule learning-towards a better understanding of covering algorithms. Mach. Learn. **58**(1), 39–77 (2005)
8. Zimmermann, A., Raedt, L.D.: Cluster-grouping: from subgroup discovery to clustering. Mach. Learn. **77**(1), 125–159 (2009)
9. Powers, D.M.W.: Evaluation: from precision, recall and F-factor to ROC, informedness, markedness and correlation. Number SIE-07-001. Flinders University, Adelaide, Australia
10. Ratcliffe, D., Taylor, K.: Closed-world concept induction for learning in OWL knowledge bases. In: Janowicz, K., Schlobach, S., Lambrix, P., Hyvönen, E. (eds.) EKAW 2014. LNCS (LNAI), vol. 8876, pp. 429–440. Springer, Cham (2014). https://doi.org/10.1007/978-3-319-13704-9_33
11. Debnath, A.K., Lopez de Compadre, R.L., Debnath, G., Shusterman, A.J., Hansch, C.: Structure-activity relationship of mutagenic aromatic and heteroaromatic nitro compounds. Correlation with molecular orbital energies and hydrophobicity. J. Med. Chem. **34**(2), 786–797 (1991)
12. Lodhi, H., Muggleton, S.: Is mutagenesis still challenging? In: Proceedings of the 15th International Conference on Inductive Logic Programming, ILP 2005, Late-Breaking Papers, pp. 35–40 (2005)
13. Ng, K.S.: Learning comprehensible theories from structured data. Ph.D. thesis, Research School of Information Sciences and Engineering and The Australian National University, October 2005
14. Taylor, K.: Generalization by absorption of definite clauses. J. Logic Program. **40**(2–3), 127–157 (1999)
15. Badea, L., Nienhuys-Cheng, S.-H.: A refinement operator for description logics. In: Cussens, J., Frisch, A. (eds.) ILP 2000. LNCS (LNAI), vol. 1866, pp. 40–59. Springer, Heidelberg (2000). https://doi.org/10.1007/3-540-44960-4_3. ACM ID: 742934
16. Lehmann, J., Haase, C.: Ideal downward refinement in the $\mathcal{EL}$ description logic. In: De Raedt, L. (ed.) ILP 2009. LNCS (LNAI), vol. 5989, pp. 73–87. Springer, Heidelberg (2010). https://doi.org/10.1007/978-3-642-13840-9_8

17. Lehmann, J.: DL-Learner: learning concepts in description logics. J. Mach. Learn. Res. (JMLR) **10**, 2639–2642 (2009)
18. Fanizzi, N., d'Amato, C., Esposito, F.: DL-FOIL concept learning in description logics. In: Železný, F., Lavrač, N. (eds.) ILP 2008. LNCS (LNAI), vol. 5194, pp. 107–121. Springer, Heidelberg (2008). https://doi.org/10.1007/978-3-540-85928-4_12
19. Quinlan, J.R.: Learning logical definitions from relations. Mach. Learn. **5**, 239–266 (1990)
20. Iannone, L., Palmisano, I., Fanizzi, N.: An algorithm based on counterfactuals for concept learning in the Semantic Web. Appl. Intell. **26**(2), 139–159 (2007)
21. Ławrynowicz, A., Potoniec, J.: Fr-ONT: an algorithm for frequent concept mining with formal ontologies. In: Kryszkiewicz, M., Rybinski, H., Skowron, A., Raś, Z.W. (eds.) ISMIS 2011. LNCS (LNAI), vol. 6804, pp. 428–437. Springer, Heidelberg (2011). https://doi.org/10.1007/978-3-642-21916-0_46

# Ontology Reasoning

# Reasoning on Context-Dependent Domain Models

Stephan Böhme[1]([⊠]) and Thomas Kühn[2]([⊠])

[1] Chair of Automata Theory, TU Dresden, Dresden, Germany
stephan.boehme@posteo.de
[2] Software Technology Group, TU Dresden, Dresden, Germany
thomas.kuehn3@tu-dresden.de

**Abstract.** Modelling context-dependent domains is hard, as capturing multiple context-dependent concepts and constraints easily leads to inconsistent models or unintended restrictions. However, current semantic technologies not yet support reasoning on context-dependent domains. To remedy this, we introduced ConDL, a set of novel description logics tailored to reason on contextual knowledge, as well as JConHT, a dedicated reasoner for ConDL ontologies. ConDL enables reasoning on the consistency and satisfiability of context-dependent domain models, e.g., Compartment Role Object Models (CROM). We evaluate the suitability and efficiency of our approach by reasoning on a modelled banking application and measuring the performance on randomly generated models.

## 1 Introduction

Modelling current information systems is hard, as they are characterised by increased context-dependence. They not only require domain analysts to capture multiple context-dependent concepts, but also to specify the particular constraints and requirements found in each context. The latter, however, can easily lead to an inconsistent model or unintended restrictions. Thus, it becomes imperative for domain analysts to reason on context-dependent domain models to uncover implicit or unsatisfiable specifications. While reasoning on classical domain modelling languages, e.g., ER [9] and UML [21], is possible, as [1,4,25] have shown, both lack the formal semantics required to make them suitable for formal reasoning. More importantly, classical domain modelling languages are unable to capture context-dependent concepts and constraints, hence, researchers have focused on more advanced modelling languages, e.g., [10,12–14,18] (see [16] for detailed surveys). We focus on the *Compartment Role Object Model* (CROM) [17] that directly supports the formal specification of context-dependent domains. Although CROM provides a formal model, due to a lack of tool support this model is not amenable for reasoning. Thus, we aim at transforming a CROM to a viable logical formalism. *Description Logics* (DLs) are a well-known family of knowledge representation formalisms that have a formal

S. Böhme and T. Kühn—Supported by the DFG in the RTG 1907 (RoSI).

Z. Wang et al. (Eds.): JIST 2017, LNCS 10675, pp. 69–85, 2017.
https://doi.org/10.1007/978-3-319-70682-5_5

semantics and allow for defining a variety of reasoning services. DLs can model application domains in a well-structured way. Yet, classical DLs lack expressive means to formalise context-dependent domains, i.e., express context dependent knowledge. To overcome their deficits, *Contextualised DLs* (ConDLs) were introduced in [6], a set of novel description logics especially tailored for reasoning on contextual knowledge. This paper utilizes ConDL to reason on CROMs. In particular, we describe a mapping from CROM to ConDL that preserves the semantics of the modelled domain. Moreover, we introduce the first reasoner dedicated to contextualised DL ontologies, JConHT. JConHT can check the consistency and satisfiability of context-dependent domain models. As a result, we show that ConDL is not only suitable to encode CROM domain models, but also allows for efficient checking of inconsistencies and unintended restrictions. We demonstrate this by reasoning on a modelled banking application and measuring the reasoner's performance on randomly generated models.

The paper is structured accordingly. Section 2 introduces a running example and formal definition of both CROM and ConDL. Afterwards, Sect. 3 presents our approach to reasoning on CROMs utilizing ConDL. Accordingly, Sect. 4 introduces JConHT, the first reasoner for ConDLs. To evaluate our approach, Sect. 5 showcases its suitability and measures its performance. In conclusion, Sect. 6 discusses related approaches and Sect. 7 summarises our results.

## 2     Basic Notions

**Running Example.** Before diving into the definitions of CROM and ConDL, we model a small banking application, extracted from [17]. Figure 1 depicts an example model of a *Bank* that employs at least one *Consultant* and provides banking services to *Customers*, who own *CheckingAccounts* and *SavingsAccounts*. They can *issue* money transferals (denoted *MoneyTransfer*), such that each transferal belongs to one customer and a customer issued an initial transferal. *Consultants advise* one or more customers. However, the advises relationship is constrained to be *irreflexive*, to prohibit self advising consultants. Besides that, *Transactions* are specified to orchestrate the transfer of money between exactly two *Accounts* by means of the roles *Source* and *Target*, such that there is a unique *Target* counterpart for each *Source*. This is ensured by the one-to-one cardinality of the *trans* relation. Additionally, the *Participants* role group with 1..1 cardinality enforces that one account cannot be *Source* and *Target* in the same *Transaction*. Finally, *Persons* can play the roles *Consultant* and *Customer*; *Companies* only *Customer*. Similarly, *Accounts* either the role of *CheckingAccount* or *SavingsAccount* in the context of a bank, as well as *Source* and *Target* in the context of transactions. Henceforth, this domain model will serve as our running example.

**CROM in a Nutshell.** The *Compartment Role Object Model* (CROM) was introduced in [17] to model dynamic, context-dependent domains. It introduces *compartment types* to represent a reified context, i.e., containing *role types* and *relationship types*. *Natural types*, in turn, fulfil role types in multiple compartment types.[1]

---

[1] A detailed ontological foundation of these kinds is provided in [17].

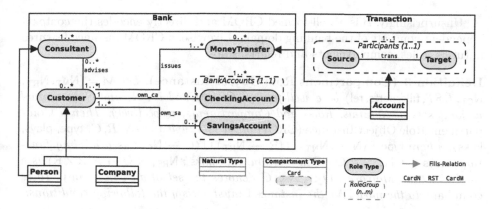

**Fig. 1.** Bank example

**Definition 1 (Compartment Role Object Model).** *Let* $N_{NT}$, $N_{RT}$, $N_{CT}$, *and* $N_{RST}$ *be mutual disjoint sets of Natural Types, Role Types, Compartment Types, and Relationship Types, respectively. Then,* $\mathcal{M} = (N_{NT}, N_{RT}, N_{CT}, N_{RST},$ fills, parts, rel) *is a* Compartment Role Object Model (CROM), *where* fills $\subseteq$ $(N_{NT} \cup N_{CT}) \times N_{RT}$ *is a relation,* parts $: N_{CT} \to \mathcal{P}N_{RT}$ *and* rel $: N_{RST} \to (N_{RT} \times N_{RT})$ *are total functions. A CROM is* well-formed *if it holds that:*

$$\forall RT \in N_{RT} \ \exists T \in (N_{NT} \cup N_{CT}) : (T, RT) \in \text{fills} \tag{1}$$

$$\forall CT \in N_{CT} : \text{parts}(CT) \neq \emptyset \tag{2}$$

$$\forall RT \in N_{RT} \ \exists! CT \in N_{CT} : RT \in \text{parts}(CT) \tag{3}$$

$$\forall RST \in N_{RST} : \text{rel}(RST) = (RT_1, RT_2) \land RT_1 \neq RT_2 \tag{4}$$

$$\forall RST \in N_{RST} \ \exists CT \in N_{CT} : \text{rel}(RST) = (RT_1, RT_2) \land RT_1, RT_2 \in \text{parts}(CT) \tag{5}$$

In detail, fills denotes that rigid types can play roles of a certain role type, parts is a partition of the set of role types wrt. the compartment type they participate in, and rel captures the two role types at the respective ends of each relationship type. The well-formedness rules ensure that the fills relation is surjective (1); each compartment type has a nonempty, disjoint set of role types as its parts (2, 3); and rel maps each relationship type to exactly two distinct role types of the same compartment type (4, 5). Accordingly, a CROM can be constructed for the banking application, depicted in Fig. 1.

*Example 2 (Compartment Role Object Model).* Let $\mathcal{B} = (N_{NT}, N_{RT}, N_{CT}, N_{RST},$ fills, parts, rel) be the model of the bank, where the components are defined as:[2]

$N_{NT} := \{\text{Person}, \text{Company}, \text{Account}\}$  $N_{RT} := \{\text{Customer}, \text{CA}, \text{SA}, \text{Source}, \dots\}$

$N_{CT} := \{\text{Bank}, \text{Transaction}\}$   $N_{RST} := \{\text{own\_ca}, \text{own\_sa}, \text{advises}, \text{issues}, \text{trans}\}$

fills $:= \{(\text{Person}, \text{Customer}), (\text{Account}, \text{Source}), (\text{Transaction}, \text{MoneyTransfer}), \dots\}$

parts $:= \{\text{Bank} \to \{\text{Consultant}, \text{Customer}, \text{CA}, \text{SA}, \text{MoneyTransfer}\}, \dots\}$

rel $:= \{\text{trans} \to (\text{Source}, \text{Target}), \text{own\_ca} \to (\text{Customer}, \text{CA}), \dots\}$

---

[2] SA and CA are abbreviations for *SavingsAccount* and *CheckingAccount*, respectively.

Unsurprisingly, $\mathcal{B}$ is a well-formed CROM and directly encodes the context-dependent concepts of the banking domain. Likewise, a CROM instance features naturals, roles, compartments and relationships.

**Definition 3 (Compartment Role Object Instance).** *Let* $\mathcal{M} = (\mathsf{N_{NT}}, \mathsf{N_{RT}},$ $\mathsf{N_{CT}}, RST, \mathsf{fills}, \mathsf{parts}, \mathsf{rel})$ *be a well-formed CROM and* $N$, $R$, *and* $C$ *be mutual disjoint sets of Naturals, Roles and Compartments, respectively. Then a Compartment Role Object Instance (CROI) of* $\mathcal{M}$ *is a tuple* $\mathfrak{i} = (N, R, C, \mathsf{type}, \mathsf{plays},$ $\mathsf{links})$, *where* $\mathsf{type} : (N \to \mathsf{N_{NT}}) \sqcup (R \to \mathsf{N_{RT}}) \cup (C \to \mathsf{N_{CT}})$ *is a labeling function,* $\mathsf{plays} \subseteq (N \cup C) \times C \times R$ *a relation, and* $\mathsf{links} : \mathsf{N_{RST}} \times C \to \mathcal{P}(R \times R)$ *is a total function. Moreover,* $O := N \cup C$ *denotes the set of all objects in* $\mathfrak{i}$. *To be compliant to the model* $\mathcal{M}$ *the instance* $\mathfrak{i}$ *must satisfy the following conditions:*

$$\forall (o, c, r) \in \mathsf{plays} : (\mathsf{type}(o), \mathsf{type}(r)) \in \mathsf{fills} \wedge \mathsf{type}(r) \in \mathsf{parts}(\mathsf{type}(c)) \tag{6}$$

$$\forall (o, c, r), (o, c, r') \in \mathsf{plays} : r \neq r' \Rightarrow \mathsf{type}(r) \neq \mathsf{type}(r') \tag{7}$$

$$\forall r \in R \; \exists! o \in O \; \exists! c \in C : (o, c, r) \in \mathsf{plays} \tag{8}$$

$$\forall RST \in \mathsf{N_{RST}} \; \forall c \in C \; \forall (r_1, r_2) \in \mathsf{links}(RST, c) : (\_, c, r_1), (\_, c, r_2) \in \mathsf{plays} \tag{9}$$

$$\forall RST \in \mathsf{N_{RST}} \; \forall c \in C \; \forall (r_1, r_2) \in \mathsf{links}(RST, c) : \mathsf{rel}(RST) = (\mathsf{type}(r_1), \mathsf{type}(r_2)) \tag{10}$$

The $\mathsf{type}$ function assigns a distinct type to each instance, $\mathsf{plays}$ identifies the objects (either natural or compartment) playing a certain role in a specific compartment, and $\mathsf{links}$ captures the roles currently linked by a relationship type in a certain compartment. A compliant CROI guarantees the consistency of both the $\mathsf{plays}$ relation and the $\mathsf{links}$ function with the model $\mathcal{M}$.[3] Axiom (6), (7) and (8) restrict the $\mathsf{plays}$ relation, such that it is consistent to the types defined in the $\mathsf{fills}$ relation and the $\mathsf{parts}$ function, an object is prohibited to play instances of the same role type multiple times in the same compartment, and each role has one distinct player in one distinct compartment, respectively. In contrast, Axiom (9) and (10) ensure that the $\mathsf{links}$ function only contains those roles, which participate in the same compartment $c$ as the relationship and whose types are consistent to the relationship's definition in the $\mathsf{rel}$ function.

Admittedly, neither Definitions 1 nor 3 captures the context-dependent constraints, showcased in Fig. 1. Hence, we introduce three context-dependent constraints, i.e., *role groups*, *occurrence constraints* and *relationship cardinalities*.

**Definition 4 (Syntax of Role Groups).** *The* set of Role Groups $\mathbb{RG}$ *is the smallest set, such that (i) every role type* $RT \in \mathsf{N_{RT}}$ *is a role group, and (ii) if* $B$ *is a role group and* $m..n \in \mathsf{Card}$, *then* $(B, n..m)$ *is also a role group, where* $\mathsf{Card} \subset \mathbb{N} \times (\mathbb{N} \cup \{\infty\})$ *with* $i \leq j$ *(elements are written as* $i..j$). *A role group contained in another one is denoted* nested.

**Definition 5 (Semantics of Role Groups).** *Let* $\mathfrak{i} = (N, R, C, \mathsf{type}, \mathsf{plays},$ $\mathsf{links})$ *a CROI compliant to* $\mathcal{M}$, $c \in C$ *a compartment, and* $o \in O$ *an object. The semantics is defined by the evaluation function* $(\cdot)^{\mathcal{I}_o^c} : \mathbb{RG} \to \{0, 1\}$: $a^{\mathcal{I}_o^c} = 1$ *iff* $a \in \mathsf{N_{RT}} \wedge \exists (o, c, r) \in \mathsf{plays}: \mathsf{type}(r) = a$ *or* $a \equiv (B, n..m) \wedge n \leq \sum_{b \in B} b^{\mathcal{I}_o^c} \leq m$.

---

[3] In contrast to [17], our definition excludes empty counter roles $\varepsilon$.

Role groups constrain the set of roles an object $o$ is allowed to play simultaneously in a certain compartment $c$. In case $a$ is a role type, $rt^{\mathcal{I}^c_o}$ checks whether $o$ plays a role of type $rt$ in $c$. If $a$ is a role group $(B, n..m)$, it checks whether the sum of the evaluations for all $b \in B$ is between $n$ and $m$. Accordingly, the following role groups directly correspond to their graphical representation in Fig. 1:

$$\text{BankAccounts} := (\{\text{CA}, \text{SA}\}, 1..1) \quad \text{Participants} := (\{\text{Source}, \text{Target}\}, 1..1)$$

Next, the *Constraint Model* is defined to collect all constraints imposed on a particular CROM $\mathcal{M}$.

**Definition 6 (Constraint Model).** *Let* $\mathcal{M} = (\mathsf{N_{NT}}, \mathsf{N_{RT}}, \mathsf{N_{CT}}, \mathsf{N_{RST}}, \mathsf{fills},$ $\mathsf{parts}, \mathsf{rel})$ *be a well-formed CROM. Then* $\mathcal{C} = (\mathsf{rolec}, \mathsf{card})$ *is a* Constraint Model *over* $\mathcal{M}$, *where* $\mathsf{rolec} : \mathsf{N_{CT}} \to \mathcal{P}\,(\text{Card} \times \mathbb{RG})$ *and* $\mathsf{card} : \mathsf{N_{RST}} \to (\text{Card} \times \text{Card})$ *are total functions.*

In detail, rolec collects the set of root role groups for each compartment type combined with a cardinality limiting the occurrence of role groups in each compartment. Moreover, card assigns a cardinality to each relationship type. Notably, all these constraints are defined context-dependent, i.e., no constraint crosses the boundary of a compartment type. Similar to the CROM $\mathcal{B}$, the corresponding constraint model is easily derived, from Fig. 1:

*Example 7 (Constraint Model).* Let $\mathcal{B}$ be the bank model from Example 2. Then $\mathcal{C}_{\mathcal{B}} = (\mathsf{rolec}, \mathsf{card})$ is the constraint model with the following components:

$$\mathsf{rolec} := \{\text{Bank} \to \{(1..\infty, \text{Consultant}), (1..\infty, \text{BankAccounts})\},$$
$$\text{Transaction} \to \{(1..1, \text{Participants})\}\}$$
$$\mathsf{card} := \{\text{own\_ca} \to (1..1, 0..\infty), \text{own\_sa} \to (1..\infty, 0..\infty), \text{issues} \to (1..1, 1..\infty),$$
$$\text{advises} \to (0..\infty, 1..\infty), \text{trans} \to (1..1, 1..1)\}$$

Finally, the validity of a given CROI is defined wrt. a constraint model.

**Definition 8 (Validity).** *Let* $\mathcal{M} = (\mathsf{N_{NT}}, \mathsf{N_{RT}}, \mathsf{N_{CT}}, \mathsf{N_{RST}}, \mathsf{fills}, \mathsf{parts}, \mathsf{rel})$ *be a well-formed CROM,* $\mathcal{C} = (\mathsf{rolec}, \mathsf{card})$ *a constraint model on* $\mathcal{M}$, *and* $\mathbf{i} = (N, R, C,$ $\mathsf{type}, \mathsf{plays}, \mathsf{links})$ *a CROI compliant to* $\mathcal{M}$. *Then* $\mathbf{i}$ *is valid with respect to* $\mathcal{C}$ *iff the following conditions hold:*

$$\forall c \in C \; \forall (i..j, a) \in \mathsf{rolec}(\mathsf{type}(CT)) : i \leq \left( \sum_{o \in O^c} a^{\mathcal{I}^c_o} \right) \leq j \tag{11}$$

$$\forall (o, c, r) \in \mathsf{plays} \; \forall (\_, a) \in \mathsf{rolec}(\mathsf{type}(c)) : \mathsf{type}(r) \in \mathsf{atoms}(a) \Rightarrow a^{\mathcal{I}^c_o} = 1 \tag{12}$$

$$\forall c \in C \; \forall RST \in \mathsf{N_{RST}} : \mathsf{rel}(RST) = (RT_1, RT_2) \wedge \mathsf{card}(RST) = (i..j, k..l) \wedge$$
$$(\forall r_2 \in R^c_{RT_2} : i \leq |\mathsf{pred}(RST, c, r_2)| \leq j) \wedge \tag{13}$$
$$(\forall r_1 \in R^c_{RT_1} : k \leq |\mathsf{succ}(RST, c, r_1)| \leq l)$$

*Here,* $\mathsf{atoms} : \mathbb{RG} \to \mathcal{P}(\mathsf{N_{RT}})$ *recursively computes all role types within a given role group. Moreover,* $R^c_{RT} := \{r \in R \mid (o, c, r) \in \mathsf{plays} \wedge \mathsf{type}(r) = RT\}$ *denotes the set of roles of type* $RT$ *played in a compartment* $c$. *Furthermore,* $\mathsf{pred}(RST, c, r) := \{r' \mid (r', r) \in \mathsf{links}(RST, c)\}$ *and* $\mathsf{succ}(RST, c, r) := \{r' \mid (r, r') \in \mathsf{links}(RST, c)\}$ *collects all predecessors respectively successors of a given role* $r$ *w.r.t. a given RST.*

**Table 1.** Syntax and Semantics of $\mathcal{SHOIQ}[\mathcal{SHOIQ}]$

| | syntax | semantics | |
|---|---|---|---|
| inverse object property | $R^-$ | $\{(e,d) \in \Delta \times \Delta \mid (d,e) \in R^{\mathcal{I}_c}\}$ | |
| object negation | $\neg C$ | $\Delta \setminus C^{\mathcal{I}_c}$ | |
| object conjunction | $C \sqcap D$ | $C^{\mathcal{I}_c} \cap D^{\mathcal{I}_c}$ | |
| obj. existential restriction | $\exists R.C$ | $\{d \in \Delta \mid$ there is some $e \in C^{\mathcal{I}_c}$ with $(d,e) \in R^{\mathcal{I}_c}\}$ | |
| object nominal | $\{a\}$ | $\{a^{\mathcal{I}_c}\}$ | |
| object at-most restriction | $\leqslant_n S.C$ | $\{d \in \Delta \mid \sharp\{e \in C^{\mathcal{I}_c} \mid (d,e) \in S^{\mathcal{I}_c}\} \leq n\}$ | |
| object concept inclusion | $C \sqsubseteq D$ | $C^{\mathcal{I}_c} \subseteq D^{\mathcal{I}_c}$ | $\left.\vphantom{\begin{array}{c}a\\a\\a\\a\end{array}}\right\}\alpha$ |
| object concept assertion | $C(a)$ | $a^{\mathcal{I}_c} \in C^{\mathcal{I}_c}$ | |
| object property assertion | $R(a,b)$ | $(a^{\mathcal{I}_c}, b^{\mathcal{I}_c}) \in R^{\mathcal{I}_c}$ | |
| object property inclusion | $R \sqsubseteq S$ | $R^{\mathcal{I}_c} \subseteq S^{\mathcal{I}_c}$ | $\left.\vphantom{\begin{array}{c}a\\a\end{array}}\right\}\mathcal{R}_\mathsf{O}$ |
| object transitivity axiom | $\mathsf{Trans}(R)$ | $R^{\mathcal{I}_c}$ is transitive. | |
| inverse meta property | $P^-$ | $\{(e,d) \in \mathbb{C} \times \mathbb{C} \mid (d,e) \in P^{\mathcal{J}}\}$ | |
| meta negation | $\neg E$ | $\mathbb{C} \setminus E^{\mathcal{J}}$ | |
| meta conjunction | $E \sqcap F$ | $E^{\mathcal{J}} \cap F^{\mathcal{J}}$ | |
| meta existential restriction | $\exists P.E$ | $\{d \in \mathbb{C} \mid$ there is some $e \in E^{\mathcal{J}}$ with $(d,e) \in P^{\mathcal{J}}\}$ | |
| meta nominal | $\{u\}$ | $\{u^{\mathcal{J}}\}$ | |
| meta at-most restriction | $\leqslant_n Q.E$ | $\{d \in \mathbb{C} \mid \sharp\{e \in E^{\mathcal{J}} \mid (d,e) \in Q^{\mathcal{J}}\} \leq n\}$ | |
| referring concept | $[\![\alpha]\!]$ | $\{d \in \mathbb{C} \mid \mathcal{I}_d \models \alpha\}$ | |
| meta concept inclusion | $E \sqsubseteq F$ | $E^{\mathcal{J}} \subseteq F^{\mathcal{J}}$ | $\left.\vphantom{\begin{array}{c}a\\a\\a\end{array}}\right\}\mathcal{B}$ |
| meta concept assertion | $E(u)$ | $u^{\mathcal{J}} \in E^{\mathcal{J}}$ | |
| meta property assertion | $T(u,v)$ | $(u^{\mathcal{J}}, v^{\mathcal{J}}) \in T^{\mathcal{J}}$ | |
| meta property inclusion | $R \sqsubseteq S$ | $R^{\mathcal{I}_c} \subseteq S^{\mathcal{I}_c}$ | $\left.\vphantom{\begin{array}{c}a\\a\end{array}}\right\}\mathcal{R}_\mathsf{M}$ |
| meta transitivity axiom | $\mathsf{Trans}(R)$ | $R^{\mathcal{I}_c}$ is transitive. | |

Each axiom verifies a particular set of constraints. Axiom (11) and (12) validate the occurrence and fulfilment of role groups, respectively. In essence, only those objects (naturals or compartments) are checked that play a corresponding role in the constrained compartment, and there are enough of such objects in that compartment. In contrast, (13) checks whether relationships respect the imposed cardinality constraints. In conclusion, the formal model easily captures the context-dependent concepts and constraints. Moreover, it allows for checking the well-formedness of CROMs and validity of CROIs. Yet, due to a lack of tool support, this formal model is not viable for verifying the consistency of a constrained CROM, thus, requiring a more suitable formalism for reasoning.

**Contextualised Description Logics.** The ConDLs we use in this paper were first introduced in [6]. We shortly recall the relevant definitions and refer the reader to [2] for a thorough introduction to DLs. ConDLs consist of two levels. On the meta level knowledge *about* contexts can be represented, e.g. their relation to each other, on object level knowledge *within* contexts can be stated.

**Definition 9 (Syntax of $\mathcal{SHOIQ}[\mathcal{SHOIQ}]$).** *Let* $O_C, O_P, O_I, M_C, M_P, M_I,$ *be non-empty, pairwise disjoint sets of* concept names, property names *and* individual names *of the* object level *and the* meta level, *respectively.*

*An* object property[4] *is either some* $R \in \mathsf{O_P}$ *or an* inverse object property $R^-$ *for* $R \in \mathsf{O_P}$. *An object RBox* $\mathcal{R_O}$ *is a finite set of property inclusion axioms* $R \sqsubseteq S$ *and transitivity axioms* $\mathsf{Trans}(R)$, *where R and S are object properties. For* $R \in \mathsf{O_P}$, *we define* $\mathsf{Inv}(R) := R^-$ *and* $\mathsf{Inv}(R^-) := R$, *and assume that* $R \sqsubseteq S \in$ $\mathcal{R_O}$ *iff* $\mathsf{Inv}(R) \sqsubseteq \mathsf{Inv}(S) \in \mathcal{R_O}$ *and that* $\mathsf{Trans}(R) \in \mathcal{R_O}$ *iff* $\mathsf{Trans}(\mathsf{Inv}(R)) \in \mathcal{R_O}$. *An object property R is called* simple *if* $\mathsf{Trans}(S) \notin \mathcal{R_O}$ *for each* $S \stackrel{*}{\sqsubseteq} R$, *where* $\stackrel{*}{\sqsubseteq}$ *is the reflexive-transitive closure of* $\sqsubseteq$. *The set of* object concepts *is inductively defined starting from object concept names* $A \in \mathsf{O_C}$, *using the constructors in the first part of Table 1, where* $a, b \in \mathsf{O_I}$, $n \in \mathbb{N}$, *R is a object property, S is a simple object property and C, D are object concepts. The second part of Table 1 shows how* object axioms *are defined.*

*A* meta property, *the* meta RBox $\mathcal{R_M}$ *and to be called* simple *are defined analogously to the object level. The set of* meta concepts *is inductively defined starting from meta property names* $P \in \mathsf{M_P}$ *and meta concept names* $B \in \mathsf{M_C}$, *using the constructors in the third part of Table 1, where* $u, v \in \mathsf{M_I}$, $n \in \mathbb{N}$, *PQ is a meta property, Q is a simple meta property, E, F are meta concepts and* $\alpha$ *is an object concept inclusion, concept assertion or property assertion. The fourth part of Table 1 shows how* meta axioms *are defined.*

*A* $\mathcal{SHOIQ}[\mathcal{SHOIQ}]$ *ontology* $\mathcal{O}$ *is a triple* $\mathcal{O} = (\mathcal{B}, \mathcal{R_M}, \mathcal{R_O})$, *where* $\mathcal{B}$ *is a finite set of meta concept inclusions, concept assertions or property assertions,* $\mathcal{R_M}$ *is a meta RBox and* $\mathcal{R_O}$ *is an object RBox.*

We use the usual abbreviations for the object level: $C \sqcup D$ (disjunction) for $\neg(\neg C \sqcap \neg D)$, $\top$ (top concept) for $A \sqcup \neg A$, where $A \in \mathsf{O_C}$ is arbitrary but fixed, $\bot$ (bottom concept) for $\neg\top$, $\forall S.C$ (value restriction) for $\neg\exists S.\neg C$, $\geqslant_n S.C$ (at-least restriction) for $\neg(\leqslant_{n-1} S.C)$, and $=_n S.C$ (exact restriction) for $(\geqslant_n S.C) \sqcap (\leqslant_n S.C)$. Abbreviations for the meta level are used analogously.

To be able to express context independent knowledge, we have the sets of *rigid concepts* $\mathsf{O_{CR}} \subseteq \mathsf{O_C}$ and *rigid properties* $\mathsf{O_{PR}} \subseteq \mathsf{O_P}$ which must be interpreted the same in all contexts. Furthermore, we employ the *constant domain assumption*, i.e. all contexts speak about the same object domain, and the *rigid individual assumption*, i.e. individuals are always the same. The semantics of ConDLs are defined in a model-theoretic way.

**Definition 10 (Semantics of** $\mathcal{SHOIQ}[\mathcal{SHOIQ}]$**).** *A* nested interpretation *is a tuple* $\mathcal{J} = (\mathbb{C}, \cdot^{\mathcal{J}}, \Delta, (\cdot^{\mathcal{I}_c})_{c \in \mathbb{C}})$, *where* $\mathbb{C}$ *and* $\Delta$ *are non-empty sets (called* contexts *and (object)* domain*),* $\cdot^{\mathcal{J}}$ *is a mapping assigning a set* $B^{\mathcal{J}} \subseteq \mathbb{C}$ *to every* $B \in \mathsf{M_C}$, *a binary relation* $P^{\mathcal{J}} \subseteq \mathbb{C} \times \mathbb{C}$ *to every* $P \in \mathsf{M_P}$ *and a context* $u^{\mathcal{J}} \in \mathbb{C}$ *to every* $u \in \mathsf{M_I}$, *and for every* $c \in \mathbb{C}$, $\cdot^{\mathcal{I}_c}$ *is a mapping assigning a set* $A^{\mathcal{I}_c} \subseteq \Delta$ *to every* $A \in \mathsf{O_C}$, *a binary relation* $R^{\mathcal{I}_c} \subseteq \Delta \times \Delta$ *to every* $R \in \mathsf{O_P}$ *and a domain element* $a^{\mathcal{I}_c} \in \Delta$ *to every* $a \in \mathsf{O_I}$ *such that for all* $c, c' \in \mathbb{C}$ *we have* $x^{\mathcal{I}_c} = x^{\mathcal{I}_{c'}}$ *for every* $x \in \mathsf{O_I} \cup \mathsf{O_{CR}} \cup \mathsf{O_{PR}}$. *The functions* $\cdot^{\mathcal{I}_c}$ *and* $\cdot^{\mathcal{J}}$ *are extended to object and meta properties and concepts, respectively, as shown in Table 1, where* $\sharp X$ *denotes the cardinality of the set X.*

---

[4] To avoid confusion with roles in CROM, the term *property* is used for binary relations instead.

*Moreover, $\mathcal{J}$ ($\mathcal{I}_c$) satisfies an meta axiom (object axiom), denoted by $\mathcal{J} \models \beta$ ($\mathcal{I}_c \models \alpha$), if the condition in the fourth (second) part of Table 1 holds, $\mathcal{J}$* satisfies *$\mathcal{B}$ ($\mathcal{R}_M$) if $\mathcal{J}$ satisfies all axioms in $\mathcal{B}$ ($\mathcal{R}_M$), $\mathcal{J}$* satisfies *$\mathcal{R}_O$ if $\mathcal{I}_c$ satisfies all axioms in $\mathcal{R}_O$ for all $c \in \mathbb{C}$, and $\mathcal{J}$* satisfies *$\mathcal{O}$ if it satisfies $\mathcal{B}$, $\mathcal{R}_M$ and $\mathcal{R}_O$. $\mathcal{O}$ is consistent if there exists a nested interpretation that satisfies $\mathcal{O}$. The consistency problem is the problem of deciding whether a given ontology is consistent.*

$\mathcal{SHOIQ}[\mathcal{SHOIQ}]$ is a suitable candidate to encode both the context-dependent concepts and constraints of CROM. It permits, for instance to encode that in every context every role must have a player, as: $\top \sqsubseteq [\![ A_{\mathsf{RT}} \sqsubseteq\, =_1 \mathsf{plays}^-.\top ]\!]$. Utilizing $\mathcal{SHOIQ}[\mathcal{SHOIQ}]$, it becomes feasible to automatically map CROM domain models to a DL ontology.

## 3   Reasoning on Role-Based Models

To verify the consistency and satisfiability of CROM domain models, it is necessary to encode both the underlying semantics as well as the context-dependent concepts and constraints in ConDL axioms. This mapping must preserve validity, i.e. the ConDL ontology is consistent if and only if there exists a CROI that is compliant with the CROM and valid w.r.t. the constraint model. Henceforth, we highlight our encoding scheme and prove that it preserves the semantics.

In general, compartment types are modelled as concepts on the meta level; whereas playing roles, relationships and constraints are modelled within a compartment on the object level. Thus, we introduce o-concepts for natural types and role types, as well as a special o-property plays. Accordingly, the fills relation is transformed into domain and range axioms for plays. Relationship types are intuitively modelled as o-properties between two played roles. Role groups are handled like roles with an additional axiom stating that "playing" a role group is equivalent to fulfilling the constraints specified in that role group. Furthermore, if an object plays an atom of a non-nested role group, that object must fulfill the role group. For occurrence constraints a fresh individual name counter and an o-property counts is introduced and each played role or fulfilled role group is connected to this counter. Thus, both occurrence constraints and relationship cardinalities can be represented as qualified number restriction. Special consideration is needed for compartments that play roles within other compartments. Even though establishing a one-to-one link between an element on the object level and one on the meta level is impossible, for consistency it is only relevant if the compartment type of the nested compartment is instantiable. Hence, we consider the o-concepts $\mathsf{N}_{\mathsf{CT}'}$, e.g. compartments which play roles on the object level, denoted as *o-compartments*, as copies of the m-concepts $\mathsf{N}_{\mathsf{CT}}$.

Ontologically, natural types would be captured as rigid concepts and their fields as rigid properties, since that information does not change within contexts. In our setting, rigidity has no influence on the consistency, and neglecting it decreases the computational complexity exponentially in the size of the input. Admittedly, the case with rigid names can be handled quite similar. In summary, we consider the following concept, property and individual names:

**Table 2.** Mapping for occurring types and the CROM $\mathcal{M}$.

$$\bigwedge_{\substack{CT_1,CT_2\in\mathsf{N}_{\mathsf{CT}}\\CT_1\neq CT_2}}$$

$$\top \sqsubseteq \bigsqcup_{CT\in\mathsf{N}_{\mathsf{CT}}} CT \tag{14}$$

$$CT_1 \sqsubseteq \neg CT_2 \tag{15}$$

$$\top \sqsubseteq [\![A_O \equiv \bigsqcup_{NT\in\mathsf{N}_{\mathsf{NT}}} NT \sqcup \bigsqcup_{CT'\in\mathsf{N}_{\mathsf{CT'}}} CT']\!] \tag{16}$$

$$\top \sqsubseteq [\![A_{RT} \equiv \bigsqcup_{RT\in\mathsf{N}_{\mathsf{RT}}} RT]\!] \tag{17}$$

$$\top \sqsubseteq \bigsqcap_{T_1,\,T_2\in\mathsf{N}_{\mathsf{NT}}\cup\mathsf{N}_{\mathsf{CT'}}\cup\mathsf{N}_{\mathsf{RT}},T_1\neq T_2} [\![T_1 \sqsubseteq \neg T_2]\!] \tag{18}$$

$$\top \sqsubseteq [\![\top \sqsubseteq A_O \sqcup A_{RT} \sqcup A_{RG} \sqcup \{\mathsf{counter}\}]\!] \tag{19}$$

$$\top \sqsubseteq [\![A_O \sqsubseteq \neg A_{RT}]\!] \sqcap [\![A_O \sqsubseteq \neg A_{RG}]\!] \sqcap [\![A_{RT} \sqsubseteq \neg A_{RG}]\!]$$
$$\sqcap [\![\neg(A_O \sqcup A_{RT} \sqcup A_{RG})(\mathsf{counter})]\!] \tag{20}$$

$$\top \sqsubseteq \bigsqcap_{RT\in\mathsf{N}_{\mathsf{RT}}} [\![A_O \sqsubseteq \,\leqslant_1\mathsf{plays}.RT]\!] \tag{21}$$

$$\top \sqsubseteq [\![A_{RT} \sqsubseteq \,=_1\mathsf{plays}^-.\top]\!] \tag{22}$$

$$\top \sqsubseteq [\![\exists\mathsf{plays}.\top \sqsubseteq A_O]\!] \tag{23}$$

$$\top \sqsubseteq [\![\top \sqsubseteq \forall\mathsf{plays}.(A_{RT} \sqcup A_{RG})]\!] \tag{24}$$

$$\bigwedge_{CT'\in\mathsf{N}_{\mathsf{CT'}}} \neg [\![CT' \sqcap \exists\mathsf{plays}.\top \sqsubseteq \bot]\!] \sqsubseteq \exists\mathsf{nested}.CT \tag{25}$$

$$\bigwedge_{CT\in\mathsf{N}_{\mathsf{CT}}}$$

$$\top \sqsubseteq \bigsqcap_{RT\in\mathsf{N}_{\mathsf{RT}}} [\![\exists\mathsf{plays}.RT \sqsubseteq (\bigsqcup_{(T,RT)\in\mathsf{fills}} T)]\!] \tag{26}$$

$$CT \sqsubseteq [\![A_{RT} \sqsubseteq \bigsqcup_{RT\in\mathsf{parts}(CT)} RT]\!] \tag{27}$$

$$\bigwedge_{RST\in\mathsf{N}_{\mathsf{RST}},\mathsf{rel}(RST)=(RT_1,RT_2)} \top \sqsubseteq [\![\exists RST.\top \sqsubseteq RT_1]\!] \sqcap [\![\top \sqsubseteq \forall RST.RT_2]\!] \tag{28}$$

- $\mathsf{N}_{\mathsf{CT}} \subseteq \mathsf{M}_\mathsf{C}$ since every compartment type is a m-concept,
- $\mathsf{nested} \in \mathsf{M}_\mathsf{P}$ to assure the existence of compartments that play roles,
- $\mathsf{N}_{\mathsf{NT}} \cup \mathsf{N}_{\mathsf{CT'}} \cup \mathsf{N}_{\mathsf{RT}} \subseteq \mathsf{O}_\mathsf{C}$ since every natural type, every o-compartment type and every role type is an o-concept,
- $\mathsf{plays} \in \mathsf{O}_\mathsf{P}$ to express the plays-relation,
- $\mathsf{N}_{\mathsf{RST}} \subseteq \mathsf{O}_\mathsf{P}$ since every relationship type is an o-property,
- $\mathsf{counter} \in \mathsf{O}_\mathsf{I}$ and $\mathsf{counts} \in \mathsf{O}_\mathsf{P}$ to express the occurrence constraints, and
- $A_O$, $A_{RT}$, $A_{RG} \in \mathsf{O}_\mathsf{C}$ for, resp., all objects eligible of playing roles, i.e. naturals and o-compartments, all played roles, and all instances of role groups.

Henceforth, the mapping is trisected, first describing how types are encoded, then how fills and rel are mapped and finally how constraints are represented.

**Encoding CROM Types.** Table 2 (first segment) summarises the encoding of the underlying semantics of a given CROM $\mathcal{M}$. On the meta level, we assure that every context belongs to exactly one compartment type (Eqs. (14), (15)). Within every context, every natural, o-compartment and role belongs to exactly one type (Eqs. (16), (17), (18)). On the object level, every element is a role, a natural, an o-compartment, a role group instance or the individual counter (Eqs. (19), (20)). Every natural or o-compartment can only play one $RT$-role in each context and each role must be played by someone (Eqs. (21), (22)). We formalise a general domain and range restriction for plays. Only naturals or o-compartments can play something, and only roles or role group instances can be played (Eqs. (23), (24)). Finally, if an o-compartment plays a role in some

**Table 3.** Mapping for constraint model $\mathcal{C}$ and for assertions.

$$\top \sqsubseteq [\![A_{\mathsf{RG}} \equiv \bigsqcup_{RG \in \mathbb{RG}(\mathcal{C})} RG]\!] \tag{29}$$

$$\top \sqsubseteq \bigsqcap_{RG_1, RG_2 \in \mathbb{RG}(\mathcal{C}), RG_1 \neq RG_2} [\![RG_1 \sqcap RG_2 \sqsubseteq \bot]\!] \tag{30}$$

$$\top \sqsubseteq [\![A_{\mathsf{RG}} \sqsubseteq \geqslant_1 \mathsf{plays}^-.\top \sqcap \leqslant_1 \mathsf{plays}^-.\top]\!] \tag{31}$$

$$\top \sqsubseteq \bigsqcap_{RG \in \mathbb{RG}(\mathcal{C})} [\![A_\mathsf{O} \sqsubseteq \leqslant_1 \mathsf{plays}.RG]\!] \tag{32}$$

$$\top \sqsubseteq \bigsqcap_{\substack{RG \in \mathbb{RG}(\mathcal{C}), \\ RG = (\{A_1, \dots, A_n\}, k, l)}} [\![\exists \mathsf{plays}.RG \equiv (\geqslant_k \mathsf{plays}.(A_1 \sqcup \dots \sqcup A_n)) \\ \sqcap (\leqslant_l \mathsf{plays}.(A_1 \sqcup \dots \sqcup A_n))]\!] \tag{33}$$

$$\top \sqsubseteq \bigsqcap_{RG \in \mathbb{RG}^\top c} [\![\exists \mathsf{plays}.(\bigsqcup_{RT \in \mathsf{atom}(RG)} RT) \sqsubseteq \exists \mathsf{plays}.RG]\!] \tag{34}$$

$$\top \sqsubseteq [\![A_{\mathsf{RT}} \sqcup A_{\mathsf{RG}} \sqsubseteq =_1 \mathsf{counts}^-.\{\mathsf{counter}\}]\!] \tag{35}$$

$$\bigwedge_{\substack{(k..l, RG) \\ \in \mathsf{occur}(CT), \\ CT \in \mathsf{N_{CT}}}} CT \sqsubseteq [\![(\geqslant_k \mathsf{counts}.RG)(\mathsf{counter})]\!] \sqcap [\![(\leqslant_l \mathsf{counts}.RG)(\mathsf{counter})]\!] \tag{36}$$

$$\top \sqsubseteq \bigsqcap_{\substack{RST \in \mathsf{N_{RST}}, \\ \mathsf{rel}(RST) = (RT_1, RT_2), \\ \mathsf{card}(RST) = (i..j, k..l)}} [\![RT_1 \sqsubseteq \geqslant_k RST.\top \sqcap \leqslant_l RST.\top]\!] \\ \sqcap [\![RT_2 \sqsubseteq \geqslant_i RST^-.\top \sqcap \leqslant_j RST^-.\top]\!] \tag{37}$$

context, the o-compartment must also exist as context (Eq. (25)). After encoding the general knowledge about types, we succinctly map a specific CROM $\mathcal{M}$ to ConDL axioms.

**Mapping CROM $\mathcal{M}$.** The fills relation restricts which natural or compartment types can play which role types. Hence, a role type has as plays predecessors only naturals or o-compartments of types which fill that role type (Eq. (26)). In conjunction with Eq. (24), we know that all plays successors of naturals or o-compartments of a specific type are either instances of a role type that are filled by that type or instances of a role group. Thus, the axiom $\top \sqsubseteq \bigsqcap_{T \in \mathsf{N_{NT}} \cup \mathsf{N_{CT}}} [\![T \sqsubseteq \forall \mathsf{plays}.(A_{\mathsf{RG}} \sqcup \bigsqcup_{(T, RT) \in \mathsf{fills}} RT)]\!]$ is entailed. Since in a compliant CROI the plays-relation respects parts, only $RT$-roles with $RT \in \mathsf{parts}(CT)$ exist in a $CT$-context (Eq. (27)). Analogous to fills restricting the domain and range of plays, the rel-function restricts these for each relationship type (Eq. (28)). Due to Eqs. (14), (17), (18) and (27) as well as the fact that parts' codomain is a partition of $\mathsf{N_{RT}}$, in any context that is not in $CT$ there are no roles of a type the participates in $CT$. Thus, the following axiom is entailed for all $CT \in \mathsf{N_{CT}}$: $\neg CT \sqsubseteq [\![\bigsqcup_{RT \in \mathsf{parts}(CT)} RT \sqsubseteq \bot]\!]$.

**Including the Constraint Model $\mathcal{C}$.** Let $\mathcal{C}$ be a constraint model, let $\mathbb{RG}(\mathcal{C})$ be the set of all complex role groups occurring in $\mathcal{C}$, and let $\mathbb{RG}^\top(\mathcal{C}) \subseteq \mathbb{RG}(\mathcal{C})$ be the subset of non-nested role groups. Analogous to roles, role groups are disjoint, every instance of a role group must be played by some object and every object can either fulfill or not fulfill a role group (Eqs. (29), (30), (31), (32)). Complex role groups are treated like role types. An object "plays" an instance of a role group if it fulfills that role group (Eq. (33)). Furthermore, if an object plays a role whose type is an atom of a non-nested role group, the object must also fulfill that role group (Eq. (34)). To capture the occurrence constraints we enforce all roles and role group instances to be connected to counter via counts and state concept assertions for counter which must hold in the respective compartment type (Eqs. (35), (36)). Cardinality constraints restrict the number of roles that are related to a role via a relationship type (Eq. (37)) (Table 3).

**Preserving Semantics.** After presenting the mapping we establish the main result of this section. For a CROM $\mathcal{M}$ and a constraint model $\mathcal{C}$, let $\mathcal{K}$ be the pair $(\mathcal{M}, \mathcal{C})$. Then, the ConDL ontology $\mathcal{O}_{\mathcal{K}}$ is the set of Axiom (14) to (37). The next theorem establishes the desired relationship between $\mathcal{K}$ and $\mathcal{O}_{\mathcal{K}}$.

**Theorem 11.** *Let $\mathcal{K}$ be the pair $(\mathcal{M}, \mathcal{C})$ with $\mathcal{M}$ being a well-formed CROM and $\mathcal{C}$ a compliant constraint model. Then, there exists a CROI that is compliant with $\mathcal{M}$ and valid w.r.t. $\mathcal{C}$ iff $\mathcal{O}_{\mathcal{K}}$ is consistent.*

The proof is a straight forward application of the axioms (see Footnote 1). Checking consistency is quite general in the sense that many other questions can be reduced to the consistency problem. With the expressive means of ConDL assertions, for instance, we can also check whether a specific compartment type is instantiable, a certain role type is playable or two roles can be linked via some relationship type. Apart from that a reasoner that is capable of processing such ontologies is also needed to use the mapping in practice.

## 4    JConHT - a $\mathcal{SHOIQ}[\mathcal{SHOIQ}]$ Reasoner

Also due to highly optimised reasoners, DLs have been successfully established. In order to reuse an existing reasoner, we convert the consistency problem in ConDL to classical reasoning tasks. In [6], a reduction into two separate decision problems is shown. Firstly, reasoning on the meta level, and secondly, checking whether the object level is consistent in each context. For several reasons we base our implementation on the hypertableau reasoner HermiT [11,19]. A model construction-based reasoner is necessary since we need information about the appearing o-axioms when reasoning on the meta level. Besides that HermiT is implemented in Java and according to the ORE Report [20] the most perfomant, model-based reasoner in the discipline *OWL DL Consistency*.

For brevity, we omit the details of the hypertableau algorithm here. The sound, complete and terminating algorithm that we construct on the basis of hypertableau is shown in Algorithm 1. Here, $\mathcal{C}$ is the set of DL clauses, $\mathcal{A}$ the ABox obtained in the clausification of $\mathcal{O}$ and $\mathcal{K}_i$ is the object ontology for the world $c_i$ which collects all o-axioms in $\mathcal{A}'$ that are asserted to hold in $c_i$ and all other o-axioms as negated axioms (see Definition 11 of [6] for details). $\mathcal{K}_{\mathrm{rig}}$ is defined analogously to $\mathcal{K}_i$, but using the renaming technique as single ontology for all worlds.

The second step of the preprocessing, i.e. the *repletion*, is necessary to ensure completeness of Algorithm 1. The hypertableau algorithm avoids the unnecessary non-determinism which is usually introduced by the GCI-rule in tableau algorithms [19]. But this optimisation disguises some implicit contradictions in the DL clauses. Consider $\mathcal{C}_{\mathrm{ex}} = \{[\neg A(a)](x) \to C(x), \top \to C(x) \lor [A \sqsubseteq \bot](x)\}$ and

---

**Algorithm 1.** Algorithm for checking consistency with hypertableau

---

**Input**   : $\mathcal{SHOIQ}[\mathcal{SHOIQ}]$-ontology $\mathcal{O}$
**Output:** true if $\mathcal{O}$ is consistent, false otherwise
Preprocessing (results in $(\mathcal{C}, \mathcal{A})$):
   1. Elimination of transitivity axioms, normalisation, clausification
   2. Repletion of DL-clauses
Let $(T, \lambda)$ be any derivation for $(\mathcal{C}, \mathcal{A})$.
$\mathfrak{A} := \{\mathcal{A}' \mid$ there exists a leaf node in $(T, \lambda)$ that is labelled with $\mathcal{A}'\}$
**for** $\mathcal{A}' \in \mathfrak{A}$ **do**
    **if** $\mathcal{A}'$ *is clash-free* **then**
       **if** $\mathcal{O}$ *contains rigid names* **then**
          **if** $\mathcal{K}_{\text{rig}} := (\mathcal{O}_{\mathcal{A}'}, \mathcal{R}'_{\text{O}})$ *is consistent* **then**
             **return** true
       **else**
          Let $\{c_1, \ldots, c_k\}$ be the individuals occurring in $\mathcal{A}'$
          **if** $\mathcal{K}_i := (\mathcal{O}_{c_i}, \mathcal{R}_{\text{O}})$ *is consistent for all* $1 \leq i \leq k$ **then**
             **return** true

    **return** *false*

---

$\mathcal{A}_{\text{ex}} = \{\neg C(s)\}$. Here, only $[\![A \sqsubseteq \bot]\!](s)$ would be derived and the ontology seems to be consistent. Let $\mathcal{J}$ be a model, then we have $s \in (\neg C)^{\mathcal{J}}$, $s \in [\![A \sqsubseteq \bot]\!]^{\mathcal{J}}$ and $s \notin [\![\neg A(a)]\!]^{\mathcal{J}}$ which, by the semantics of ConDL, implies $s \in [\![A(a)]\!]^{\mathcal{J}}$. This contradicts $[\![A \sqsubseteq \bot]\!](s)$ and $(\mathcal{C}_{\text{ex}}, \mathcal{A}_{\text{ex}})$ is indeed inconsistent. To make these implicitly negated o-axioms visible, we introduce the repletion of $\mathcal{C}$.

**Definition 12 (Repletion of DL-Clauses).** *Let $\mathcal{C}$ be a set of DL-clauses. The repletion of $\mathcal{C}$ is obtained from $\mathcal{C}$ by adding the DL-clause $\top \rightarrow [\![\alpha]\!](x) \vee [\![\neg\alpha]\!](x)$ for each o-axiom $[\![\alpha]\!]$ occurring in $\mathcal{C}$.*

A drawback of the repletion is the high amount of non-determinism it introduces, but it is only necessary if o-axioms occur in the antecedent of a DL-clause. Apparently only Ax. (25), i.e. only if compartments play roles, introduces such o-axioms in the antecedent. Therefore, only then the repletion is necessary when reasoning on CROMs. Arguably, when constraints are omitted, CROM can be mapped to a less expressive ConDL, which further reduces the reasoning time.

## 5    Case Studies

**Implicit Knowledge in the Banking Domain.** Let us consider our running example. Instead of writing down all axioms of the respective ontology $\mathcal{O}_{\text{Bank}}$, we will rather point out those inferences that uncover hidden restrictions. In detail, we first inspect the Bank compartment type and its internal role types, role groups, and relationship types. Omitting general axioms, the Axioms (38) to (43) are contained in the ontology. Consequently, in any interpretation $\mathcal{J}$ that satisfies $\mathcal{O}_{\text{Bank}}$ with $c \in \text{Bank}^{\mathcal{J}}$, Ax. (38) and (39) entail the existence of an element in $\text{CheckingAccount}^{\mathcal{I}_c}$ or $\text{SavingsAccount}^{\mathcal{I}_c}$. Due to (40) and (41), there

**Table 4.** ConDL axioms of the banking example

| | |
|---|---|
| Bank $\sqsubseteq [\![(\geqslant_1 \text{counts}.\text{BankAccounts})(\text{counter})]\!]$ | (38) |
| $\top \sqsubseteq [\![\exists \text{plays}.\text{BankAccounts} \equiv =_1 \text{plays}.(\text{CheckingAccount} \sqcup \text{SavingsAccount})]\!]$ | (39) |
| $\top \sqsubseteq [\![\text{SavingsAccount} \sqsubseteq \geqslant_1 \text{own\_sa}^-.\top]\!]$ | (40) |
| $\top \sqsubseteq [\![\text{CheckingAccount} \sqsubseteq \geqslant_1 \text{own\_ca}^-.\top]\!]$ | (41) |
| $\top \sqsubseteq [\![\exists \text{own\_sa}.\top \sqsubseteq \text{Customer}]\!]$ | (42) |
| $\top \sqsubseteq [\![\exists \text{own\_ca}.\top \sqsubseteq \text{Customer}]\!]$ | (43) |
| Bank $\sqsubseteq [\![(\geqslant_1 \text{counts}.\text{Customer})(\text{counter})]\!]$ | (44) |
| Transaction $\sqsubseteq [\![(=_1 \text{counts}.\text{Participants})(\text{counter})]\!]$ | (45) |
| $\top \sqsubseteq [\![\exists \text{plays}.\text{Participants} \equiv =_1 \text{plays}.(\text{Source} \sqcup \text{Target})]\!]$ | (46) |
| $\top \sqsubseteq [\![\text{Source} \sqsubseteq =_1 \text{trans}.\top]\!] \sqcap [\![\text{Target} \sqsubseteq =_1 \text{trans}^-.\top]\!]$ | (47) |
| $\top \sqsubseteq [\![\exists \text{trans}.\top \sqsubseteq \text{Source}]\!] \sqcap [\![\exists \text{trans}^-.\top \sqsubseteq \text{Target}]\!]$ | (48) |
| Transaction $\sqsubseteq \bot$ | (49) |
| $\top \sqsubseteq [\![\text{Customer} \sqsubseteq \geqslant_1 \text{issues}.\top]\!]$ | (50) |
| $\top \sqsubseteq [\![\exists \text{issues}^-.\top \sqsubseteq \text{MoneyTransfer}]\!]$ | (51) |
| $\top \sqsubseteq [\![\exists \text{plays}.\text{MoneyTransfer} \sqsubseteq \text{Transaction}']\!]$ | (52) |
| $\neg [\![\text{Transaction}' \sqcap \exists \text{plays}.\top \sqsubseteq \bot]\!] \sqsubseteq \exists \text{nested}.\text{Transaction}$ | (53) |

must be some element "owning a CA or SA", which, by (42) and (43), must be in Customer$^{\mathcal{I}_c}$. As a result, Ax. (44) is entailed, i.e. the occurrence constraint for Customer is essentially 1..*. Similarly, when investigating the Transaction compartment type, we infer Ax. (45) to (48). For instance, assume there is a $d \in$ Transaction$^{\mathcal{J}}$. By (45) and (46), we infer the existence of exactly one element in Source$^{\mathcal{I}_d}$ or Target$^{\mathcal{I}_d}$. However, due to (47) and (78), there must also be an element in Target$^{\mathcal{I}_d}$ or Source$^{\mathcal{I}_d}$, respectively. It follows that Participants$^{\mathcal{I}_d}$ contains two elements, which contradicts (45). In conclusion, Ax. 49 is entailed, i.e. Transaction is not instantiable, due to the occurrence constraint of the Participants role group. With this knowledge, we can further reason on the Bank compartment type. Due to (44) and (50) to (52), there must be an element in MoneyTransfer$^{\mathcal{I}_c}$ playing an element in Transaction$'^{\mathcal{I}_c}$. Thus, by (53), there must be a context $c_2$ connected to $c$ via nested with $c_2 \in$ Transaction. Yet, this contradicts (49). In consequence, the banking domain model is indeed inconsistent, due to a small modelling error in an occurrence constraints (Table 4).

**Performance Evaluation of JConHT.** To investigate the performance of our approach, we conducted a set of benchmarks to test both the translation to a contextualised DL ontology (Sect. 3) as well as the subsequent reasoning with JConHT (Sect. 4). Hence, we developed a generator for CROM to create pseudo-random domain models of increasing complexity. Then, these models are transformed to the corresponding OWL ontology, and finally tested for consistency using our reasoner. Notably though, we focus on the execution time of JConHT, as the transformation time is polynomial bounded in the input size

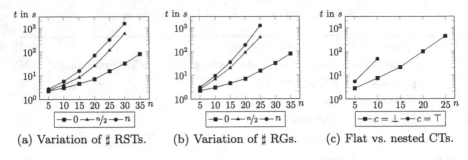

(a) Variation of ♯ RSTs.      (b) Variation of ♯ RGs.      (c) Flat vs. nested CTs.

**Fig. 2.** Average execution times of JConHT for benchmark ontologies.

(i.e. $\mathcal{O}(n^2)$) and negligible small. We investigated in the impact of three variables on the performance: (a) number of relationship types defined and constrained per compartment type, (b) number of role groups introduced per compartment type, and (c) a Boolean indicating whether compartment types can play roles. Our experiments generates random CROM models of stepwise increased complexity, varying each of these variables. The generator itself utilises a pseudo-random number generator (initialised with a given seed $s$), to create CROM models of size $n$, i.e., a domain model with $n$ natural types and $n$ compartment types with $n$ role types each, such that each role type is filled by two player types (either natural or compartment type). Additional parameters determine the number of relationship types $m$ between two distinct role types for each compartment (a), the number of role groups $k$ of two random role types for each compartment (b), and a Boolean $c$ indicating that compartment types are eligible as player types (c). The constraint model is generated accordingly, by assigning random cardinalities to the occurrence of role types and role groups, to role groups, as well as to the ends of relationships. Notably though, the set of cardinalities is limited $Card := \{0..0, 0..1, 0..\infty, 1..1, 1..\infty\}$. Utilizing this generator, we can individually test the performance impact of each of the variables.

We generated CROM domain models with $k \in \{0, n/2, n\}$ relationship types for (a) and $m \in \{0, n/2, n\}$ role groups for (b). To investigate nested compartment types (c), CROMs were created with $c \in \{\bot, \top\}$ whereas $k = n/2$ relationship types and $m = n/2$ role group. In each case, we generated, transformed, and verified CROM domain models with $n = 5, 10, 15, \ldots$ until the reasoner threw an out-of-memory exception. We repeated this process for each configuration, i.e., generating 100 models for each configuration with seed $s$, $1 \leq s \leq 100$, and calculate the average execution time of the reasoner to decide consistency.[5]

Figure 2 sums up the impact of each variable on JConHT's performance. As ConDL's reasoning time is exponential in the size of $n$, the time axis is logarithmic. In fact, Fig. 2a and b illustrates the impact of constrained relationships and role groups, respectively. The baseline for both is a configuration without relationship cardinalities and role groups, i.e., only occurrence constraints. We

---

[5] The tests were performed on a 3.3 GHz i5-2500 quad core with 12 GB heap size dedicated to an openjdk-8-jre (build 1.8.0) running on an Ubuntu 16.04.

found that the number of constrained relationships has a lower performance impact than the number of role groups. This is unsurprising, as role groups can represent arbitrary propositional logic formulae [17]. Thus, role groups become unsatisfiable easily. In turn, Fig. 2c indicates a significant performance penalty of nested compartment types are present. While reasoning on CROMs with $n/2$ role groups and relationship types is tractable, permitting compartment types to play roles quickly leads to a state space explosion, i.e. only models of size $n \leq 10$ could be checked. This is due to Axiom (25), requiring the addition of repletion clauses, and thus, introducing a large amount of non-determinism. In sum, our performance evaluation indicates that variable (c), nested compartment, has the highest performance impact, whereas the inclusion of role groups (b) and constrained relationships (a) has comparatively small impact. Besides that, we identified the heap size as a limiting factor, especially, when reasoning on more complex or nested models. When dealing with an average number of relationship types and role groups without nested compartments, however, reasoning was feasible for domain models of size $n \leq 25$.

## 6   Related Work

In the past, several approaches for formal frameworks to reason on UML arose, e.g. [1,4,7,24,25], from which we adopted some ideas, e.g. to model attributes of a UML class with DL properties and multiplicities of associations with counters [7]. In general, UML lacks expressive power to model context-dependent domains and while some approaches extended UML in this regard [10,12,23], there semantics is usually more ambiguous. In contrast, CROM has both a well-defined and formal semantics [17]. As classical DLs cannot properly formalise contextual knowledge, many different approaches and extensions of DLs have been proposed, e.g. [3,8,15,22]. Yet, many were tailored to different goals, e.g., to support context-specific reuse of ontologies. On one side, in most cases they have a different understanding of contexts, e.g. defined as set of attribute-value declarations for given *dimensions* [22]. Consequently, except a coverage relation, one can hardly express any other knowledge about contexts, such as their relational structure, rendering them insufficient to model CROMs. Similarily, in [5] a multidimensional data model is introduced with the same restrictions. On the other side, $\mathcal{ALC}_{\mathcal{ALC}}$ [15] formalizes contexts as formal objects with properties and relational structure, resulting in a similar two-dimensional DL that permits object knowledge to transcend through contexts. Yet, this leads to a double exponential time complexity and, in the presence of rigid roles, to undecidability.

## 7   Conclusion

To cope with context-dependent knowledge of today's Information Systems, advanced domain models permitting formal validation are indispensable. We enabled reasoning on context-dependent domain models by mapping CROM to the contextualized description logic $\mathcal{SHOIQ}[\mathcal{SHOIQ}]$ for which the dedicated reasoner JConHT can efficiently decide consistency. We showcased the semantically correctness, suitability and performance of our approach. In future, we will map additional constraints and further optimize both our mapping and JConHT.

# References

1. Ahmad, M.A., Nadeem, A.: Consistency checking of UML models using description logics: a critical review. In: Proceedings of the ICET 2010, pp. 310–315 (2010)
2. Baader, F., et al. (eds.): The Description Logic Handbook: Theory, Implementation, and Applications, 2nd edn. Cambridge University Press, Cambridge (2007)
3. Baader, F., et al.: Context-dependent views to axioms and consequences of semantic web ontologies. J. Web Semant. **12**, 22–40 (2012)
4. Berardi, D., Calvanese, D., De Giacomo, G.: Reasoning on UML class diagrams. Artif. Intell. **168**(1–2), 70–118 (2005)
5. Bertossi, L., Milani, M.: The ontological multidimensional data model in quality data specification and extraction. In: Calì, A., Wood, P., Martin, N., Poulovassilis, A. (eds.) BICOD 2017. LNCS, vol. 10365, pp. 126–130. Springer, Cham (2017). https://doi.org/10.1007/978-3-319-60795-5_13
6. Böhme, S., Lippmann, M.: Decidable description logics of context with rigid roles. In: Proceedings of the FroCoS 2015, pp. 17–32 (2015)
7. Calì, A., Calvanese, D., Giacomo, G., Lenzerini, M.: A formal framework for reasoning on UML class diagrams. In: Hacid, M.-S., Raś, Z.W., Zighed, D.A., Kodratoff, Y. (eds.) ISMIS 2002. LNCS (LNAI), vol. 2366, pp. 503–513. Springer, Heidelberg (2002). https://doi.org/10.1007/3-540-48050-1_54
8. Ceylan, I.I., Peñaloza, R.: The bayesian ontology language $\mathcal{BEL}$. J. Autom. Reasoning **58**(1), 67–95 (2017)
9. Chen, P.P.S.: The entity-relationship model toward a unified view of data. ACM Trans. Database Syst. (TODS) **1**(1), 9–36 (1976)
10. Genovese, V.: A meta-model for roles: introducing sessions. In: Proceedings of the Ws. on Roles and Relationships in OOP, Multiagent Systems, and Ontologies, pp. 27–38 (2007)
11. Glimm, B., Horrocks, I., Motik, B., Stoilos, G., Wang, Z.: Hermit: an OWL 2 reasoner. J. Autom. Reasoning **53**(3), 245–269 (2014)
12. Guizzardi, G., Wagner, G.: Conceptual simulation modeling with Onto-UML. In: Proceedings of the Winter Simulation Conference, p. 5 (2012)
13. Halpin, T.: Object-role modeling (ORM/NIAM). In: Bernus, P., Mertins, K., Schmidt, G. (eds.) Handbook on Architectures of Information Systems, pp. 81–103. Springer, Heidelberg (2006). https://doi.org/10.1007/978-3-662-03526-9_4
14. Hennicker, R., Klarl, A.: Foundations for ensemble modeling – the HELENA approach. In: Iida, S., Meseguer, J., Ogata, K. (eds.) Specification, Algebra, and Software. LNCS, vol. 8373, pp. 359–381. Springer, Heidelberg (2014). https://doi.org/10.1007/978-3-642-54624-2_18
15. Klarman, S., Gutiérrez-Basulto, V.: Description logics of context. J. Logic Comput. **26**(3), 817–854 (2016)
16. Kühn, T., Leuthäuser, M., Götz, S., Seidl, C., Aßmann, U.: A metamodel family for role-based modeling and programming languages. In: Combemale, B., Pearce, D.J., Barais, O., Vinju, J.J. (eds.) SLE 2014. LNCS, vol. 8706, pp. 141–160. Springer, Cham (2014). https://doi.org/10.1007/978-3-319-11245-9_8

17. Kühn, T., et al.: A combined formal model for relational context-dependent roles. In: Proceedings of the SLE 2015, pp. 113–124 (2015)
18. Liu, M., Hu, J.: Information networking model. In: Laender, A.H.F., Castano, S., Dayal, U., Casati, F., Oliveira, J.P.M. (eds.) ER 2009. LNCS, vol. 5829, pp. 131–144. Springer, Heidelberg (2009). https://doi.org/10.1007/978-3-642-04840-1_12
19. Motik, B., Shearer, R., Horrocks, I.: Hypertableau reasoning for description logics. J. Artif. Intell. Res. 36, 165–228 (2009)
20. Parsia, B., et al.: The OWL reasoner evaluation (ORE) 2015 competition report. In: Proceedings of the SSWS 2015, co-located with ISWC 2015, pp. 2–15 (2015)
21. Rumbaugh, J., Jacobson, R., Booch, G.: The Unified Modelling Language Reference Manual, 1st edn. Addison-Wesley, USA (1999)
22. Serafini, L., Homola, M.: Contextualized knowledge repositories for the semantic web. J. Web Semant. 12, 64–87 (2012)
23. Sheng, Q.Z., et al.: Contextuml: A UML-based modeling language for model-driven development of context-aware web services. In: Proceedings of the ICMB 2005, pp. 206–212 (2005)
24. Simmonds, J., Straeten, R.V.D., Jonckers, V., Mens, T.: Maintaining consistency between UML models using description logic. L'Objet 10(2–3), 231–244 (2004)
25. Simmonds, J., et al.: A tool based on DL for UML model consistency checking. Int. J. Softw. Eng. Knowl. Eng. 18(6), 713–735 (2008)

# Energy-Efficiency of OWL Reasoners—Frequency Matters

Patrick Koopmann[1], Marcus Hähnel[2], and Anni-Yasmin Turhan[1(✉)]

[1] Institute of Theoretical Computer Science, Technische Universität Dresden,
Dresden, Germany
{patrick.koopmann,anni-yasmin.turhan}@tu-dresden.de
[2] Operating Systems Group, Technische Universität Dresden, Dresden, Germany
mhaehnel@os.inf.tu-dresden.de

**Abstract.** While running times of ontology reasoners have been studied extensively, studies on energy-consumption of reasoning are scarce, and the energy-efficiency of ontology reasoning is not fully understood yet. Earlier empirical studies on the energy-consumption of ontology reasoners focused on reasoning on smart phones and used measurement methods prone to noise and side-effects. This paper presents an evaluation of the energy-efficiency of five state-of-the-art OWL reasoners on an ARM single-board computer that has built-in sensors to measure the energy consumption of CPUs and memory precisely. Using such a machine gives full control over installed and running software, active clusters and CPU frequencies, allowing for a more precise and detailed picture of the energy consumption of ontology reasoning. Besides evaluating the energy consumption of reasoning, our study further explores the relationship between computation power of the CPU, reasoning time, and energy consumption.

## 1  Introduction

Semantic technology applications often use ontologies and ontology reasoners as the core machinery to accomplish their tasks. Such applications are increasingly used on mobile devices [24]. Running times of ontology reasoning systems have been in the centre of attention of developers and users as long as these systems exist. On mobile devices, energy is a restricted resource, and as such at least as important as running times, but little is known so far about the energy consumption of ontology reasoners—although the motivation to investigate the energy consumption of reasoners is manifold.

*Selection of the hardware* for a reasoning task in an ontology-based mobile application requires knowledge on the energy consumption of carrying out this task. Reasoning could either be performed on a remote server or on the mobile device

This work is supported (in part) by the German Research Foundation (DFG) within the Collaborative Research Center SFB 912 HAEC.

directly. While reasoning on a remote machine may save energy and computation time, it can bring about problems in terms of privacy and security, and makes the service dependent on internet connectivity, as data has to be sent to another server. But if reasoning is performed on the mobile device (by reasoners such as Mini-Me developed specifically for mobile devices [17]) its energy consumption becomes relevant, as the energy available is simply limited by the battery.

*Selection of the reasoner system* for a reasoning task and ontology might regard its energy consumption. Little is known whether OWL reasoners differ as strongly in their energy consumption as they differ in the approaches they implement. Even for reasoners ported to and evaluated on mobile platforms [2,12] there is only little research on how the size, expressivity, and structure of the ontology and the performed reasoning task relate to energy consumption of the reasoner system.

*Development of energy-efficient reasoners* which use algorithms that behave energy-aware, requires detailed and reliable measurement methods for the hardware on which they are to be used. Such measurement methods should facilitate energy profiling of the different reasoning tasks—ideally for the individual components of the hardware.

*Prediction functions* for energy consumption trained by machine learning algorithms require reliable information about the energy consumption of a reasoning task at hand. While most research on prediction functions for reasoners focuses on running times, first research on predicting energy consumption has been undertaken in [6], albeit in a setup where only imprecise data on the energy consumption were available.

In this paper, we present an empirical study on the energy consumption (and running time) of OWL reasoners. We are not the first to address this topic. Motivated by the hard energy constraints of mobile devices, several research groups have evaluated the energy consumption of ontology reasoning on smartphones [6,16,23]. To the best of our knowledge, the first study on energy consumption of ontology reasoning was carried out by Patton et al. [16], who evaluated the energy consumption of answering SPARQL queries in the LUBM benchmark [8] and Schema.org [7]. They evaluated the reasoners Pellet [19], HermiT [5] and JENA rules [3] on the smartphone Samsung Galaxy S4. In order to measure the energy consumption, they replaced the battery of the phone with an external power supply that allows for power monitoring of the overall device. Based on their observations, they hypothesise that there is an almost linear relationship between execution time and power consumption of reasoners.

As the approach in [16] only works for smartphones with a replaceable battery, Valincius et al. [23] proposed an approach which uses the power management integrated circuit (PMIC) of the smartphone battery. Some of these PMICs, called *Fuel Gauge Chips* by Valincius et al., contain monitoring features that can be accessed by standard software libraries. Similarly to Patton et al., the authors evaluated SPARQL queries on the LUBM benchmark using the same set of reasoners, but on a OnePlus One smartphone. They observed that the capacity of the battery affected the measured values. The capacity was reduced significantly by the experiments, so that experiments had to be rerun in different orders to compensate for this effect. This framework was later used by Guclu et al. [6] to evaluate the ontology reasoners HermiT and TrOWL [21] on a Samsung Galaxy S6 and a Sony XPeria Z3, this time using a large set of ontologies taken from the OWL reasoner evaluation (ORE) competition from 2014 [1]. Their aim was to learn a prediction function for the energy consumption based on ontology metrics. The authors again found that the measured values differ significantly depending on the battery level of the device. This necessitated to incorporate the observed error rate in the interpretation of their measurements. The authors observed that performance and predictability of energy consumption can vary a lot depending on the hardware used. Moreover, contrary to the hypothesis by Patton et al., the execution time was not always linearly related to the energy consumption. The reason is that one of the smartphones, the Sony XPeria Z3, uses an ARM big.LITTLE architecture. This architecture allows the machine to switch operation freely between a slower, more energy-efficient cluster and a faster, less energy-efficient cluster, and makes it harder to obtain predictable measurements on energy consumption.

In conclusion, earlier studies on energy consumption of reasoners considered only a small set of available reasoning systems and ontologies (except the latter in [6]), and were only able to measure the energy consumption during reasoning for a smart phone as a whole and not for its components. All teams executed their experiments on Android smartphones, on which active background services, which have an impact on the overall energy consumption can only be controlled up to a certain point. Consequently, such measurements yield limited precision leading to uncertainties in the observed results. So far there are neither fine-grained, well-established methods of measurement nor benchmarks to assess the energy consumption of ontology reasoners.

To overcome the software and hardware related limitations that had an impact on the precision of energy measurements in these earlier evaluations, we used a different hardware setup in our experiments. More precisely, we chose a single-board computer with built-in sensors that measure power and energy consumption of various hardware components in a precise fashion. The device has a hardware architecture commonly found in Android smartphones, but gives the user full control over the hardware and software configuration. This way, we avoid the side-effects of unrelated tasks in the measurements, while obtaining results that can indicate energy consumption of mobile devices in general. The chosen hardware and architecture allows not only for more precise measurements of the energy consumption of ontology reasoning, but even for an evaluation of the energy consumption in regard of different CPU frequencies.

We used the ontologies from the ORE'15 benchmark [15] for computing ABox realisation in the OWL 2 DL and the OWL 2 EL profile. ABox realisation infers for all individuals in the data of the ontology to which of the named classes from the ontology they belong. Our choice is motivated on the one hand by the relevance reasoning about individuals has to mobile semantic technology applications, and on the other hand by the fact that there are more OWL reasoner systems available that are capable of full ABox realisation than for (full SPARQL) query answering. To understand the impact of the hardware parameters on energy efficiency, we further chose to carry out our experiments under different CPU frequencies. We observed that, since the reasoning systems do not take full advantage of the computation power available, they perform more energy-efficient on lower CPU frequencies. For example, by reducing the CPU frequency from 2.0 GHz to 1.5 GHz, the energy consumption of reasoning is reduced by 43% on average, while the reasoning time is now increased by only 19%—giving rise to our claim that frequency matters.

The paper is structured as follows. In the next section we describe our experimental setup, used systems and data in detail. Section 3 lists and discusses our observations on running times, energy and power consumption as well as on the effects of the CPU frequency. The paper ends with conclusions and pointers to future work.

## 2 Experimental Setup

We describe our experimental setup and the rationale for its design in detail. The goal of this study is to obtain detailed measurements of the energy consumption of OWL reasoners. Furthermore, we want to explore the energy consumption and running time of OWL reasoners in regard of CPU frequency.

### 2.1 Experimental Data and Systems

Our experiments used a specific type of hardware, two sets of ontologies—one for the OWL DL and one for the OWL EL profile—and a set of reasoning systems capable of performing ABox realisation in at least one of the used profiles. For readers interested in access to the reasoners and ontologies used in our experiments, we provide links and further results online.[1]

*Hardware.* We performed our experiments on the ODROID XU3 by Hardkernel, a single-board computer with inbuilt-sensors to measure energy consumption of different components, whose asymmetric ARM big.LITTLE architecture provides an interesting trade-off space for software energy efficiency [9]. The ODROID XU3 has two clusters with 4 cores each:

- an ARM Cortex-A15 quadcore with 2.0 GHz maximum frequency (the 'big cluster'), and
- an ARM Cortex-A7 quadcore with 1.4 GHz maximum frequency (the 'little cluster').

---

[1] See http://lat.inf.tu-dresden.de/~koopmann/energy-evaluation/.

**Table 1.** Metrics on the benchmark ontologies used in our evaluation.

| Profile | #Ontologies | #TBox axioms | | #ABox axioms | |
|---------|-------------|--------------|--------|--------------|--------|
| | | Average | Median | Average | Median |
| OWL DL | 150 | 1,690 | 506 | 1,208 | 494 |
| OWL EL | 109 | 31,272 | 2,718 | 47,045 | 2,279 |

It further has 2 GiB of LPDDR3 RAM with a frequency of 933MHz. Built-in sensors allow to measure the energy consumption of both clusters, the memory, and the GPU independently. As operating system, we used Arch Linux with Linux kernel version 4.9. We performed all experiments on the big cluster, while the evaluation environment was executed on the little cluster. To evaluate the impact of the CPU frequency on performance and energy consumption, we performed our experiments with CPU frequencies of 0.2 GHz, 0.5 GHz, 1.0 GHz, 1.5 GHz, and 2.0 GHz.

*Ontologies.* To get a balanced mix of ontologies with varying structures and properties, we took the ontologies from the benchmark used in the 2015 edition of the *OWL Reasoner Evaluation competition* (ORE'15) [15]. The competition evaluated OWL reasoners on reasoning time and success rates. It has different tracks for the reasoning tasks *consistency checking*, *classification* and *realisation* and provides for each of them a set of ontologies in the lightweight OWL EL profile and a set of ontologies in the expressive OWL DL profile. For both OWL profiles, we evaluated the reasoning task realisation, since it reasons over ABox data and is implemented in several reasoning systems.

At the ORE'15 ABox realisation track for the OWL DL profile, the reasoners used in our evaluation could only solve between 106 and 163 of the 264 ontologies. To obtain an experimental setup significant for the comparison of reasoners without overly many timeouts, we restricted the test set to the 150 ontologies of the track that had less than 10,000 statements. For ORE'15 ABox realisation track for the OWL EL profile, the situation was different. Here the dedicated OWL EL profile reasoner ELK could compute realisation for all but 7 of the ontologies at the competition, which is why we selected all of the 109 ontologies in the OWL EL profile of the ORE'15 realisation track for our experiments. Table 1 shows for both profiles the average and median of the number of TBox axioms and ABox axioms of the selected ontologies.

*OWL reasoning systems.* Our evaluation uses reasoners implemented in Java, as these can be executed directly on the ARM architecture of the ODROID. Reasoners not implemented in Java would require a recompilation from the sources for the architecture of the ODROID. Unfortunately, this technical constraint ruled out state of the art reasoners such as Konclude [20], Fact++ [22], PAGOdA [25] and ELepHant [18], and they are left for future work. Note that while PAGOdA itself is implemented in Java, the latest version has dependencies to the system-dependent datalog engine RDFox [14]. We used two sets of reasoners each dedicated to the respective OWL profile. Of all of these reasoners, we used the latest version available on the official websites when we initiated the experiments (status January 2017).

To obtain a set of relevant OWL DL reasoners, we picked the four reasoners implemented in Java that performed best at the OWL DL realisation track at ORE'15. Listed according their performance at ORE'15, we used the following reasoners and versions:

- HermiT 1.3.8 [5],
- TrOWL 1.5 [21],
- Pellet 2.4.0 [19], and
- JFact 5.0.2.[2]

The reasoner TrOWL differs from the other reasoners in that for the OWL DL profile, it deliberately sacrifices completeness for performance, i.e. it does not guarantee to compute all instance relationships.

For the OWL EL realisation track of ORE'15, the four best performing reasoners were ELK [13], TrOWL, JFact, and Pellet. Since the latter two reasoners are optimised for more expressive ontology languages and perform significantly worse on OWL EL ontologies, we restricted our evaluation to

- ELK 0.5.0 [13] and
- TrOWL 1.5 [21].

ELK and TrOWL implement reasoning algorithms that are complete only for the OWL EL profile. ELK is the only reasoner that uses a dedicated multithreading implementation and thus implements parallel reasoning.

## 2.2 Setup of the Experiments

Our experiments are designed to investigate mainly the energy consumption and not so much the running times of the OWL reasoners. For this reason we used a higher timeout of 10 min than the 3 min timeout used at ORE'15, also to accommodate for the lower computation power of the CPU at lower frequencies. We ran each ABox realisation for the two OWL profiles for the respective set of ontologies and OWL reasoning systems. For each run, we logged the following information:

- histories of the different energy sensors,
- running time,
- CPU utilisation,
- number of CPU instructions, and
- cache references and cache misses per instruction.

---

[2] http://jfact.sourceforge.net/.

The low-level information from the system was collected to gain insight on the causes for differing energy consumptions and reasoning times. Since all reasoners parsed the ontologies via the OWL API [11], they would use the same time for this task. We measured the time and energy used by parsing the ontologies once (for all) and excluded it in the measurements for the individual ABox realisation runs of the reasoners.

At ORE'15, the number of successful computations within the timeout was prioritised in the ranking, and computation time was only taken into account if the number of successful runs of all reasoners was the same for that ontology. Our focus here is on energy consumption of ontology reasoning, about which the number of timeouts hardly gives any insights. For this reason, and to allow for a meaningful comparison, we excluded those ontologies from the comparison which caused a timeout or an error for any reasoner at any frequency.[3] Excluding ontologies from the comparison that caused a timeout or an error for any reasoner at any frequency left us with 82 ontologies of 150 for the OWL DL profile and with 75 ontologies of 109 for the OWL EL profile. In the following, unless stated otherwise, all numbers refer to these sets of ontologies. For the interested reader, we provided corresponding numbers for the complete set of ontologies (including failed/incomplete computations), on our aforementioned webpage.

## 3    Observations

We report on our measurements carried out for the set of 82 OWL DL ontologies and 75 OWL EL ontologies for which ABox realisation was performed on the hardware and by the reasoning systems described in the last section. We start reporting on running times for ABox reasoning, because these values are important to put the measured values on energy consumption into perspective. We then turn our attention to power and energy consumption in Sect. 3.2, and discuss possible reasons for our observations in Sect. 3.3.

### 3.1    Running Times

The overall time required for ABox realisation is composed of the time needed for parsing the ontology and the time needed for the actual reasoning. To distinguish better the reasoning times of the different reasoners, we consider them separately.

*Running times for parsing* via the OWL API were measured for ontologies from both profiles. The obtained values are displayed in Table 2. The table relates to each frequency of the CPU the average, the standard deviation, and the median of running times for parsing the ontologies of the two profiles. As one could expect, the average running time, as well as the median of the running time, decreases as the frequency of the CPU increases. However, note that this increase is not proportional to the frequency of the CPU, as loading and parsing

---

[3] Note that this can lead to a different ranking as the one obtained at ORE'15.

**Table 2.** Execution time for parsing in seconds.

| Frequency | OWL DL | | | OWL EL | | |
|---|---|---|---|---|---|---|
| | Average | Standrd. dev. | Median | Average | Standrd. dev. | Median |
| 0.2 GHz | 91.05 | 35.20 | 97.61 | 143.10 | 56.00 | 186.41 |
| 0.5 GHz | 53.15 | 14.60 | 74.73 | 104.92 | 25.50 | 143.92 |
| 1.0 GHz | 61.84 | 8.00 | 82.81 | 92.50 | 14.00 | 156.55 |
| 1.5 GHz | 45.70 | 6.00 | 61.03 | 84.58 | 10.50 | 150.08 |
| 2.0 GHz | 37.60 | 5.00 | 50.06 | 76.30 | 9.00 | 137.34 |

**Table 3.** Reasoning time in seconds for the OWL DL profile.

| Frequency | HermiT | | | JFact | | |
|---|---|---|---|---|---|---|
| | Average | Standrd. dev. | Median | Average | Standrd. dev. | Median |
| 0.2 GHz | 84.85 | 205.16 | 11.58 | 115.82 | 282.06 | 14.32 |
| 0.5 GHz | 34.45 | 82.59 | 4.75 | 48.79 | 116.42 | 6.05 |
| 1.0 GHz | 18.20 | 43.33 | 2.61 | 27.42 | 64.35 | 3.41 |
| 1.5 GHz | 12.98 | 30.55 | 1.92 | 20.72 | 47.73 | 2.66 |
| 2.0 GHz | 10.33 | 24.06 | 1.58 | 17.40 | 39.36 | 2.26 |
| **Frequency** | **Pellet** | | | **TrOWL** | | |
| | Average | Standrd. dev. | Median | Average | Standrd. dev. | Median |
| 0.2 GHz | 31.16 | 64.12 | 11.43 | 4.98 | 7.76 | 2.97 |
| 0.5 GHz | 13.74 | 30.79 | 4.74 | 2.12 | 3.41 | 1.26 |
| 1.0 GHz | 7.80 | 18.00 | 2.63 | 1.20 | 1.98 | 0.69 |
| 1.5 GHz | 5.92 | 13.85 | 1.97 | 0.91 | 1.56 | 0.51 |
| 2.0 GHz | 5.02 | 11.91 | 1.63 | 0.77 | 1.33 | 0.43 |

requires frequent accesses to both hard drive and RAM. In general, the parsing times for OWL EL ontologies were higher than the ones for the OWL DL profile. This was to be expected, as the OWL EL ontologies were larger on average.

*Running times for ABox realisation in the OWL DL profile* were measured for the reasoners HermiT, JFact, Pellet, and TrOWL. The data obtained for running times is displayed in Table 3. Clearly, the running times of all reasoners decreased as the CPU frequencies increased. Here the difference between the reasoners in regard of the average and the standard deviation of reasoning times compared to the median is prominent. While HermiT and Pellet differ strongly on the average, their running times for the median are surprisingly close at all used CPU frequencies. TrOWL has the lowest running times, but recall that it implements an incomplete reasoning method and might miss inferences. The biggest decrease of running times consistently occurred when using a CPU frequency of 0.5 GHz instead of 0.2 GHz.

**Table 4.** Reasoning time in seconds for the OWL EL profile.

| Frequency | ELK | | | TrOWL | | |
|-----------|---------|--------------|--------|---------|--------------|--------|
|           | Average | Standrd. dev. | Median | Average | Standrd. dev. | Median |
| 0.2 GHz   | 19.77   | 22.79        | 11.51  | 30.51   | 114.45       | 2.99   |
| 0.5 GHz   | 8.36    | 10.19        | 4.85   | 13.13   | 51.51        | 1.24   |
| 1.0 GHz   | 4.78    | 6.37         | 2.70   | 7.42    | 29.92        | 0.75   |
| 1.5 GHz   | 3.74    | 5.34         | 2.03   | 5.76    | 24.39        | 0.51   |
| 2.0 GHz   | 3.23    | 4.66         | 1.68   | 4.78    | 20.70        | 0.47   |

*Running times for ABox realisation in the OWL EL profile* were measured for the reasoners ELK and TrOWL and are displayed in Table 4. Here the difference between average and median is even stronger—while for ELK, the average is roughly the double of the median, for TrOWL, the average is almost an order of magnitude higher than its median.

When comparing the running times for parsing with those for computing ABox realisation, it is apparent that the overall running times are dominated by the parsing of the ontologies.

## 3.2   Energy and Power Consumption

*Energy consumption of parsing* for both profiles is displayed in Table 5. The energy consumption of both reasoners first decreased when the CPU frequency is increased up to 1.0 GHz. But the energy consumption increased again for frequencies higher than 1.0 GHz—unlike the running times. We will see a similar effect when regarding the energy consumption of ABox realisation, and give some explanations for this in Sect. 3.3. Similarly to the running times, the parsing of ontologies from the OWL EL profile consumed more energy, which can again be explained by the larger size of the ontologies.

**Table 5.** Energy consumption for parsing in Joule.

| Frequency | OWL DL | | | OWL EL | | |
|-----------|---------|--------------|--------|---------|--------------|--------|
|           | Average | Standrd. dev. | Median | Average | Standrd. dev. | Median |
| 0.2 GHz   | 7.31    | 6.33         | 2.47   | 50.98   | 11.27        | 94.58  |
| 0.5 GHz   | 3.97    | 3.18         | 1.53   | 37.48   | 6.56         | 85.10  |
| 1.0 GHz   | 3.07    | 2.45         | 1.43   | 37.18   | 5.22         | 86.87  |
| 1.5 GHz   | 3.60    | 2.50         | 2.11   | 56.53   | 6.89         | 128.38 |
| 2.0 GHz   | 5.59    | 4.84         | 3.72   | 106.77  | 10.82        | 250.05 |

*Energy consumption for ABox realisation.* In the OWL DL profile, there were 82 ontologies that could be realised by all reasoners in all frequency settings. The results of our energy measurements are shown in Table 6. In the OWL EL profile, there were 75 ontologies that could be realised by all the selected OWL EL reasoners in all frequencies. The measured values are displayed in Table 7. Note that there was generally a high standard deviation, and the average and median values often differed significantly, For example, while TrOWL had a higher average energy consumption in the OWL EL profile than ELK, the median of the energy consumption of TrOWL was significantly lower than for ELK, indicating that TrOWL usually consumed less energy than ELK. In fact, if we consider all 90 ontologies for which ELK and TrOWL successfully performed realisation at 2.0 GHz, we see that for 70 of those, TrOWL consumed less energy than ELK. This indicates that choosing a reasoner specifically for a given ontology can provide significant savings in energy and reasoning time. Mobile applications that require simpler ontologies can therefore save significant energy and reasoning time by carefully selecting an appropriate reasoner for the ontology at hand.

Our measurements seem to confirm the hypothesis by Patton et al. that there is an almost linear relationship between energy consumption and reasoning time. Figure 1 plots for each reasoner the energy consumption against the running time for reasoning in the OWL DL ontologies at 1.0 GHz. For the other frequencies and the OWL EL profile, the picture looks similar.

However, when considering the average power consumption (i.e., energy per time) during realisation, one can note differences between reasoners as well as between profiles. At 2.0 GHz, in the OWL DL profile, the lowest power consumption of the CPU was by TrOWL with 2.37 W, followed by Pellet with 2.52 W, with

**Table 6.** Energy consumption in Joule for reasoning in the OWL DL profile.

| Frequency | HermiT | | | JFact | | |
|---|---|---|---|---|---|---|
| | Average | Standrd. dev. | Median | Average | Standrd. dev. | Median |
| 0.2 GHz | 31.46 | 74.60 | 4.82 | 42.94 | 102.62 | 5.96 |
| 0.5 GHz | 18.66 | 42.99 | 3.33 | 25.92 | 60.00 | 3.82 |
| 1.0 GHz | 16.79 | 37.53 | 3.49 | 24.18 | 53.92 | 4.22 |
| 1.5 GHz | 20.67 | 44.60 | 4.60 | 30.28 | 65.30 | 5.67 |
| 2.0 GHz | 34.17 | 73.08 | 8.28 | 53.33 | 111.86 | 9.72 |
| Frequency | Pellet | | | TrOWL | | |
| | Average | Standrd. dev. | Median | Average | Standrd. dev. | Median |
| 0.2 GHz | 11.98 | 23.58 | 4.82 | 2.30 | 2.95 | 1.61 |
| 0.5 GHz | 7.82 | 15.94 | 3.29 | 1.67 | 1.88 | 1.11 |
| 1.0 GHz | 7.64 | 15.07 | 3.54 | 1.92 | 1.72 | 1.61 |
| 1.5 GHz | 9.36 | 18.57 | 4.23 | 2.73 | 2.21 | 2.55 |
| 2.0 GHz | 17.18 | 32.66 | 8.22 | 4.72 | 3.92 | 3.23 |

**Table 7.** Energy consumption in Joule for reasoning in the OWL EL profile.

| Frequency | ELK | | | TrOWL | | |
|---|---|---|---|---|---|---|
| | Average | Standrd. dev. | Median | Average | Standrd. dev. | Median |
| 0.2 GHz | 8.69 | 10.04 | 5.14 | 12.47 | 46.45 | 1.64 |
| 0.5 GHz | 5.76 | 6.76 | 3.57 | 8.03 | 29.89 | 1.16 |
| 1.0 GHz | 5.98 | 7.25 | 3.83 | 7.79 | 28.22 | 1.70 |
| 1.5 GHz | 7.76 | 10.29 | 4.70 | 10.12 | 36.33 | 2.49 |
| 2.0 GHz | 14.41 | 18.73 | 9.12 | 17.34 | 61.87 | 4.87 |

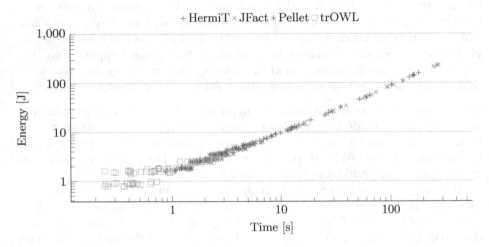

**Fig. 1.** Energy consumption in Joule in relation to reasoning time in seconds for the OWL DL profile (at CPU frequency of 1.0 GHz).

Hermit and JFact having the highest consumption of 2.62 W. In the OWL EL profile, the CPU consumed more power in general, with at 2.0 GHz, TrOWL consuming 2.60 W on average, and ELK 3.18 W. At the lower frequencies, though less power was consumed by the CPU, we obtain the same ranking among the reasoners. There were almost no differences in power consumption in the other components (memory, little cluster, GPU), which together accounted for between 0.26 and 0.28 W on average for all reasoners, profiles and CPU frequencies.

## 3.3   Impact of the CPU Frequency

We observed that the power consumption of the CPU grew exponentially with the selected frequency. While at 2.0 GHz, reasoning consumed 2.69 W on average for all reasoners and profiles, this value halves with each lower frequency used:

- 1.20 W at 1.5 GHz,
- 0.65 W at 1.0 GHz,

– 0.31 W at 0.5 GHz, and
– 0.16 W at 0.2 GHz.

In contrast, reasoning times were less than linearly affected by the CPU frequency, which is why the reasoning systems consumed significantly more energy at 2.0 GHz than at the lower frequencies.

The non-linear behaviour of the energy consumption in relation to the system frequency is to be expected for our system. The power that a system draws during execution is usually calculated using the following formula:

$$P = c \cdot f \cdot V^2,$$

where $c$ is the capacitance of the system, $f$ is the frequency and $V$ is the core voltage. Since voltage has a quadratic influence on power, it is also the dominating factor for power efficiency, i.e., for power consumption in relation to work. The capacitance is a value specific to the system and very hard to determine for a specific instance of a system in practice. Many factors can change the capacitance of the system during runtime, because they disable unused parts of the chip. In general, the execution time decreases with increasing frequency, while power consumption increases.

Energy usage corresponds to execution time multiplied by average power used. To save energy, the decrease in execution time due to increased frequency has to outweigh the increase in power draw. Especially memory intensive workloads, such as those caused by reasoning systems, may benefit less from higher frequencies, because most of the cycles are spent waiting for memory, which is significantly slower than the CPU. While this waiting does not consume as much power as executing work, it still is not as efficient compared to a lower clocked chip that can run at a lower voltage.

We observed exactly this behaviour in our experiments. Interestingly, different reasoners benefited differently from the computation power available. Comparing reasoning times at 0.2 GHz with reasoning times at 2.0 GHz, we find for the OWL DL profile:

– HermiT taking 13.5% of the time,
– Pellet 14.5% of the time,
– TrOWL 14.8% of the time, and
– JFact 15.6% of the time

at the higher frequency. For performing realisation on the OWL EL ontologies,

– TrOWL took 14.5% of the time and
– ELK 15.1% of the time

at 2.0 GHz than it took at the frequency of 0.2 GHz. These values contrast the fact that the CPU is actually ten times faster at 2.0 GHz than it is at 0.2 GHz.

Based on this observation, it is no surprise that the lowest energy consumption occurred at the lower CPU frequencies, as can already be seen in Tables 6 and 7. Figure 2 shows, for the OWL DL profile, at which frequencies how many

ontologies caused the lowest energy consumption of the CPU. Figure 3 displays
the same kind of information, but for the energy consumption of all measured
components combined. Indeed, by using a lower CPU frequency, the energy consumption can be reduced significantly, with comparatively small increases in
running time. For example, at 1.5 GHz, on average, realisation used only 57% of
the energy used at the maximum frequency of 2.0 GHz, while taking only 119%
of the time.

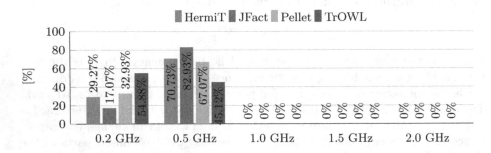

**Fig. 2.** Proportion of ontologies that caused the lowest energy consumption of the
CPU. (E.g. with HermiT for almost 71% of the ontologies the CPU consumed least
energy at 0.5 GHz.)

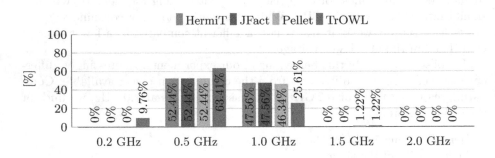

**Fig. 3.** Proportion of ontologies that caused the lowest energy consumption at a given
frequency of all measured components. (E.g. with HermiT for almost 48% of the ontologies the measured components consumed least energy at 1.0 GHz.)

To get a clearer understanding of what happens at the lower frequencies,
we first looked at the CPU utilisation, that is, the average number of active
cores, during reasoning. We observed that higher CPU frequencies often caused
a decrease in CPU utilisation, which means that on average, less cores were used
at the higher frequencies. For ELK, which had the highest CPU utilisation, the
average CPU utilisation changed from 1.63 at 0.2 GHz to 1.42 at 2.0 GHz. For
TrOWL, in both profiles, it even went from 1.21 at 0.2 GHz down to 0.85 at

2.0 GHz. For the other reasoners, the CPU utilisation was less affected by the CPU frequency, and generally lower than for ELK, which can be explained by the less dedicated use of multithreading. Note that using less cores means using less of the computation power available, and therefore has a negative effect on the reasoning time. The numbers also indicate that at higher frequencies, threads were often idle, and having a quadcore system was most useful when the CPU frequency was low.

The expected reason for why the reasoning systems did not make full use of the computation power available is that they are *memory-intensive* applications, and therefore slowed down by frequent accesses to the RAM. To confirm this hypothesis, and to understand better why different reasoners benefited differently from the CPU frequency, we examined the number of cache misses per CPU instruction measured during the experiments. A cache miss occurs when data accessed is not available in the cache, and has to be transferred from the (much slower) RAM to the cache. Comparing the number of cache misses per CPU instruction, we found significant differences between the reasoning systems, which partly reflect the above observations. The lowest number of cache misses per CPU instruction was by HermiT (1.06%) and TrOWL (1.01% in both profiles), the highest was by JFact (1.28%) and ELK (1.34%).

Since ontology reasoning is memory-intensive by nature, higher numbers of cache misses cannot be avoided in general. However, the observed differences between the reasoning systems indicate that it might be possible to minimise the number of cache misses by a careful implementation with dedicated optimisations for cache-access. Optimisations like this have for instance already been applied fruitfully in the area of SAT-solvers [4,10]. Such optimisations to alleviate the memory bottle neck could potentially also improve performance of future OWL reasoning systems.

## 4 Conclusion

We evaluated energy consumption of computing ABox realisation in two OWL profiles with five state-of-the-art OWL reasoners that are implemented in Java. The experiments were run on a computer with built-in energy sensors, which allowed for more exact energy measurements than previously used methods for evaluating energy consumption of OWL reasoning.

Our empirical results confirm the hypothesis originally stated by Patton et al. that there is an almost linear relation between energy consumption and running time of OWL reasoning. However, different reasoners vary in the amount of energy they consume per second. Even though the energy consumption of reasoning is dominated by the energy consumption of the CPU, the performance of the reasoners is still affected significantly by the memory. The reason is that, as memory-intensive applications, reasoning systems regularly spend time waiting for data to be transferred from the RAM to the cache. As a result, they benefit less from higher CPU frequencies, while they consume significantly less energy at lower frequencies, and OWL reasoning turns out to be more energy-efficient on lower frequencies. The impact memory access have on reasoning times differs however from reasoning system to reasoning system. Determining (close to) optimal configurations for energy-efficiency and saving energy needs further investigations.

In the future, we would like to use our measurement framework to investigate the energy consumption of non-Java reasoners, and examine the energy consumption of query answering. We also see quite some potential in using our measurements to improve the results on predicting energy consumption of OWL reasoning obtained in [6]. In order to generate high quality prediction functions, one would need to relate the ontology metrics to the energy consumption of the OWL reasoners in regard of the hardware setup and the reasoner used.

# References

1. Bail, S., Glimm, B., Jiménez-Ruiz, E., Matentzoglu, N., Parsia, B., Steigmiller, A. (eds.): Informal Proceedings of the 3rd International Workshop on OWL Reasoner Evaluation, ORE, CEUR Workshop Proceedings, vol. 1207. CEUR-WS.org (2014)
2. Bobed, C., Yus, R., Bobillo, F., Mena, E.: Semantic reasoning on mobile devices: do androids dream of efficient reasoners? Web Semant. Sci. Serv. Agents World Wide Web **35**, 167–183 (2015)
3. Carroll, J.J., Dickinson, I., Dollin, C., Reynolds, D., Seaborne, A., Wilkinson, K.: Jena: implementing the semantic web recommendations. In: Proceedings of the 13th International World Wide Web Conference (Alternate track papers & posters), pp. 74–83. ACM (2004)
4. Elffers, J., Johannsen, J., Lauria, M., Magnard, T., Nordström, J., Vinyals, M.: Trade-offs between time and memory in a tighter model of CDCL SAT solvers. In: Creignou, N., Le Berre, D. (eds.) SAT ·2016. LNCS, vol. 9710, pp. 160–176. Springer, Cham (2016). https://doi.org/10.1007/978-3-319-40970-2_11
5. Glimm, B., Horrocks, I., Motik, B., Stoilos, G., Wang, Z.: HermiT: an OWL 2 reasoner. J. Autom. Reasoning **53**(3), 245–269 (2014)
6. Guclu, I., Li, Y.-F., Pan, J.Z., Kollingbaum, M.J.: Predicting energy consumption of ontology reasoning over mobile devices. In: Groth, P., Simperl, E., Gray, A., Sabou, M., Krötzsch, M., Lecue, F., Flöck, F., Gil, Y. (eds.) ISWC 2016. LNCS, vol. 9981, pp. 289–304. Springer, Cham (2016). https://doi.org/10.1007/978-3-319-46523-4_18
7. Guha, R.V., Brickley, D., Macbeth, S.: Schema.org: evolution of structured data on the web. Commun. ACM **59**(2), 44–51 (2016)
8. Guo, Y., Pan, Z., Heflin, J.: LUBM: a benchmark for OWL knowledge base systems. Web Semant. Sci. Serv. Agents World Wide Web **3**(2), 158–182 (2005)
9. Hähnel, M., Härtig, H.: Heterogeneity by the numbers: a study of the ODROID XU+E big.LITTLE platform. In: 6th Workshop on Power-Aware Computing and Systems (HotPower 2014). USENIX Association (2014)
10. Hölldobler, S., Manthey, N., Saptawijaya, A.: Improving resource-unaware SAT solvers. In: Fermüller, C.G., Voronkov, A. (eds.) LPAR 2010. LNCS, vol. 6397, pp. 519–534. Springer, Heidelberg (2010). https://doi.org/10.1007/978-3-642-16242-8_37
11. Horridge, M., Bechhofer, S.: The OWL API: a Java API for OWL ontologies. Semant. Web **2**(1), 11–21 (2011)

12. Kazakov, Y., Klinov, P.: Experimenting with ELK reasoner on Android. In: Proceedings of the ORE 2013, pp. 68–74 (2013)
13. Kazakov, Y., Krötzsch, M., Simančík, F.: The incredible ELK. J. Autom. Reasoning **53**(1), 1–61 (2014)
14. Motik, B., Nenov, Y., Piro, R., Horrocks, I.: Combining rewriting and incremental materialisation maintenance for datalog programs with equality. In: IJCAI, pp. 3127–3133 (2015)
15. Parsia, B., Matentzoglu, N., Gonçalves, R.S., Glimm, B., Steigmiller, A.: The OWL reasoner evaluation (ORE) 2015 competition report. J. Autom. Reasoning, 1–28 (2015)
16. Patton, E.W., McGuinness, D.L.: A power consumption benchmark for reasoners on mobile devices. In: Mika, P., et al. (eds.) ISWC 2014. LNCS, vol. 8796, pp. 409–424. Springer, Cham (2014). https://doi.org/10.1007/978-3-319-11964-9_26
17. Scioscia, F., Ruta, M., Loseto, G., Gramegna, F., Ieva, S., Pinto, A., Di Sciascio, E.: A mobile matchmaker for the ubiquitous semantic web. In: Mobile Computing and Wireless Networks: Concepts, Methodologies, Tools, and Applications, pp. 994–1017. IGI Global (2016)
18. Sertkaya, B.: The ELepHant reasoner system description. In: Proceedings of the ORE 2013, pp. 87–93 (2013)
19. Sirin, E., Parsia, B., Grau, B.C., Kalyanpur, A., Katz, Y.: Pellet: a practical OWL-DL reasoner. Web Semant. Sci. Serv. Agents World Wide Web **5**(2), 51–53 (2007)
20. Steigmiller, A., Liebig, T., Glimm, B.: Konclude: system description. J. Web Semant. Sci. Serv. Agents World Wide Web **27**, 78–85 (2014)
21. Thomas, E., Pan, J.Z., Ren, Y.: TrOWL: tractable OWL 2 reasoning infrastructure. In: Aroyo, L., Antoniou, G., Hyvönen, E., ten Teije, A., Stuckenschmidt, H., Cabral, L., Tudorache, T. (eds.) ESWC 2010. LNCS, vol. 6089, pp. 431–435. Springer, Heidelberg (2010). https://doi.org/10.1007/978-3-642-13489-0_38
22. Tsarkov, D., Horrocks, I.: FaCT++ description logic reasoner: system description. In: Furbach, U., Shankar, N. (eds.) IJCAR 2006. LNCS (LNAI), vol. 4130, pp. 292–297. Springer, Heidelberg (2006). https://doi.org/10.1007/11814771_26
23. Valincius, E., Nguyen, H.H., Pan, J.Z.: A power consumption benchmark framework for ontology reasoning on Android devices. In: Proceedings of the ORE 2015, pp. 80–86 (2015)
24. Yus, R., Pappachan, P.: Are apps going semantic? A systematic review of semantic mobile applications. In: MoDeST@ ISWC, pp. 2–13 (2015)
25. Zhou, Y., Cuenca Grau, B., Nenov, Y., Kaminski, M., Horrocks, I.: PAGOdA: pay-as-you-go ontology query answering using a datalog reasoner. J. Artif. Intell. Res. **54**, 309–367 (2015)

# The Identity Problem in Description Logic Ontologies and Its Application to View-Based Information Hiding

Franz Baader$^{(\boxtimes)}$, Daniel Borchmann, and Adrian Nuradiansyah

Theoretical Computer Science, TU Dresden, Dresden, Germany
{franz.baader,daniel.borchmann,adrian.nuradiansyah}@tu-dresden.de

**Abstract.** The work in this paper is motivated by a privacy scenario in which the identity of certain persons (represented as anonymous individuals) should be hidden. We assume that factual information about known individuals (i.e., individuals whose identity is known) and anonymous individuals is stored in an ABox and general background information is expressed in a TBox, where both the TBox and the ABox are publicly accessible. The identity problem then asks whether one can deduce from the TBox and the ABox that a given anonymous individual is equal to a known one. Since this would reveal the identity of the anonymous individual, such a situation needs to be avoided. We first observe that not all Description Logics (DLs) are able to derive any such equalities between individuals, and thus the identity problem is trivial in these DLs. We then consider DLs with nominals, number restrictions, or function dependencies, in which the identity problem is non-trivial. We show that in these DLs the identity problem has the same complexity as the instance problem. Finally, we consider an extended scenario in which users with different rôles can access different parts of the TBox and ABox, and we want to check whether, by a sequence of rôle changes and queries asked in each rôle, one can deduce the identity of an anonymous individual.

## 1 Introduction

In order to illustrate the privacy scenario sketched in the abstract, assume that you are asked to perform a survey regarding the satisfaction of employees with the management of a company. Since the boss of the company is known not to respond well to criticism, the employees insist that you perform the survey such that the identity of persons voicing criticism cannot be deduced by the boss. Thus, you let the employees use a pseudonym when answering the survey. However, the survey does ask some personal data from the participants, and you are concerned that the boss can use the provided answers, in combination with the employee database and general knowledge about how things work in the

---

A. Nuradiansyah—Funded by DFG within the Research Training Group 1907 "RoSI".

Z. Wang et al. (Eds.): JIST 2017, LNCS 10675, pp. 102–117, 2017.
https://doi.org/10.1007/978-3-319-70682-5_7

company, to deduce that a certain pseudonym corresponds to a specific employee. For example, assume that in the survey the anonymous individual $x$ states that she is female and has expertise in logic and privacy. The boss knows that all employees with expertise logic belong to the formal verification task force and all employees with expertise privacy belong to the security task force. In addition, the employee database contains the information that the members of the first task force are John, Linda, Paul, Pattie and of the second Jim, John, Linda, Pamela. Since Linda is the only female employee belonging to both task forces, the boss can deduce that Linda hides behind the pseudonym $x$. The question is now whether you can use an automated system to check whether such a breach of privacy can occur in your survey.

The purpose of this paper is to show that ontology reasoners can in principle be used for this purpose. We assume that both the information provided in the survey and the employee database are represented in a DL ABox $\mathcal{A}$, where the employees from the database are represented as known individuals in $\mathcal{A}$ and the pseudonyms used in the survey are represented as anonymous individuals in $\mathcal{A}$. Background information (such as disjointness of the concepts Male and Female, or the connection between expertise and task forces) are represented in a DL TBox $\mathcal{T}$. In order to detect a breach of privacy, we then need to check whether the ontology $\mathcal{O}$ consisting of $\mathcal{T}$ and $\mathcal{A}$ implies an identity between some anonymous individual $x$ and a known individual $a$. We call the underlying reasoning task the *identity problem* for $\mathcal{O}$, $x$, and $a$.

In Sect. 2 we formally introduce the identity problem and show that, for a large class of DLs, this problem is trivial in the sense that no identities between distinct individuals can be deduced from consistent ontologies formulated in these DLs. Not surprisingly, this class consists of the DLs that are fragments of first-order logic without equality. In Sect. 3, we introduce three DLs for which the identity problem is non-trivial, i.e., the DL $\mathcal{ALCO}$ [14], where nominals allow us to derive identities; $\mathcal{ALCQ}$ [11], where number restrictions allow us to derive identities; and $\mathcal{CFD}_{nc}$ [22], where functional dependencies allow us to derive identities. In Sect. 4 we show that the identity problem can be reduced in polynomial time to the instance problem, and that for the three DLs mentioned above this actually yields an optimal procedure w.r.t. worst-case complexity. Section 5 considers the identity problem in the context of rôle-based access control [13] to ontologies. Basically, we assume that a user rôle $\hat{r}$ is associated with access to a subset $\mathcal{O}_{\hat{r}}$ of the ontology.[1] While having rôle $\hat{r}$, the user can access $\mathcal{O}_{\hat{r}}$ through queries, and can then store the result in a view $V_{\hat{r}}$. In a setting where rôles can dynamically change, the user may have collected (and stored) a sequence of views for different rôles. The question is then whether it is possible to derive the identity of an anonymous individual with a known one using these views. We will show that answering this question can be reduced to the identity problem investigated in the previous sections.

---

[1] To distinguish user rôles from DL roles, we write them with "ô" and also denote specific such rôles with letters with a hat.

Similar privacy scenarios have been considered for databases [4], but also in the context of ontology-based data access [7,8,18]. In particular, [18] introduces a setting with sub-ontologies and views that is similar to what we consider in Sect. 5. However, the main difference between these works and ours is that we concentrate on hiding the *identity* of an anonymous individual with a known one. In contrast, the other works are trying to hide *properties* of known individuals, i.e., the membership of an individual (or a tuple of individuals) in the answers to certain queries.

A preliminary version of this paper was presented at the Description Logic Workshop [1].

## 2   The Identity Problem

We assume that the reader is familiar with the basic notions of Description Logics, as e.g. introduced in [2,3]. We denote the set of concept names by $N_C$, the set of role names by $N_R$, and the set of individual names by $N_I$. Using the constructors of the given DL, one can then build concept descriptions over a given set of concept and role names. TBoxes and ABoxes are assumed to be defined in the standard way: TBoxes are finite sets of general concept inclusions (GCIs) of the form $C \sqsubseteq D$, for concept descriptions $C, D$; and ABoxes are finite sets of assertions of the form $C(a)$ and $r(a, b)$, where $C$ is a concept description, $r$ is a role name, and $a, b$ are individual names. An ontology is of the form $\mathfrak{O} = (\mathcal{T}, \mathcal{A})$ where $\mathcal{T}$ is a TBox and $\mathcal{A}$ is an ABox. The semantics of DLs is defined as usual by considering first-order interpretations $\mathcal{I}$, which consist of a non-empty domain $\Delta^{\mathcal{I}}$ and assign concept names $A$ with subsets $A^{\mathcal{I}}$ of $\Delta^{\mathcal{I}}$, role names $r$ with binary relations $r^{\mathcal{I}}$ on $\Delta^{\mathcal{I}}$, and individual names $a$ with elements $a^{\mathcal{I}}$ of $\Delta^{\mathcal{I}}$. Note, however, that we do *not* make the unique name assumption (UNA) for individual names, i.e., different individual names may be interpreted by the same element of the interpretation domain. The semantics of the concept constructors then allows us to assign subsets $C^{\mathcal{I}}$ of $\Delta^{\mathcal{I}}$ to concept descriptions. The interpretation $\mathcal{I}$ is a model of the ontology $\mathfrak{O} = (\mathcal{T}, \mathcal{A})$ if it satisfies all the GCIs in $\mathcal{T}$ and assertions in $\mathcal{A}$, i.e., $C^{\mathcal{I}} \subseteq D^{\mathcal{I}}$ for all $C \sqsubseteq D$ in $\mathcal{T}$, $a^{\mathcal{I}} \in C^{\mathcal{I}}$ for all $C(a)$ in $\mathcal{A}$, and $(a^{\mathcal{I}}, b^{\mathcal{I}}) \in r^{\mathcal{I}}$ for all $r(a, b)$ in $\mathcal{A}$. The ontology $\mathfrak{O}$ is consistent if it has a model.

The *identity problem* asks whether two individuals are equal w.r.t. a given ontology. Since anything (also identities) follows from an inconsistent ontology, we consider this problem only for the case where the ontology is consistent.

**Definition 1.** *Let $a, b \in N_I$ be distinct individual names and $\mathfrak{O}$ a consistent ontology. Then $a$ is* equal *to $b$ w.r.t. $\mathfrak{O}$ (denoted by $\mathfrak{O} \models a \doteq b$) iff $a^{\mathcal{I}} = b^{\mathcal{I}}$ for all models $\mathcal{I}$ of $\mathfrak{O}$. The* identity problem *for $\mathfrak{O}, a, b$ asks whether $\mathfrak{O} \models a \doteq b$.*

Not all DLs are able to derive equality of individuals. We call those that can *DLs with equality power.*

**Definition 2.** $\mathcal{L}$ *is a* description logic without equality power *if there is no consistent ontology $\mathfrak{O}$ formulated in $\mathcal{L}$ and two distinct individual names $a, b \in N_I$ such that $\mathfrak{O} \models a \doteq b$. Otherwise we say that $\mathcal{L}$ has equality power.*

It is well-known (see, e.g., Chap. 6 in [2]) that many DLs can be translated into first-order predicate logic (FOL). Basically, concept names and role names are translated into unary and binary predicates, respectively, and complex concept descriptions are translated into FOL formulas with one free variable. Individual names are translated into constant symbols and TBoxes and ABoxes into closed formulas. For the translation of some DLs, FOL without equality is sufficient whereas for others equality is needed.

**Theorem 1.** *If the DL $\mathcal{L}$ can be translated into FOL without equality, then it is a DL without equality power.*

*Proof.* Let $\mathfrak{O} = (\mathcal{T}, \mathcal{A})$ be a consistent ontology of $\mathcal{L}$ and $a, b \in N_I$ be two distinct individual names. We must show that $\mathfrak{O} \not\models a \doteq b$. According to our assumption on $\mathcal{L}$, there is an FOL formula $\phi$ not containing the equality symbol that is equivalent to $\mathfrak{O}$. Consequently, it is sufficient to show that $\phi \not\models a = b$ according to the semantics of FOL, where the equality symbol $=$ is interpreted as equality. Since $\mathfrak{O}$ is consistent, the formula $\phi$ is satisfiable.

Using well-known approaches and results regarding FOL [6], we can transform $\phi$ into a formula $\phi'$ in Skolem form containing additional function symbols such that (i) $\phi$ is satisfiable iff $\phi'$ is satisfiable, and (ii) any model of $\phi'$ is a model of $\phi$. Thus, $\phi'$ is satisfiable and since it is in Skolem form it has a Herbrand model $\mathcal{I}_H$. Since $\phi'$ does not contain equality, distinct terms (and thus in particular distinct constants) are interpreted by distinct elements in $\mathcal{I}_H$. Finally, we know that $\mathcal{I}_H$ is also a model of $\phi$, which shows that there is a model of $\phi$ in which $a$ and $b$ are not interpreted by the same domain element. This proves $\phi \not\models a = b$.     □

As a consequence of this theorem, we conclude that the basic DL $\mathcal{ALC}$ [16] (see below) and its fragments, but also more expressive DLs such as $\mathcal{SRI}$ (see the Appendix in [3]), do not have equality power, and thus the identity problem is trivial for these DLs.

## 3   Three DLs with Equality Power

In this section, we introduce three DLs that are able to derive equalities between individuals, and for which thus the identity problem is non-trivial. The first two DLs are $\mathcal{ALCO}$, which extends $\mathcal{ALC}$ by *nominals*, and $\mathcal{ALCQ}$, which extends $\mathcal{ALC}$ by qualified number restrictions. Recall that concept descriptions $C, D$ of $\mathcal{ALC}$ are defined using the following syntax rules:

$$C, D ::= \bot \mid \top \mid A \mid \neg C \mid C \sqcap D \mid \forall r.C \mid \exists r.C,$$

where $A \in N_C$ and $r \in N_R$. The semantics of these concept constructors are defined as follows:

- $\bot^{\mathcal{I}} := \emptyset, \top^{\mathcal{I}} := \Delta^{\mathcal{I}};$
- $(C \sqcap D)^{\mathcal{I}} := C^{\mathcal{I}} \cap D^{\mathcal{I}}, (C \sqcup D)^{\mathcal{I}} := C^{\mathcal{I}} \cup D^{\mathcal{I}}, (\neg C)^{\mathcal{I}} := \Delta^{\mathcal{I}} \setminus C^{\mathcal{I}};$
- $(\forall r.C)^{\mathcal{I}} := \{d \in \Delta^{\mathcal{I}} \mid \forall e \in \Delta^{\mathcal{I}}.(d,e) \in r^{\mathcal{I}} \Rightarrow e \in C^{\mathcal{I}}\};$
- $(\exists r.C)^{\mathcal{I}} := \{d \in \Delta^{\mathcal{I}} \mid \exists e \in \Delta^{\mathcal{I}}.(d,e) \in r^{\mathcal{I}} \wedge e \in C^{\mathcal{I}}\}.$

Nominals can be used to generate singleton concepts from individual names: if $a \in N_I$, then $\{a\}$ is a concept description of $\mathcal{ALCO}$, whose semantics is defined as $\{a\}^{\mathcal{I}} := \{a^{\mathcal{I}}\}$. Qualified number restrictions are of the form $\geqslant n\,r.C$ and $\leqslant n\,r.C$, with associated semantics

- $(\geqslant n\,r.C)^{\mathcal{I}} = \{d \in \Delta^{\mathcal{I}} \mid$ there are at least $n$ elements $e \in \Delta^{\mathcal{I}}$ with $(d,e) \in r^{\mathcal{I}}$ and $e \in C^{\mathcal{I}}\};$
- $(\leqslant n\,r.C)^{\mathcal{I}} = \{d \in \Delta^{\mathcal{I}} \mid$ there are at most $n$ elements $e \in \Delta^{\mathcal{I}}$ with $(d,e) \in r^{\mathcal{I}}$ and $e \in C^{\mathcal{I}}\}.$

The third DL, called $\mathcal{CFD}_{nc}$ [22], derives its equality power from so-called functional dependencies. Instead of roles, this logic uses attributes, which are interpreted as total functions. We use the symbol $N_A$ to denote the set of all attributes, replacing the set $N_R$. Concept descriptions $C, D$ of $\mathcal{CFD}_{nc}$ are defined using the following syntax rules:

$$C, D ::= A \mid \neg A \mid C \sqcap D \mid \forall \mathsf{Pf}.C \mid A : \mathsf{Pf}_1, \ldots, \mathsf{Pf}_k \to \mathsf{Pf},$$

where $A \in N_C$, $k \geq 1$, and the *path functions* $\mathsf{Pf}, \mathsf{Pf}_i$ are words in $N_A^*$ with the convention that the empty word is denoted by $id$. A concept description of the form $A : \mathsf{Pf}_1, \ldots, \mathsf{Pf}_k \to \mathsf{Pf}$ is called a *path functional dependency* (PFD). In $\mathcal{CFD}_{nc}$ there is an additional restriction on PFDs to ensure that reasoning in this logic is polynomial: for any PFD of the form above there is an $i, 1 \leq i \leq k$ such that

1. $\mathsf{Pf}$ is a prefix of $\mathsf{Pf}_i$, or
2. $\mathsf{Pf} = \mathsf{Pf}'f$ for $f \in N_A$ and $\mathsf{Pf}'$ is a prefix of $\mathsf{Pf}_i$.

Note that PFDs whose right-hand side $\mathsf{Pf}$ has length $\leq 1$ trivially satisfy this restriction.

The interpretation of attributes as total functions is extended to path functions by using composition of functions and interpreting $id$ as the identity function. The semantics of atomic negation $(\neg A)$ and conjunction $(C \sqcap D)$ is defined in the same way as in $\mathcal{ALC}$. For the constructors involving path functions, it is defined as follows:

$$(\forall \mathsf{Pf}.C)^{\mathcal{I}} := \{d \in \Delta^{\mathcal{I}} \mid \mathsf{Pf}^{\mathcal{I}}(d) \in C^{\mathcal{I}}\},$$

$(A : \mathsf{Pf}_1, \ldots, \mathsf{Pf}_k \to \mathsf{Pf})^{\mathcal{I}} :=$

$$\{d \in \Delta^{\mathcal{I}} \mid \forall e \in \Delta^{\mathcal{I}}. \left( \bigwedge_{1 \leq i \leq k} \mathsf{Pf}_i^{\mathcal{I}}(d) = \mathsf{Pf}_i^{\mathcal{I}}(e) \right) \Rightarrow \mathsf{Pf}^{\mathcal{I}}(d) = \mathsf{Pf}^{\mathcal{I}}(e)\}$$

A TBox $\mathcal{T}$ in $\mathcal{CFD}_{nc}$ consists of a finite set of inclusion dependencies $A \sqsubseteq C$, and an ABox $\mathcal{A}$ consists of a finite set of concept assertions $A(a)$ and path function assertions $\mathtt{Pf}_1(a) = \mathtt{Pf}_2(b)$, where $A \in N_C$, $C$ is a $\mathcal{CFD}_{nc}$ concept description, $a, b \in N_I$, and $\mathtt{Pf}_i \in N_A^*$.

**Theorem 2.** *The DLs $\mathcal{ALCO}$, $\mathcal{ALCQ}$, and $\mathcal{CFD}_{nc}$ have equality power.*

This theorem is an immediate consequence of the following three examples, which each shows for the respective DL that it can derive equality between different individuals.

*Example 1.* Here we formulate the example from the introduction in the DL $\mathcal{ALCO}$. Let $\mathfrak{O} = (\mathcal{T}, \mathcal{A})$ where

$$\mathcal{T} := \{\exists \mathsf{expertise}.\{\mathsf{LOGIC}\} \sqsubseteq \mathsf{VerTF}, \quad \exists \mathsf{expertise}.\{\mathsf{PRIVACY}\} \sqsubseteq \mathsf{SecTF},$$
$$\mathsf{VerTF} \sqsubseteq \{\mathsf{JOHN}\} \sqcup \{\mathsf{LINDA}\} \sqcup \{\mathsf{PAUL}\} \sqcup \{\mathsf{PATTIE}\},$$
$$\mathsf{SecTF} \sqsubseteq \{\mathsf{JIM}\} \cup \{\mathsf{JOHN}\} \cup \{\mathsf{LINDA}\} \cup \{\mathsf{PAMELA}\}, \quad \mathsf{Female} \sqsubseteq \neg\mathsf{Male}\},$$
$$\mathcal{A} := \{\mathsf{Female}(x), \mathsf{expertise}(x, \mathsf{LOGIC}), \mathsf{expertise}(x, \mathsf{PRIVACY}),$$
$$\mathsf{Female}(\mathsf{LINDA}), \mathsf{Female}(\mathsf{PATTIE}), \mathsf{Female}(\mathsf{PAMELA}),$$
$$\mathsf{Male}(\mathsf{JOHN}), \mathsf{Male}(\mathsf{JIM}), \mathsf{Male}(\mathsf{PAUL})\}.$$

It is easy to see that $\mathfrak{O} \models x \doteq \mathsf{LINDA}$ since $x$'s expertise implies that she belongs to both the verification and the security task force, but the only female employee belonging to both is Linda.

For the sake of brevity, we use abstract examples to show that $\mathcal{ALCQ}$ and $\mathcal{CFD}_{nc}$ have equality power. It would, however, be easy to provide intuitive examples also for these two DLs.

*Example 2.* Consider the $\mathcal{ALCQ}$ ontology $\mathfrak{O} = (\mathcal{T}, \mathcal{A})$ where

$$\mathcal{T} := \{A \sqsubseteq\, \leqslant 1\, r.B\} \quad \text{and} \quad \mathcal{A} := \{A(a), r(a, b), r(a, x), B(a), B(x)\}.$$

Obviously, we have $\mathfrak{O} \models x \doteq b$.

*Example 3.* Consider the $\mathcal{CFD}_{nc}$ ontology $\mathfrak{O} = (\mathcal{T}, \mathcal{A})$ where

$$\mathcal{T} := \{A \sqsubseteq A : f \rightarrow id\} \quad \text{and} \quad \mathcal{A} := \{A(a), f(a) = b, A(x), f(x) = b\}.$$

Since both $x$ and $a$ belong to $A$ and have the same value $b$ for the path function $f$, the path functional dependency in $\mathcal{T}$ implies that they must be equal, i.e., we have $\mathfrak{O} \models x \doteq a$.

Theorem 2 together with Theorem 1 implies that the DLs $\mathcal{ALCO}$, $\mathcal{ALCQ}$, and $\mathcal{CFD}_{nc}$ cannot be expressed in FOL without equality. We leave it to the reader to come up with translations of nominals, qualified number restrictions, and path functional dependencies into FOL with equality.

## 4    The Complexity of the Identity Problem

In this section, we first show that the identity problem can be polynomially reduced to the *instance problem* for all DLs with equality power. Note that the instance problem is one of the basic inference problems for DLs, and thus instance checking facilities are available in most DL reasoners. Given an ontology $\mathfrak{O}$, a concept description $C$, and an individual name $a$, we say that $a$ is an *instance* of $C$ w.r.t. $\mathfrak{O}$ (written $\mathfrak{O} \models C(a)$) if $a^{\mathcal{I}} \in C^{\mathcal{I}}$ holds for all models $\mathcal{I}$ of $\mathfrak{O}$.

**Lemma 1.** *Let $\mathcal{L}$ be a DL with equality power, $\mathfrak{O} = (\mathcal{T}, \mathcal{A})$ an $\mathcal{L}$ ontology and $a, b$ two distinct individual names. If $B$ is a concept name not occurring in $\mathfrak{O}$, then we have*

$$\mathfrak{O} \models a \doteq b \ \ iff \ \ (\mathcal{T}, \mathcal{A} \cup \{B(a)\}) \models B(b).$$

*Proof.* The direction from left to right is trivial. We show the other direction by contraposition. Thus, assume that $\mathfrak{O} \not\models a \doteq b$. Let $\mathcal{I}$ be a model of $\mathfrak{O}$ such that $a^{\mathcal{I}} \neq b^{\mathcal{I}}$. Let $\mathcal{I}'$ be the interpretation that coincides with $\mathcal{I}$ on all role names, individual names, and concept names different from $B$. For $B$ we define $B^{\mathcal{I}'} := \{a^{\mathcal{I}}\}$. Since $B$ does not occur in $\mathfrak{O}$, the interpretation $\mathcal{I}'$ is still a model of $\mathcal{T}$ and $\mathcal{A}$, and it satisfies $B(a)$ by our definition of $B^{\mathcal{I}'}$. However, it does not satisfy $B(b)$ since $b^{\mathcal{I}'} = b^{\mathcal{I}} \neq a^{\mathcal{I}}$ does not belong to $B^{\mathcal{I}'}$.    □

This lemma shows that the identity problem is at most as complex as the instance problem for all DLs with equality power that allow instance assertions for concept names in the ABox. Since the instance problem is polynomial for $\mathcal{CFD}_{nc}$ [22], this implies that also the identity problem is polynomial for this DL. In [22] it is mentioned that P-hardness of the consistency problem for $\mathcal{CFD}_{nc}$ ontologies is an easy consequence of P-hardness of satisfiability of propositional Horn formulas [5]. We now show that the same is true also for the identity problem.

**Theorem 3.** *The identity problem is P-complete for $\mathcal{CFD}_{nc}$ ontologies.*

*Proof.* We already know that the problem is in P. To show P-hardness, we reduce Horn-SAT to the identity problem. Recall that a Horn-formula $\phi$ is a finite set of clauses of the form

(a) $p_1 \wedge \ldots \wedge p_n \rightarrow p_0$ where $n > 0$ and $p_0, \ldots, p_n$ are propositional variables;
(b) $\rightarrow p_0$, which states that the propositional variable $p_0$ must be true;
(c) $p_1 \wedge \ldots \wedge p_n \rightarrow$ for $n > 0$ propositional variables $p_1, \ldots, p_n$, which states that $p_1, \ldots, p_n$ cannot be true at the same time.

Given $\phi$, we construct a $\mathcal{CFD}_{nc}$-ontology $\mathfrak{O}_\phi = (\mathcal{T}_\phi, \mathcal{A}_\phi)$ as follows. For every propositional variable $p$ occurring in $\phi$ we introduce a functional role $f_p$ as well as individuals $c_p, d_p$. In addition, we introduce the functional role $f_\perp$ and the individuals $c_\perp, d_\perp$ and $a, b$. Intuitively, we encode truth of the propositional variable $p$ as equality of the individuals $c_p$ and $d_p$, and inconsistency as equality of $c_\perp$ and $d_\perp$. Clauses of the form (a) and (c) are encoded using path functional dependencies in the TBox and clauses of the form (b) as path function assertions in the ABox. To be more precise, we define:

$$\mathcal{T}_\phi := \{A \sqsubseteq A : f_{p_1}, \ldots f_{p_n} \to f_{p_0} \mid p_1 \wedge \ldots \wedge p_n \to p_0 \in \phi\}$$
$$\{A \sqsubseteq A : f_{p_1}, \ldots f_{p_n} \to f_\perp \mid p_1 \wedge \ldots \wedge p_n \to \ \in \phi\}$$
$$\mathcal{A} := \{A(a), A(b)\} \cup$$
$$\{f_p(a) = c_p, f_p(b) = d_p \mid p \in \mathsf{var}(\phi)\} \cup \{f_\perp(a) = c_\perp, f_\perp(b) = d_\perp\} \cup$$
$$\{c_{p_0} = d_{p_0} \mid \to p_0 \in \phi\}.$$

where $A$ is a concept name. The ontology constructed this way satisfies the syntactic restrictions on $\mathcal{CFD}_{nc}$ ontologies. Moreover, it can be constructed in logarithmic space since it can simply be read off the representation of $\phi$.

By definition, the assertions $c_{p_0} = d_{p_0}$ enforce equality of these individuals iff $\phi$ contains a clause $\to p_0$ of the form (b). The path functional dependencies in $\mathcal{T}_\phi$ can then be used to derive further equalities according to the clauses of the form (a) and (c) in $\phi$. It is thus easy to see that equality of the individuals $c_p$ and $d_p$ can be derived from $\mathcal{O}$ iff $\phi$ implies that the propositional variable $p$ must be set to true. Consequently, deriving an equality of the individuals $c_\perp$ and $d_\perp$ indicates that a clause $p_1 \wedge \ldots \wedge p_n \to$ of the form (c) in $\phi$ is violated. In fact, deriving $c_\perp = d_\perp$ is only possible if there is such a clause of the form (c) in $\phi$ and the equalities $c_{p_i} = d_{p_i}$ $(i = 1, \ldots, n)$ have already been derived.

Using this intuition, it is then easy to prove the following claim:

$$\phi \text{ is unsatisfiable iff } \mathcal{O} \models c_\perp \dot{=} d_\perp,$$

which states correctness of our reduction, and thus establishes P-hardness of the identity problem in $\mathcal{CFD}_{nc}$. □

For $\mathcal{ALCO}$ and $\mathcal{ALCQ}$, the instance problem is ExpTime-complete [14,21]. Thus, we obtain exponential-time upper bounds for the identity problem in these DLs. To show that these upper bounds are optimal, we basically prove that there are polynomial-time reductions of the instance problem in $\mathcal{ALC}$ to the identity problem in these logics. In fact, the instance problem is already ExpTime-hard for the common sub-logic $\mathcal{ALC}$ of $\mathcal{ALCO}$ and $\mathcal{ALCQ}$ [15]. Before introducing these reductions and proving that they are correct, we have to deal with a subtlety that shows up in these proofs.

Note that, in $\mathcal{ALC}$, we can assume without loss of generality that any instance relationship that does not follow from an ontology can be refuted by a model of cardinality greater than 1.

**Lemma 2.** Let $\mathcal{O} = (\mathcal{T}, \mathcal{A})$ be an $\mathcal{ALC}$ ontology, $C$ an $\mathcal{ALC}$ concept description, and $a$ an individual name. If $\mathcal{O} \not\models C(a)$, then there is a model $\mathcal{I}$ of $\mathcal{O}$ such that $a^\mathcal{I} \notin C^\mathcal{I}$ and $|\Delta^\mathcal{I}| \geq 2$.

*Proof.* This follows from the fact that models of $\mathcal{ALC}$ ontologies are closed under disjoint union (see [3], Theorem 3.8). In fact, if $\mathcal{O} \not\models C(a)$, then there is a model $\mathcal{I}$ of $\mathcal{O}$ such that $a^\mathcal{I} \notin C^\mathcal{I}$. However, this model could have cardinality 1. If we take the disjoint union $\mathcal{J} = \mathcal{I}_1 \uplus \mathcal{I}_2$ of $\mathcal{I}$ with itself, then the cardinality of $\Delta^\mathcal{J}$

is twice the cardinality of $\Delta^{\mathcal{I}}$, and thus at least 2. Theorem 3.8 in [3] says that $\mathcal{J}$ is a model of $\mathcal{T}$. Regarding the ABox, we assume that all individual names occurring in $\mathcal{A}$ are interpreted in $\mathcal{J}$ by their interpretation in the renaming $\mathcal{I}_1$ of $\mathcal{I}$. Using Lemma 3.7 in [3], it is easy to see that this ensures that $\mathcal{J}$ is also a model of $\mathcal{A}$. □

Note that this lemma does not hold for $\mathcal{ALCO}$ ontologies. For example, $\mathfrak{O} = (\{\top \sqsubseteq \{a\}\}, \emptyset)$ has only models of size 1, and $\mathfrak{O} \not\models A(a)$. This is the reason why we use the DL $\mathcal{ALC}$ rather than the more expressive logics $\mathcal{ALCO}$ or $\mathcal{ALCQ}$ in our reductions.

**Lemma 3.** *Let $\mathcal{L} \in \{\mathcal{ALCO}, \mathcal{ALCQ}\}$, $\mathfrak{O}$ be an $\mathcal{ALC}$ ontology, $C$ an $\mathcal{ALC}$ concept description, and $a$ an individual name. Then we can construct in polynomial time an $\mathcal{L}$ ontology $\mathfrak{O}'$ and individuals $a', b'$ such that*

$$\mathfrak{O} \models C(a) \quad \text{iff} \quad \mathfrak{O}' \models a' \doteq b'.$$

*Proof.* Let $\mathfrak{O} = (\mathcal{T}, \mathcal{A})$. We consider the two DLs separately.

(1.) $\mathcal{L} = \mathcal{ALCO}$:
We define $\mathfrak{O}' := (\mathcal{T} \cup \{C \sqsubseteq \forall r.\{b'\}\}, \mathcal{A} \cup \{r(a, a'), r(a, b')\})$, where $a', b'$ are distinct individual names and $r$ is a role name such that $a', b', r$ do not occur in $\mathfrak{O}$. The direction from left to right is again trivial. The other direction is shown by contraposition. Let $\mathcal{I}$ be a model of $\mathfrak{O}$ such that $a^{\mathcal{I}} \notin C^{\mathcal{I}}$. By Lemma 2, we can assume without loss of generality that the domain of $\mathcal{I}$ contains at least two distinct elements $d_1 \neq d_2$. We construct an interpretation $\mathcal{I}'$ that coincides with $\mathcal{I}$ on all concept, role, and individual names occurring in $\mathfrak{O}$, and thus is also a model of $\mathfrak{O}$. In addition, $\mathcal{I}'$ interprets $r$ as $r^{\mathcal{I}'} := \{(a^{\mathcal{I}}, d_1), (a^{\mathcal{I}}, d_2)\}$ and the new individual names as $a'^{\mathcal{I}'} := d_1$ and $b'^{\mathcal{I}'} := d_2$. By construction, $\mathcal{I}'$ satisfies the assertional part of $\mathfrak{O}'$. To see that it also satisfies the GCI $C \sqsubseteq \forall r.\{b'\}$, note that $a^{\mathcal{I}} = a^{\mathcal{I}'}$ is the only element of $\mathcal{I}'$ that has successors w.r.t. the role $r$. Since it does not belong to $C^{\mathcal{I}} = C^{\mathcal{I}'}$, the elements of $C^{\mathcal{I}'}$ trivially satisfy the value restriction $\forall r.\{b'\}$. Thus, $\mathcal{I}'$ is a model of $\mathfrak{O}'$ in which the individuals $a', b'$ are interpreted by different elements, which shows $\mathfrak{O}' \not\models a' \doteq b'$.

(2.) $\mathcal{L} = \mathcal{ALCQ}$:
We define $\mathfrak{O}' := (\mathcal{T} \cup \{C \sqsubseteq \, \leqslant 1\, r.\top\}, \mathcal{A} \cup \{r(a, a'), r(a, b')\})$, where $a', b'$ are distinct new individuals and $r$ is a new role name not occurring in $\mathfrak{O}$. The direction from left to right is again trivial. To show the other direction, assume that $\mathcal{I}$ is a model of $\mathfrak{O}$ such that $a^{\mathcal{I}} \notin C^{\mathcal{I}}$. Again, we assume without loss of generality that the domain of $\mathcal{I}$ contains at least two distinct elements $d_1 \neq d_2$. We construct an interpretation $\mathcal{I}'$ in the same way as in case 1. above. Also, the argument why $\mathcal{I}'$ is a model of $\mathfrak{O}'$ in which $a', b'$ are interpreted by different elements is identical to the one above. □

As an easy consequence of Lemmas 1 and 3 we obtain the exact complexity of the identity problem in $\mathcal{ALCO}$ and $\mathcal{ALCQ}$. In fact, Lemma 1 yields ExpTime upper bounds. To show that Lemma 3 indeed yields ExpTime lower bounds, we need to take into account the fact that we have defined the identity problem with only consistent ontologies as possible input. In fact, since the consistency problem can be reduced to the instance problem in $\mathcal{ALC}$, it could potentially be the case that the reason for the ExpTime-hardness of the instance problem comes from the hardness of consistency only. However, we will show now that this is not the case, i.e., we show that ExpTime-hardness of the instance problem in $\mathcal{ALC}$ also holds if we consider the instance problem only for consistent $\mathcal{ALC}$ ontologies $\mathfrak{O}$.

**Lemma 4.** *The instance problem w.r.t. consistent $\mathcal{ALC}$ ontologies is ExpTime-hard.*

*Proof.* We show this by a reduction of the (un)satisfiability problem for $\mathcal{ALC}$-concepts w.r.t. TBoxes, which is also known to be ExpTime-complete ([3], Theorem 5.13). Recall that $C$ is satisfiable w.r.t. $\mathcal{T}$ iff there is a model $\mathcal{I}$ of $\mathcal{T}$ satisfying $C^{\mathcal{I}} \neq \emptyset$.

Thus, let $C$ be an $\mathcal{ALC}$ concept description and $\mathcal{T}$ an $\mathcal{ALC}$ TBox. We can assume without loss of generality that $\mathcal{T}$ consists of a single GCI $\top \sqsubseteq D$ for an $\mathcal{ALC}$ concept description $D$ (see [3], page 117). Note that $\mathcal{T}$ may actually be inconsistent.

Given $C$ and $D$, we now construct a consistent $\mathcal{ALC}$ ontology $\mathfrak{O}_{C,D} = (\mathcal{T}_{C,D}, \emptyset)$ as follows:

$$\mathcal{T}_{C,D} := \{B \sqsubseteq \exists r.(C \sqcap A), A \sqsubseteq D\} \cup \{A \sqsubseteq \forall s.A \mid s \text{ occurs in } C, D\},$$

where $A, B$ are concept names not occurring in $C, D$ and $r$ is a role name not occurring in $C, D$. It is easy to see that $\mathfrak{O}_{C,D}$ is consistent. In fact, any interpretation $\mathcal{I}$ with $A^{\mathcal{I}} = B^{\mathcal{I}} = \emptyset$ is obviously a model of $\mathcal{T}_{C,D}$. Thus, to prove the lemma it is sufficient to show that the following holds (for an arbitrary individual name $a$):

$$C \text{ is satisfiable w.r.t. } \{\top \sqsubseteq D\} \text{ iff } \mathfrak{O}_{C,D} \not\models \neg B(a).$$

First, assume that $\mathfrak{O}_{C,D} \not\models \neg B(a)$. This means that there is a model $\mathcal{I}$ of $\mathfrak{O}_{C,D}$ that interprets $B$ as a non-empty set. Then the first GCI ensures that there is an element $d_0$ of $A$ that also belongs to $C$. In addition, all the elements connected via roles occurring in $C, D$ with $d_0$ also belong to $A$, and thus to $D$ because of the second GCI. Consequently, if we restrict $\mathcal{I}$ to these elements, we obtain a model of $\top \sqsubseteq D$ in which $d_0$ belongs to $C$. This shows that $C$ is satisfiable w.r.t. $\{\top \sqsubseteq D\}$.

Conversely, assume that $\mathcal{I}$ is a model of $\{\top \sqsubseteq D\}$ with $d_0 \in C^{\mathcal{I}}$. Then $\mathcal{I}$ can easily be extended to a model of $\mathcal{T}_{C,D}$ in which $a$ belongs to $B$ by (i) introducing an additional element $d$ belonging to $B$, (ii) interpreting $a$ as $d$, (iii) interpreting $r$ as $\{(d, d_0)\}$, and (iv) putting $d_0$ as well as all the elements reachable from it into $A$. ☐

In addition, if $\mathfrak{O}$ is a consistent $\mathcal{ALC}$ ontology, then so are the ontologies $\mathfrak{O}'$ constructed from it in the proof of Lemma 3. Thus, Lemma 3 together with

Lemma 4 yields the matching ExpTime lower bounds for the identity problem in $\mathcal{ALCO}$ and $\mathcal{ALCQ}$.

**Theorem 4.** *The identity problem is ExpTime-complete for $\mathcal{ALCO}$ and $\mathcal{ALCQ}$ ontologies.*

For the three DLs with equality power considered in this paper, the identity problem has the same complexity as the instance problem. A natural question to ask is whether this is always the case. A simple example shows that the answer to this question is negative. In fact, let $\mathcal{ALC}^=$ be the DL $\mathcal{ALC}$, with the only difference that $\mathcal{ALC}^=$ ABoxes may contain equality assertions $a \doteq b$ between individual names. It is easy to see that the identity problem in this DL is non-trivial, but it can be solved in polynomial time. In fact, to check whether a consistent $\mathcal{ALC}^=$ ontology implies an equality $a \doteq b$, we only need to construct the reflexive, transitive, and symmetric closure of the explicitly stated equalities. However, since $\mathcal{ALC}$ is a sub-logic of $\mathcal{ALC}^=$, the instance problem in this DL is ExpTime-hard (and it is easy to show that it is also in ExpTime).

One may also wonder whether the complexity of the instance problem can be transferred to the identity problem also for DLs where the instance problem has a higher complexity than ExpTime. For example, the DL $\mathcal{ALCOIQ}$, which extends both $\mathcal{ALCO}$ and $\mathcal{ALCQ}$ and additionally allows the use of inverse roles, has a NExpTime-complete satisfiability problem [20], even w.r.t. the empty TBox. This implies that the instance problem w.r.t. consistent $\mathcal{ALCOIQ}$ ontologies is coNExpTime-complete. In fact, the $\mathcal{ALCOIQ}$ concept description $C$ is unsatisfiable iff $(\emptyset, \emptyset) \models \neg C(a)$ (for a new individual name $a$), which shows coNExpTime-hardness also w.r.t. consistent ontologies. The complexity upper bound follows from the NExpTime upper bound of satisfiability in $C^2$, i.e., two-variable fragment of first-order logic with counting quantifiers [12].

Since $\mathcal{ALCOIQ}$ contains $\mathcal{ALCO}$, it has equality power and can force models to have cardinality 1. Lemma 1 implies that the identity problem in $\mathcal{ALCOIQ}$ is in coNExpTime. Regarding hardness, the reductions employed in the proof of Lemma 3 can in principle both be used since the constructors employed in them are available in $\mathcal{ALCOIQ}$. However, Lemma 3 uses an $\mathcal{ALC}$ ontology $\mathcal{O}$ in the reduction, which yields only an ExpTime lower bound. Simply using an $\mathcal{ALCOIQ}$ ontology instead does not work since the proof depends on the fact that $\mathcal{O}$ has models refuting the instance relation of cardinality at least 2. However, by looking at the NExpTime-hardness proof for satisfiability in $\mathcal{ALCOIQ}$ in [20], it is easy to see that the following modified instance problem is also coNExpTime-hard for consistent $\mathcal{ALCOIQ}$ ontologies: is $a$ an instance of $C$ in all models of $\mathcal{O}$ of cardinality $\geq 2$? Thus, one can without loss of generality restrict the attention to models of cardinality $\geq 2$ when reducing the instance problem for $\mathcal{ALCOIQ}$ to the identity problem for this logic.

**Theorem 5.** *The identity problem is coNExpTime-complete for $\mathcal{ALCOIQ}$ ontologies.*

# 5   The View-Based Identity Problem

In this section, we will adapt the approach of [17,18] for view-based information hiding such that it can formalize the rôle-based access control scenario sketched in the introduction. We assume that ontologies are written using some DL $\mathcal{L}$ with equality power.

To define what kind of information is to be hidden, we divide the set of individual names into the disjoint sets $N_{AI}$ and $N_{KI}$ consisting of anonymous and known individuals, respectively. As before, we do not make the unique name assumption for these individuals. Given an anonymous individual $x \in N_{AI}$ and an ontology $\mathfrak{O}$, we define the *identity* of $x$ w.r.t. $\mathfrak{O}$ as

$$idn(x, \mathfrak{O}) := \{b \in N_{KI} \mid \mathfrak{O} \models x \doteq b\}.$$

Note that $b, b' \in idn(x, \mathfrak{O})$ implies that $\mathfrak{O} \models b' \doteq b$. Thus, if the cardinality of $idn(x, \mathfrak{O})$ is greater 1, this does not mean that $x$ is equal to one of these individuals, but rather that it is equal to all of them (and thus that all of them are equal). We say that $x$ is a *hidden* if $idn(x, \mathfrak{O}) = \emptyset$.

In the rôle-based access control scenario we assume that there is a "large" *input ontology* $\mathfrak{O}_I$ that is always consistent, but users can only see a part of it depending on which rôle they currently have. More formally, we assume that there is a finite set of *user rôles* $\mathfrak{R}$, and that playing the rôle $\hat{r} \in \mathfrak{R}$ gives access to a subset $\mathfrak{O}_{\hat{r}} \subseteq \mathfrak{O}_I$ of the input ontology. Here "access" does not mean that a user with rôle $\hat{r}$ can download the ontology $\mathfrak{O}_{\hat{r}}$. Instead, the users can ask queries to $\mathfrak{O}_{\hat{r}}$, where a *subsumption query* is of the form $C \sqsubseteq D$ for concept descriptions $C, D$ and a *retrieval query* is of the form $C$ for concept descriptions $C$ or $r$ for role names $r$.

**Definition 3.** *Let $\mathfrak{O}_I$ be the input ontology, $\mathfrak{O}_{\hat{r}} \subseteq \mathfrak{O}_I$ the ontology accessible by users with rôle $\hat{r} \in \mathfrak{R}$, and $q$ be a query. The answer to $q$ w.r.t. $\hat{r}$, denoted by $ans(q, \hat{r})$, is defined as follows:*

- *$ans(q, \hat{r}) := \{\mathbf{true}\}$, if $q = C \sqsubseteq D$ and $\mathfrak{O}_{\hat{r}} \models C \sqsubseteq D$,*
- *$ans(q, \hat{r}) := \emptyset$, if $q = C \sqsubseteq D$ and $\mathfrak{O}_{\hat{r}} \not\models C \sqsubseteq D$,*
- *$ans(q, \hat{r}) := \{a \in N_I \mid \mathfrak{O}_{\hat{r}} \models C(a)\}$, if $q = C$,*
- *$ans(q, \hat{r}) := \{(a, b) \in N_I \times N_I \mid \mathfrak{O}_{\hat{r}} \models r(a, b)\}$, if $q = r$.*

Since $\mathfrak{O}_{\hat{r}} \subseteq \mathfrak{O}_I$, positive answers to queries, i.e., $ans(C \sqsubseteq D, \hat{r}) = \{\mathbf{true}\}$, $a \in ans(C, \hat{r})$, or $(a, b) \in ans(r, \hat{r})$, imply that this subsumption, instance, or role relationship also holds in $\mathfrak{O}_I$. In contrast, negative answers do not tell us anything about what holds in $\mathfrak{O}_I$ since the inclusion may be strict. Answers to queries w.r.t. rôle $\hat{r}$ can be stored in a view.

**Definition 4.** *A view is a total function $V : dom(V) \to 2^{N_I} \cup 2^{N_I \times N_I} \cup \{\{\mathbf{true}\}\}$ where the view definition $dom(V)$ is a finite set of queries and $V(q)$ is a finite set for all $q \in dom(V)$. This view is a view for $\hat{r} \in \mathfrak{R}$ (written $\hat{r} \models V$) if $V(q) = ans(q, \hat{r})$ holds for all $q \in dom(V)$. The size of the view $V$ is defined as $\sum_{q \in dom(V)} size(q) \cdot |V(q)|$, where the size of a query $q$ is obtained in an obvious way from the sizes of the concepts/roles defining it.*

In a setting where user rôles can dynamically change, a user may successively play rôles $\hat{r}_1, \hat{r}_2, \ldots, \hat{r}_k$, in each rôle $\hat{r}_i$ generating (and storing) a view $V_{\hat{r}_i}$ for $\hat{r}_i$ by asking queries. The question is now whether these views can be used to find out the identity of a given anonymous individual $x \in N_{AI}$. Assume that the user wants to know whether there is a $b \in N_{KI}$ such that $b \in idn(x, \mathfrak{O}_I)$. However, the user cannot access $\mathfrak{O}_I$ as a whole, all she knows is that the positive answers to the queries in the views $V_{\hat{r}_i}$ are justified by subsets of $\mathfrak{O}_I$. Consequently, instead of one (unknown) ontology $\mathfrak{O}_I$, the user needs to consider all possible ontologies, i.e., all ontologies that are compatible with the positive answers in the views.

**Definition 5.** *The ontology $\mathfrak{P}$ is a* possible ontology *for the sequence of views $V_{\hat{r}_1}, \ldots, V_{\hat{r}_k}$ if $\mathfrak{P}$ is consistent and compatible with all positive answers in these views, where $\mathfrak{P}$ is* compatible *with*

- $V_{\hat{r}_i}(C \sqsubseteq D) = \{\mathbf{true}\}$ *if* $\mathfrak{P} \models C \sqsubseteq D$,
- $a \in V_{\hat{r}_i}(C)$ *if* $\mathfrak{P} \models C(a)$, *and* $(a, b) \in V_{\hat{r}_i}(r)$ *if* $\mathfrak{P} \models r(a, b)$.

*We denote the* set *of all possible ontologies for $V_{\hat{r}_1}, \ldots, V_{\hat{r}_k}$ with $Poss(V_{\hat{r}_1}, \ldots, V_{\hat{r}_k})$. The* certain identity *of $x$ w.r.t. $V_{\hat{r}_1}, \ldots, V_{\hat{r}_k}$ is defined as*

$$cert\_idn(x, V_{\hat{r}_1}, \ldots, V_{\hat{r}_k}) := \bigcap_{\mathfrak{P} \in Poss(V_{\hat{r}_1}, \ldots, V_{\hat{r}_k})} idn(x, \mathfrak{P}).$$

*We say that $x$ is* hidden *w.r.t. $V_{\hat{r}_1}, \ldots, V_{\hat{r}_k}$ if $cert\_idn(x, V_{\hat{r}_1}, \ldots, V_{\hat{r}_k}) = \emptyset$.*

Since $\mathfrak{O}_I \in Poss(V_{\hat{r}_1}, \ldots, V_{\hat{r}_k})$, we know that $b \in cert\_idn(x, V_{\hat{r}_1}, \ldots, V_{\hat{r}_k})$ implies that $b \in idn(x, \mathfrak{O}_I)$. Thus, if $cert\_idn(x, V_{\hat{r}_1}, \ldots, V_{\hat{r}_k}) \neq \emptyset$, the identity of $x$ in $\mathfrak{O}_I$ is no longer hidden. Conversely, if $cert\_idn(x, V_{\hat{r}_1}, \ldots, V_{\hat{r}_k}) = \emptyset$, then for all $b \in N_{KI}$ there is a $\mathfrak{P} \in Poss(V_{\hat{r}_1}, \ldots, V_{\hat{r}_k})$ such that $\mathfrak{P} \not\models x \doteq b$. Since, according to the information available to the user, $\mathfrak{O}_I$ could be this $\mathfrak{P}$, she cannot conclude for any $b \in N_{KI}$ that $\mathfrak{O}_I \models x \doteq b$. This shows that $cert\_idn(x, V_{\hat{r}_1}, \ldots, V_{\hat{r}_k}) = \emptyset$ indeed corresponds to the fact that the views $V_{\hat{r}_1}, \ldots, V_{\hat{r}_k}$ do not disclose the identity of $x$.

Since the set $Poss(V_{\hat{r}_1}, \ldots, V_{\hat{r}_k})$ consists of infinitely many ontologies, the definition of $cert\_idn(x, V_{\hat{r}_1}, \ldots, V_{\hat{r}_k})$ does not directly yield an approach for computing this set. We will now show that we can reduce this computation to the identity problem for the *canonical ontology* of $V_{\hat{r}_1}, \ldots, V_{\hat{r}_k}$. Basically, this ontology consists of the GCIs, concept assertions, and role assertions obtained from the positive answers in the views.

**Definition 6.** *The* canonical ontology $\mathcal{C}(V_{\hat{r}_1}, \ldots, V_{\hat{r}_k})$ *of $V_{\hat{r}_1}, \ldots, V_{\hat{r}_k}$ is defined as $\mathcal{C}(V_{\hat{r}_1}, \ldots, V_{\hat{r}_k}) := (\mathcal{T}, \mathcal{A})$ where*

$$\mathcal{T} := \{C \sqsubseteq D \mid V_{\hat{r}_i}(C \sqsubseteq D) = \{\mathbf{true}\} \text{ for some } i, 1 \leq i \leq k\}$$
$$\mathcal{A} := \{C(a) \mid a \in V_{\hat{r}_i}(C) \text{ for some } i, 1 \leq i \leq k\} \cup$$
$$\{r(a, b) \mid (a, b) \in V_{\hat{r}_i}(r) \text{ for some } i, 1 \leq i \leq k\}.$$

Note that the size of $\mathcal{C}(V_{\hat{r}_1}, \ldots, V_{\hat{r}_k})$ is linear in the sum of the sizes of the views $V_{\hat{r}_1}, \ldots, V_{\hat{r}_k}$.

Since $\mathcal{C}(V_{\hat{r}_1}, \ldots, V_{\hat{r}_k})$ consists of all positive answers in the views $V_{\hat{r}_1}, \ldots, V_{\hat{r}_k}$, it clearly implies them, and thus $\mathcal{C}(V_{\hat{r}_1}, \ldots, V_{\hat{r}_k}) \in \mathsf{Poss}(V_{\hat{r}_1}, \ldots, V_{\hat{r}_k})$. Conversely, every ontology $\mathfrak{P} \in \mathsf{Poss}(V_{\hat{r}_1}, \ldots, V_{\hat{r}_k})$ implies all these positive answers, and thus all the GCIs, concept assertions, and role assertions in $\mathcal{C}(V_{\hat{r}_1}, \ldots, V_{\hat{r}_k})$. This implies that every consequence of $\mathcal{C}(V_{\hat{r}_1}, \ldots, V_{\hat{r}_k})$ is also a consequence of $\mathfrak{P}$.

**Theorem 6.** *Given views $V_{\hat{r}_1}, \ldots, V_{\hat{r}_k}$ and an anonymous individual $x \in N_{AI}$, we have* $cert\_idn(x, V_{\hat{r}_1}, \ldots, V_{\hat{r}_k}) = idn(x, \mathcal{C}(V_{\hat{r}_1}, \ldots, V_{\hat{r}_k}))$.

*Proof.* First assume that $b \in cert\_idn(x, V_{\hat{r}_1}, \ldots, V_{\hat{r}_k})$. Then we have $\mathfrak{P} \models x \doteq b$ for all $\mathfrak{P} \in \mathsf{Poss}(V_{\hat{r}_1}, \ldots, V_{\hat{r}_k})$. Since $\mathcal{C}(V_{\hat{r}_1}, \ldots, V_{\hat{r}_k}) \in \mathsf{Poss}(V_{\hat{r}_1}, \ldots, V_{\hat{r}_k})$, this yields $\mathcal{C}(V_{\hat{r}_1}, \ldots, V_{\hat{r}_k}) \models x \doteq b$, and thus $b \in idn(x, \mathcal{C}(V_{\hat{r}_1}, \ldots, V_{\hat{r}_k}))$.

Conversely, assume $b \in idn(x, \mathcal{C}(V_{\hat{r}_1}, \ldots, V_{\hat{r}_k}))$, and thus $\mathcal{C}(V_{\hat{r}_1}, \ldots, V_{\hat{r}_k}) \models x \doteq b$. We must show that, for all $\mathfrak{P} \in \mathsf{Poss}(V_{\hat{r}_1}, \ldots, V_{\hat{r}_k})$, we have $\mathfrak{P} \models x \doteq b$. This is an immediate consequence of the fact that all the consequences of $\mathcal{C}(V_{\hat{r}_1}, \ldots, V_{\hat{r}_k})$ are also consequences of $\mathfrak{P}$. □

This theorem shows that, to check whether $x$ is *hidden* w.r.t. $V_{\hat{r}_1}, \ldots, V_{\hat{r}_k}$, it is sufficient to compute $idn(x, \mathcal{C}(V_{\hat{r}_1}, \ldots, V_{\hat{r}_k}))$. In case the employed ontology language $\mathcal{L}$ allows for unrestricted GCIs, concept assertions, and role assertions, the set $idn(x, \mathcal{C}(V_{\hat{r}_1}, \ldots, V_{\hat{r}_k}))$ can clearly be computed using an algorithm that solves the identity problem for $\mathcal{L}$ ontologies a polynomial number of times. Note that this applies to the DLs $\mathcal{ALCO}$, $\mathcal{ALCQ}$, and $\mathcal{ALCOIQ}$ considered in the previous sections, but not to $\mathcal{CFD}_{nc}$ since there GCIs and concept assertions need to satisfy certain restrictions.

**Corollary 1.** *For $\mathcal{L} \in \{\mathcal{ALCO}, \mathcal{ALCQ}\}$ we can check in exponential time whether an anonymous individual $x$ is hidden w.r.t. views $V_{\hat{r}_1}, \ldots, V_{\hat{r}_k}$. For $\mathcal{L} = \mathcal{ALCOIQ}$, this problem can be solved in NExpTime.*

The ExpTime upper bound for $\mathcal{ALCO}$ and $\mathcal{ALCQ}$ is obvious. For $\mathcal{ALCOIQ}$, one considers all the (polynomially many) known individuals $a_1, \ldots, a_p$. Using a NExpTime procedure for the complement of the identity problem, one then checks whether $x$ is not identical to $a_1$. The non-successful paths of this nondeterministic computation stop with failure whereas the successful ones continue with the same test for $a_2$, etc. It is easy to see that this yields the desired NExpTime procedure. In fact, any path of this procedure has only exponential length, and a successful path indicates that inequality with $x$ holds for all known individuals.

## 6  Conclusions and Future Work

In this paper, we have provided some initial definitions and results regarding the identity problem in DL ontologies, i.e., the question whether the ontology implies that a given anonymous individual is equal to a known individual.

We have also considered a more involved rôle-based access control scenario where users can access parts of the ontology depending on their rôle. In a setting where users can change rôles dynamically, the question is then whether, by changing rôles and asking queries in these rôles, the user can find out the identity of an anonymous individual although this may not be possible for a single rôle. We have shown how to use the identity problem to address this question.

Until now, we have only investigated how to find out whether the identity of an anonymous individual is disclosed in a certain situation. We have not considered what to do when this is the case. One possibility would be to additionally anonymize the available information, e.g., by replacing some of the known individuals in assertions by new anonymous ones, similar to what is done in [9].

Another direction for future research could be to look at $k$-anonymity [19] rather than identity. In principle, our identity problem is concerned with 1-anonymity, i.e., we want to avoid that one can deduce from the given information that an anonymous individual belongs to a singleton set consisting of only one known individual. In many applications, one also wants to ensure that the set of known individuals to which the anonymous one is known to belong has a large enough cardinality, i.e., one $> k$. Of course, in this setting additional anonymization (as mentioned above) is also relevant in cases where $k$-anonymity is not given.

Finally, we intend to consider cases where the information about known and anonymous individuals holds only with a certain probability, e.g., using ontologies with subjective probability as introduced in [10]. In this setting, equality can also only be derived with a certain probability, and one might want to keep the probability of derived identities low enough.

# References

1. Baader, F., Borchmann, D., Nuradiansyah, A.: Preliminary results on the identity problem in description logic ontologies. In: Proceedings of the 30th International Workshop on Description Logics (2017)
2. Baader, F., Calvanese, D., McGuinness, D.L., Patel-Schneider, P.F., Nardi, D. (eds.): The Description Logic Handbook: Theory, Implementation, and Applications. Cambridge University Press, Cambridge (2003)
3. Baader, F., Horrocks, I., Sattler, U.: An Introduction to Description Logic. Cambridge University Press, Cambridge (2017)
4. Biskup, J., Bonatti, P.A.: Controlled query evaluation for enforcing confidentiality in complete information systems. Int. J. Inf. Sec. 3(1), 14–27 (2004)
5. Cook, S., Nguyen, P.: Logical Foundations of Proof Complexity, 1st edn. Cambridge University Press, New York (2010)
6. Gallier, J.: Logic for Computer Science: Foundations of Automatic Theorem Proving, 2nd edn. Dover (2015)
7. Grau, B.C.: Privacy in ontology-based information systems: a pending matter. Semant. Web 1, 137–141 (2010)
8. Grau, B.C., Horrocks, I.: Privacy-preserving query answering in logic-based information systems. In: Proceedings of the 18th European Conference on Artificial Intelligence, pp. 40–44 (2008)

9. Grau, B.C., Kostylev, E.V.: Logical foundations of privacy-preserving publishing of linked data. In: Proceedings of the Thirtieth AAAI Conference on Artificial Intelligence, pp. 943–949. AAAI Press (2016)
10. Gutiérrez-Basulto, V., Jung, J.C., Lutz, C., Schröder, L.: Probabilistic description logics for subjective uncertainty. J. Artif. Intell. Res. (JAIR) **58**, 1–66 (2017)
11. Hollunder, B., Baader, F.: Qualifying number restrictions in concept languages. In: Proceedings of the 2nd International Conference on the Principles of Knowledge Representation and Reasoning (KR 1991), pp. 335–346 (1991)
12. Pratt-Hartmann, I.: Complexity of the two-variable fragment with counting quantifiers. J. Logic Lang. Inform. **14**(3), 369–395 (2005)
13. Sandhu, R.S., Coyne, E.J., Feinstein, H.L., Youman, C.E.: Role-based access control models. Computer **29**(2), 38–47 (1996)
14. Schaerf, A.: Reasoning with individuals in concept languages. In: Torasso, P. (ed.) AI*IA 1993. LNCS, vol. 728, pp. 108–119. Springer, Heidelberg (1993). https://doi.org/10.1007/3-540-57292-9_49
15. Schild, K.: A correspondence theory for terminological logics: preliminary report. In: Proceedings of the 12th International Joint Conference on Artificial Intelligence (IJCAI 1991), pp. 466–471 (1991)
16. Schmidt-Schauß, M., Smolka, G.: Attributive concept descriptions with complements. Artif. Intell. **48**(1), 1–26 (1991)
17. Stouppa, P., Studer, T.: A formal model of data privacy. In: Virbitskaite, I., Voronkov, A. (eds.) PSI 2006. LNCS, vol. 4378, pp. 400–408. Springer, Heidelberg (2007). https://doi.org/10.1007/978-3-540-70881-0_34
18. Stouppa, P., Studer, T.: Data privacy for $\mathcal{ALC}$ knowledge bases. In: Artemov, S., Nerode, A. (eds.) LFCS 2009. LNCS, vol. 5407, pp. 409–421. Springer, Heidelberg (2008). https://doi.org/10.1007/978-3-540-92687-0_28
19. Sweeney, L.: K-anonymity: a model for protecting privacy. Int. J. Uncertain. Fuzziness Knowl.-Based Syst. **10**(5), 557–570 (2002)
20. Tobies, S.: The complexity of reasoning with cardinality restrictions and nominals in expressive description logics. J. Artif. Intell. Res. **12**, 199–217 (2000)
21. Tobies, S.: Complexity results and practical algorithms for logics in knowledge representation. CoRR, cs.LO/0106031, PhD thesis, RWTH Aachen (2001)
22. Toman, D., Weddell, G.: Conjunctive query answering in $\mathcal{CFD}_{nc}$: a PTIME description logic with functional constraints and disjointness. In: Cranefield, S., Nayak, A. (eds.) AI 2013. LNCS (LNAI), vol. 8272, pp. 350–361. Springer, Cham (2013). https://doi.org/10.1007/978-3-319-03680-9_36

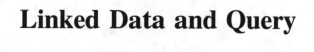

# Linked Data and Query

# Resolving Range Violations in DBpedia

Piyawat Lertvittayakumjorn[1,2](✉), Natthawut Kertkeidkachorn[2,3],
and Ryutaro Ichise[2,3]

[1] Department of Computing, Imperial College London, London SW7 2AZ, UK
pl1515@imperial.ac.uk
[2] National Institute of Informatics, Tokyo 101-8430, Japan
{natthawut,ichise}@nii.ac.jp
[3] SOKENDAI (The Graduate University for Advanced Studies),
Tokyo 101-8430, Japan

**Abstract.** DBpedia, a large-scale multi-disciplinary knowledge graph extracted from structured data in Wikipedia, is an essential part of the Linked Open Data (LOD). However, several previous works report many types of errors existing in DBpedia. The crucial one is *a range violation error* – a problem when an object of a triple does not have a type required by the range of the triple's predicate. This inconsistency could undermine the effectiveness of any applications using DBpedia. In this paper, we aim to correct these erroneous triples by finding correct objects with the required type to replace the incorrect objects. Our approach is based on graph analysis and keyword matching. It also exploits information from the incorrect objects because, despite their incorrectness, they contain useful clues to find the correct objects. The results from eight different datasets show that our proposed approach outperforms various baseline methods, including entity search (e.g., Soft-TFIDF and DBpedia Lookup) and knowledge graph completion (TransE and AMIE+).

**Keywords:** DBpedia · Linked data · Data quality · Error correction · Range violation error · Knowledge graph refinement

## 1 Introduction

DBpedia is a large knowledge graph extracted from structured data in Wikipedia such as infobox templates, categorization information, images, and disambiguation pages [12]. The graph structure of DBpedia is so powerful that it can answer many complex queries, based on Wikipedia data. The current English version of DBpedia contains more than 4.6 million things (entities) and more than 100 million facts (triples) [6]. Due to its large-scale size and the multi-disciplinary characteristic inherited from Wikipedia, DBpedia becomes a nucleus of the Linked Open Data (LOD) project to which numerous linked data resources connect [2].

It is obvious that the information extraction process, from tons of Wikipedia articles, to construct DBpedia must be done automatically. Although the extraction framework was carefully designed, the resulting knowledge graph is not

© Springer International Publishing AG 2017
Z. Wang et al. (Eds.): JIST 2017, LNCS 10675, pp. 121–137, 2017.
https://doi.org/10.1007/978-3-319-70682-5_8

free of errors for many reasons such as human errors and inconsistencies within Wikipedia. One major type of errors in DBpedia is a problem when an object of a triple does not have a type required by the range of the triple's predicate [7]. We call this error type as *a range violation error* (RVE). Currently, 24.6% of DBpedia triples whose predicate is a mapping-based object property with a defined range are suffering from this kind of error. For example, the triple <dbr[1]:Sedo, dbo[2]:locationCountry, dbr:Cologne> in DBpedia is erroneous because the predicate dbo:locationCountry requires an object with the type dbo:Country, which dbr:Cologne is devoid of since Cologne is a city, not a country. This inconsistency could undermine the effectiveness of any applications using DBpedia. To correct this error, the object dbr:Cologne should be replaced by dbr:Germany, the country where Cologne is located.

In this paper, we aim to automatically solve the range violation error (of mapping-based object properties) by finding a correct object to replace the incorrect object for each of the erroneous triples. An underlying idea of our approach is that the clues as to the correct object could be found in the profile of the subject and the incorrect object. Even though there are other strategies to fix the range violation error, depending on its root cause, such as adding the missing type to the object and refining the range indicated in the ontology, our paper still makes a significant impact on the research field because (1) it complements existing works on knowledge graph refinement which follow other strategies [16,18,20] and (2) based on our investigation, fixing by replacing objects is applicable to more than 51.7% of all range violation errors in DBpedia.

Overall, the main contributions of this paper are as follows:

- A novel approach finding correct replacements for incorrect objects of DBpedia's range violation triples with mapping-based object properties.
- An evaluation of the proposed approach showing promising results compared to several baseline methods.
- Eight public datasets manually constructed as a gold standard for the range violation problem which can be used in further research.

## 2    Range Violation Errors in DBpedia

In the latest version of DBpedia[3], there are 1,103 object properties (owl:Object Property) each of which requires an entity as its object and 1,746 datatype properties (owl:DataTypeProperty) each of which requires a literal as its object. Every datatype property is provided with the corresponding range, i.e., the required data type of the literal object. For instance, the range of dbo:salary is xsd:double, while the range of dbo:filename is xsd:string. In contrast, there are only 828 object properties whose range is indicated in the ontology. The range of each object property specifies a required (DBpedia ontology) class that the object entity must have as its rdf:type. For example, the

---

[1] dbr: http://dbpedia.org/resource/.
[2] dbo: http://dbpedia.org/ontology/.
[3] http://wiki.dbpedia.org/downloads-2016-04 (accessed: 6th June 2017).

range of dbo:commander is dbo:Person, whereas the range of dbo:residence is dbo:Place. In this paper, we focus on solving the range violation errors for these 828 object properties.

The easiest way to do so is just removing the erroneous triples from DBpedia. However, this is not a good practice because we may lose some distorted but useful information stored in those removed triples. A better way is fixing these erroneous triples in order to recover as many accurate facts as possible. Since the range violation error is a conflict between the *property*'s *range* and the *object*'s *type*, we can fix this error by (one or more than one of) the four strategies below.

1. **Change the property** – If there is a more appropriate property that preserves the triple's knowledge but does not cause the range violation error, we could change the triple's property. For example, the triple <dbr:John_ Harvard_(statue), dbo:museum, dbr:Harvard_Yard> suffers the range violation error because the property dbo:museum requires its object to have the type dbo:Museum, but dbr:Harvard_Yard is not a museum. Actually, the knowledge this triple stores is that the statue of John Havard is located in the Havard Yard. So, a more appropriate property for this triple could be dbo:location because *(i)* it is more semantically fitting and *(ii)* its range is dbo:Place and dbr:Harvard_Yard has this type. Therefore, a proper fix for this erroneous triple is replacing dbo:museum with dbo:location.

2. **Refine the property's range** – The ranges of some object properties in the ontology cannot support all possible facts of those properties. For example, the DBpedia ontology specifies that the range of dbo:compiler (the person or entity responsible for selecting the album's track listing) is dbo:Person. This leads to several range-violation triples such as <dbr:Love_Rocks, dbo:compiler, dbr:Human_Rights_Campaign> and <dbr:Journeys_by_DJ, dbo:compiler, dbr:Coldcut> whose objects are a dbo:Organisation and a dbo:Band respectively, not a dbo:Person. Therefore, an appropriate way to resolve these conflicts is refining the range of dbo:compiler from dbo:Person to dbo:Agent which is a superclass of not only dbo:Person but also dbo:Band and dbo:Organisation.

3. **Change the object** – Some range violation errors exist because the object entities are incorrect. These situations could be attributed to faults in Wikipedia articles, in extraction templates, or in an entity resolution process. For instance, the triple <dbr:The_Fighting_Devil_Dogs, dbo:language, dbr:John_English_(director)> is incorrect because John English is a person, not a language as required by the dbo:language property. This error happens due to the wrong entity resolution which misinterpreted the English language as John English. So, we should fix this triple by changing its object from dbr:John_English_(director) to dbr:English_language.

4. **Refine the object's type(s)** – In this case, the property and the object together enunciate the accurate fact, but the missing or incorrect type of the object leads to range violation. For example, Albert Ernest Doyle (A. E. Doyle) in the triple <dbr:Butler_Bank, dbo:architect, dbr:A._E. _Doyle> was a veritable architect; however, the required type dbo:Architect

is missing from the entity dbr:A._E._Doyle. The best way to fix this inconsistency is adding the type dbo:Architect to dbr:A._E._Doyle.

Among the four strategies, only refining the property's range alters the knowledge graph schema and this task has been being studied under the topic *ontology learning and refinement.* Meanwhile, advanced research in *entity type prediction* can address refining the object's type(s). However, few earlier works suggest the replacements of literal, entity, or property to fix errors found in knowledge graphs [17]. So, our proposed approach aims to fill this niche by automatically changing the incorrect objects that inflict the range violation.

Let us now consider a repair by changing objects. Usually, we can replace the incorrect object with only one correct object to fix the error. For example, in the previous case, dbr:English_language is the one and only perfect choice to replace dbr:John_English_(director). However, in some cases, we have to replace the incorrect object with more than one correct object. This is evident in the case of <dbr:KidsCo, dbo:locationCountry, dbr:Europe>. KidsCo was a brand launched in many countries of Europe. So, we must replace dbr:Europe with all European countries where KidsCo was launched such as Poland, Romania, Hungary, etc. To reduce the complexity of the problem, in this work, we deal only with cases in which the incorrect object should be replaced by only one correct object. Formally, our problem formulation is

- **Given** an erroneous triple $t = \langle s, p, o \rangle$ in DBpedia where
  - $p$ is a DBpedia object property with the range $r_p$
  - $o$ is an incorrect object which has at least one DBpedia ontology class (dbo) as its type but does not have $r_p$ as its type
- **Find** a semantically correct object $o'$ that has the type $r_p$ and best replaces the incorrect object $o$ in $\langle s, p, o \rangle$.

The condition "$o$ has at least one DBpedia ontology class as its type" helps us filter out a triple whose object could already be the correct one but has not been classified to any class yet. So, in a realistic scenario, the triple should be handled by refining the object's type first. If we later find that the object is certainly not an instance of $r_p$, we can apply our approach to change the object.

## 3   Related Work

Correcting range violation errors is a way to improve the *quality* of knowledge graphs. Up to now, many dimensions of knowledge graph quality have been studied, assessed, and improved. However, our work is closely related to three of them – *completeness, correctness,* and *consistency.*

Knowledge graph completion tries to increase the completeness of knowledge graphs by predicting missing types for entities or missing relations between entities. It can be employed to refine an object's type(s) for resolving range violations (see Strategy 4 in Sect. 2). Previously, various methods have been applied to complete missing types and/or relations in knowledge graphs such as association rule mining [8], natural language processing [10], statistical relational learning [3], etc. As regards our task, changing the incorrect objects of range violation triples may be separated into two steps – *(i)* removing the problematic links followed by *(ii)* constructing the correct links from the subjects and the predicates. In principle, knowledge graph completion techniques can address the second step after the first step is managed. Nonetheless, adopting this idea neglects useful information stored in the incorrect objects. So, our approach attempts to exploit all components of the erroneous triples including their incorrect objects. This makes our work distinct from other knowledge graph completion techniques.

Regarding the other two dimensions, knowledge graph correctness and consistency are related concepts, but not equivalent. While correctness concerns whether the data stored in a knowledge graph correctly represents the real-world facts or not, consistency corresponds to the absence of contradictions among triples, axioms, and the ontology of the knowledge graph. Literally, the range violation error is a kind of consistency problems; however, some correctness problems can also lead to range violation such as the use of incorrect properties or objects (see Strategy 1 and 3 in Sect. 2). Many research papers aim to detect correctness errors in knowledge graphs by different means such as using outlier detection methods [22], using learned axioms [15] or axioms from an external ontology [19], and cross-checking with other knowledge graphs [14]. However, far too little attention has been paid to automatic correction of detected errors [17] which makes our contribution novel and significant.

More recently, some publications turn to address inconsistencies between a knowledge graph and its ontology including the domain and range violations. For example, Tonon et al. found that some of these violations occur because the properties are used in multiple contexts. So, after identifying these properties, they created a new sub-property with the proper domain and range for each context and used it instead of the old property [20]. Paulheim identified and manually fixed the mapping statements which contributed to each group of inconsistencies in DBpedia he automatically detected. Also, he started discussion threads about changing some domains and ranges in the ontology to fix the violations [16] (see Strategy 2 in Sect. 2). In contrast, Dimou et al. transformed the mappings into RML mapping language before assessing and refining them (semi-)automatically to prevent several OWL axiom violations [7]. Currently, DBpedia uses their work for domain validation[4]. Nevertheless, fixing mappings and re-running them as the last two papers did can solve only the errors due to incorrect mappings, but fail to handle the errors due to mistakes in the original data source. Unlike them, our approach working on the triples directly can deal with errors originated from both mappings and data source.

## 4   Our Approach

For an erroneous triple $t = \langle s, p, o \rangle$, our approach finds a semantically correct object $o'$ that has the type $r_p$ and best replaces the incorrect object $o$. It is clear

---

[4] http://mappings.dbpedia.org/validation/index.html.

that only objects with the type $r_p$ could be the correct answer, so the complete search space of this problem $(S_p)$ is a set of all entities with the type $r_p$. However, $S_p$ is enormous for some properties $p$. So, our approach *(i)* constructs a reduced search space $(S_t)$ that contains only the entities related to the erroneous triple $t$ and then *(ii)* calculates scores of the entities in $S_t$. An object with a higher score is more likely to be the correct object.

## 4.1  Constructing a Reduced Search Space

To obtain the reduced search space of triple $t$ $(S_t)$, we follow the pseudocode in Fig. 1. First, we create the complete search space $S_p$ using a SPARQL query in line 2. Then we find related properties of $p$ (line 3). Formally, $p'$ is a related property of $p$ if and only if there exists at least one $(x, y)$, $y \in S_p$, such that both $\langle x, p, y \rangle$ and $\langle x, p', y \rangle$ are in the knowledge base. To illustrate, dbo:headquarter is a related property of dbo:targetAirport because headquarters of some airlines are also their target airports. After that, in line 7–11, we create the first portion of $S_t$ named $S_{t,1}$ which stores all entities in $S_p$ that are linked to $s$ by at least one related property of $p$ (written as $p'$). In some cases, $s$ may have $p'$ links to objects (entities or literals) that lack the type $r_p$ and, therefore, they are not included in $S_{t,1}$. However, we do not completely ignore these objects because they may give us some hints to the correct object. So, if the conditional probability $P(\langle x, p, y \rangle | \langle x, p', y \rangle, y \in S_p)$ is larger than a threshold $\tau$, we transform the objects $e$ into clue texts stored in $C_t$ (line 12–14) together with the label of the incorrect object $o$. Note that the notation lab($x$) means the label of $x$ when $x$ is an entity, but when $x$ is a literal lab($x$) is $x$ itself.

**Input**    $t = \langle s, p, o \rangle$ : an erroneous triple
$r_p$ : the range of property $p$
$\tau$ : a threshold for selecting clue texts
**Output**  $S_t$ : a search space for triple $t$

1.  $S_{t,1}, S_{t,2}, S_{t,3}$ = empty set
2.  $S_p$ = SPARQLq($\langle ?e, \text{rdf:type}, r_p \rangle$)
3.  $RelatedProps$ = findRelatedPropOf($p$)
4.  $C_t$ = [lab($o$)]
5.  $K$ = dict()
6.  $AllKw$ = []
7.  **for** $p'$ **in** $RelatedProps$:
8.      $E1$ = SPARQLq($\langle s, p', ?e \rangle$)
9.      **for** $e$ **in** $E1$:
10.        **if** $e \in S_p$:
11.          $S_{t,1}$.add($e$)
12.        **else**:
13.          **if** conditionalProb($p, p', S_p$) > $\tau$:
14.            $C_t$.append(lab($e$))
15.  **for** $c$ **in** $C_t$:
16.      $K[c]$ = getKeywordSet($c$)
17.      $AllKw$.union($K[c]$)
18.  $AllKw$ = removeGeneral($AllKw$)
19.  **for** $e$ **in** $S_p$:
20.      **if** $\exists w \in AllKw[w$ is in $abs(e)]$:
21.        $S_{t,2}$.add($e$)
22.  $E2$ = SPARQLq($\langle o, ?r, ?e \rangle$)
23.  $E2$.union(SPARQLq($\langle ?e, ?r, o \rangle$))
24.  **for** $(e, r)$ **in** $E2$:
25.      **if** $e \in S_p$:
26.        $S_{t,3}$.add($e$)
27.  $S_t$ = merge($S_{t,1}, S_{t,2}, S_{t,3}$)

**Fig. 1.** Pseudocode for constructing the reduced search space $S_t$

Next, for each clue text $c \in C_t$, create a set of keywords $K_c$ (written as $K[c]$ in Fig. 1) by *(i)* tokenizing $c$ into a set of keywords excluding stop words and digits *(ii)* enlarging the set with stems of existing keywords and then *(iii)* removing elements in the set whose substring is already in the set. All of these steps are executed in function getKeywordSet (line 16). Then gather keywords from all clue texts of $t$ and keep them in a set called $AllKw$ (line 17). We will use $AllKw$ to search for candidate objects, so we consider only the keywords whose length is at least four to prevent matches by accident. Moreover, we exclude the label of $r_p$ from the set; otherwise, it might match a lot of objects in $S_p$ with no relations to the correct object. For example, when $p$ is dbo:targetAirport, "airport" ($lab(r_p)$) should not be in $AllKw$ because it matches almost all entities in $S_p$. The function removeGeneral in line 18 performs this task. When our keyword set is ready, we create the second portion of $S_t$ named $S_{t,2}$ storing all entities in $S_p$ whose abstract contains at least one keyword in $AllKw$ (line 19–21). Please note that the term $abs(e)$ returns an English abstract of $e$, i.e. $\langle e, \text{dbo:abstract}, abs(e) \rangle$. If the abstract does not exist, we use $lab(e)$ instead.

Last but not least, we create the third portion of $S_t$ named $S_{t,3}$ which collects all entities in $S_p$ that connect immediately to the incorrect object $o$ in any direction (line 22–26). Finally, we union (merge) the three portions $(S_{t,1}, S_{t,2}, S_{t,3})$ to be $S_t$ in line 27.

## 4.2 Calculating Scores

We have designed three novel scoring methods to evaluate the likelihood that a candidate object $e \in S_t$ is the correct object of the triple $t$. The first and second methods are based on graph analysis and keyword matching respectively, while the last one averages out the normalized scores of a candidate object given by the first two methods. In any case, the object with a higher score is more likely to be the correct object.

**Method 1: Graph Method.** Intuitively, the correct object $o'$ should be strongly related to $o$ compared to other entities in $S_t$, and the more related two entities are, the more objects connect both entities. For example, compared to dbr:Japan, dbr:Australia is more likely to be a correct dbo:Country object for replacing dbr:Sydney due to a lot of objects which connect to both dbr:Sydney and dbr:Australia (e.g., dbo:Organisation entities which have dbo:city links to dbr:Sydney and dbo:country links to dbr:Australia and dbo:Person entities with dbr:deathPlace links to both dbr:Sydney and dbr:Australia). In contrast, a few objects connect to both dbr:Japan and dbr:Sydney in this manner. Moreover, immediate links between dbr:Sydney and dbr:Australia (e.g., <dbr:Australia, dbo:largestCity, dbr:Sydney>) also emphasize the strong relationship between both entities. Therefore, we define a scoring function based on graph analysis, $g : S_t \to \mathbb{R}$, as

$$g(e) = |A(o, e)| + b(e) \tag{1}$$

$$A(o, e) = \{x : \text{isURI}(x) \land (\langle o, r_1, x \rangle \lor \langle x, r_1, o \rangle) \land \tag{2}$$
$$(\langle e, r_2, x \rangle \lor \langle x, r_2, e \rangle) \land r_1, r_2 \neq \text{rdf:type}\}$$

and

$$b(e) = \begin{cases} 1 \text{ if } e \in S_{t,3}; \\ 0 \text{ otherwise.} \end{cases} \tag{3}$$

In essence, $A(o, e)$ is a set of entities that have direct links to both $o$ and $e$ regardless of the links' direction, and $b(e)$ indicates whether $e$ links immediately to the incorrect object $o$ or not. However, SPARQL takes a long time to compute the function isURI, which checks if $x$ is an entity, so our implementation substitutes isURI with a semantically equivalent condition $\langle x, \text{rdf:type}, \text{owl:Thing} \rangle$ to optimize runtime performance.

**Method 2: Keyword Method.** The clue texts we obtain while constructing the search space $S_t$ may be able to help us find the correct object. From the previous example of dbr:Sydney, "Sydney" (the label of the incorrect object) becomes a clue text. So, entities in $S_p$ whose abstracts contain the clue text "Sydney" (such as dbr:Australia and dbr:Electoral_district_of_City_of_Sydney) are likely to be the correct object in this case. With this intuition, we develop the keyword matching method to find $e \in S_t$ which the clue texts support. The scoring function of this method, $m : S_t \to \mathbb{R}$, is defined as

$$m(e) = \sum_{c \in C_t} \frac{|\{w \in K_c : w \text{ is in } abs(e) \land cap(w) \land w \text{ is in } prof(o)\}| + 1}{|K_c| + 1} + r(e). \tag{4}$$

where

$$prof(o) = uri(o) + \sum_{(r,x) \in P_o} (\text{``} - \text{''} = + lab(r) + \text{`` : ''} + lab(x)) \tag{5}$$

$$P_o = \{(r, x) : \langle o, r, x \rangle \lor \langle x, r, o \rangle\} \tag{6}$$

and

$$r(e) = \begin{cases} 1 \text{ if } \exists p'[\langle s, p', e \rangle \land P(\langle x, p, y \rangle | \langle x, p', y \rangle, y \in S_p) > \tau] \text{ and} \\ \quad e \text{ is in } prof(o); \\ 0 \text{ otherwise.} \end{cases} \tag{7}$$

The score of an entity $e$ is calculated by aggregating scores of $e$ with respect to each clue text $c \in C_t$. The score with respect to $c$ reflects a proportion of keywords in $c$ which are found in the *abstract* of $e$. To ensure that the keywords refer to a named-entity which guides us to the replacement of $o$, we count only keywords $w$ that begin with a capital letter ($cap(w)$) and are related to $o$ ($w$ is in the *profile* of $o$). Equation 5 explains how to construct the profile string of an entity $o$ ($prof(o)$). Please note that, in this equation, the operation $+$ and $\sum$ denote string concatenation. $P_o$ in Eq. 6 stores all pairs of predicates and objects connecting to the entity $o$ in any direction. Additionally, the term $r(e)$ is a bonus point which will be 1 only if the conditions indicated in Eq. 7 are satisfied. This is because satisfying the conditions is equivalent to the match of one clue text where the clue is an entity in the search space $S_t$.

**Method 3: Combined method.** The reason behind this method is that the correct object $o'$ should gain relatively high scores from both graph and keyword methods. So, we define the scoring function $f : S_t \to \mathbb{R}$ as

$$f(e) = \frac{1}{2} \left( \frac{g(e)}{\max_{x \in S_t}(g(x))} + \frac{m(e)}{\max_{x \in S_t}(m(x))} \right). \tag{8}$$

Essentially, the combined method returns an average of the normalized scores from the graph method and the keyword method.

**Ranking Candidate Objects.** After calculating the scores by one of the proposed methods, we can rank the candidate objects and select one with the highest score to replace $o$ in $t$. In case the scores are equal, we have two more criteria to prioritize candidate objects. The first one is the number of portions of $S_t$ that contain the candidate, i.e., $|\{i : e \in S_{t,i}\}|$. If an entity has more reasons to be a candidate object for this triple, it is more likely to be the correct object. In case of the second tie, the number of $p$ in-links to the candidate will be used as a final criterion since it reflects the possibility of becoming the object of $p$ in $t$.

## 5  Experiments

### 5.1  Datasets

Since our problem formulation is different from previous studies, to the best of our knowledge, there is no publicly available dataset that fits with this problem. So, we decided to manually create datasets for evaluation. From the most recent version of DBpedia (see Footnote 3), we searched for all object properties of which more than 200 object entities contain at least one DBpedia ontology class but lack the required class. With this criteria, we found 92 properties substantially suffering from the range violation problem. Some of them can be repaired by replacing incorrect objects as we aim to do, while the rest needs other fixing strategies as discussed in Sect. 2. Considering only the former group, we selected eight properties with different characteristics to create gold standard datasets as summarized in Table 1. Each of the datasets contains one hundred erroneous triples $\langle s, p, o \rangle$ of a particular object property $p$ together with their correct objects $o'$.

The selected properties are dissimilar in terms of the required class $r_p$, the complete search space size $|S_p|$, and other interesting aspects. For example,

- Incorrect objects of some properties (e.g., language and college) usually have similar labels to their correct objects such as (dbr:Slovenia, dbr:Slovene_language) and (dbr:Arizona_Wildcats, dbr:University_of_Arizona). However, this feature does not exist in other properties.

**Table 1.** Eights different datasets used in the experiments (The datasets are available to download at http://ri-www.nii.ac.jp/FixRVE/Dataset8)

| Property $p$ (dbo) | Range $r_p$ (dbo) | $|S_p|$ |
|---|---|---|
| locationCountry | Country | 3, 424 |
| language | Language | 9, 215 |
| targetAirport | Airport | 14, 813 |
| routeEnd | RouteOfTransportation | 30, 782 |
| formerTeam | SportsTeam | 33, 614 |
| college | EducationalInstitution | 55, 302 |
| employer | Organisation | 280, 482 |
| birthPlace[a] | Place | 839, 987 |

[a]Normally, when our approach needs to use entities' abstracts, it queries the data from DBpedia sparql endpoint, but due to a large search space size of birth-Place dataset, the endpoint cannot return abstracts of all entities in the search space $S_p$. So, we use the abstracts from an abstract dataset provided on the DBpedia official website (see Footnote 3) instead.

- Incorrect objects of some properties are clearly related to *only one* object in $S_p$. For instance, in the locationCountry dataset, the incorrect objects are mostly places and each of them is always located in only one country. In contrast, for the targetAirport property, each incorrect object (city, state, or country) could be related to more than one airport in $S_p$.

A variety of error characteristics from different properties like these helps us prevent bias in the experiments.

## 5.2   Experiment 1: Evaluating the Search Space

If the correct object does not lie in the constructed search space $S_t$, our approach will fail to find it. So, in this experiment, we analyzed the quality of the search space $S_t$ to see if it successfully included the correct object $o'$. For each dataset, we calculated the average search space size and the number of cases whose $S_t$ contains $o'$. We also examined the distribution of correct answers among the three portions of $S_t - S_{t,1}$, $S_{t,2}$, and $S_{t,3}$. Please note that, in this (and the next) experiment, the parameter $\tau$ for selecting clue texts was set at 0.9 and we used Snowball stemmer[5] to stem the keywords of the clue texts.

**Results.** The results of this experiment are presented in Table 2. The first and second columns list the eight properties and the size of their complete search space $S_p$. The third column indicates the average size of the constructed search space $S_t$, whereas its proportion to the complete search space size is calculated

---

[5] http://snowball.tartarus.org/texts/introduction.html.

as a percentage in the fourth column. The fifth column shows the proportion
of cases whose $S_t$ contains the correct object $o'$. Finally, the last three columns
present the proportion of the correct objects in $S_{t,1}$, $S_{t,2}$, and $S_{t,3}$, respectively.

Table 2. The analysis results of the constructed search space $S_t$

| Property $p$ (dbo) | $|S_p|$ | Avg. $|S_t|$ | % | $o' \in S_t$ | $o' \in S_{t,1}$ | $o' \in S_{t,2}$ | $o' \in S_{t,3}$ |
|---|---|---|---|---|---|---|---|
| locationCountry | 3,424 | 87 | 2.53 | 0.99 | 0.61 | 0.43 | **0.92** |
| language | 9,215 | 686 | 7.44 | 0.96 | 0.28 | 0.87 | **0.89** |
| targetAirport | 14,813 | 242 | 1.64 | 1.00 | 0.40 | **0.97** | 0.94 |
| routeEnd | 30,782 | 1,582 | 5.14 | 1.00 | 0.66 | **0.90** | 0.88 |
| formerTeam | 33,614 | 9,935 | 29.56 | 0.98 | 0.78 | **0.92** | **0.92** |
| college | 55,302 | 3,690 | 6.67 | 1.00 | 0.48 | 0.98 | **1.00** |
| employer | 280,482 | 10,088 | 3.60 | 0.97 | 0.09 | 0.60 | **0.96** |
| birthPlace | 839,987 | 16,103 | 1.92 | 0.96 | 0.51 | **0.90** | 0.43 |
| Average | | | 7.31 | 0.98 | 0.48 | 0.82 | **0.87** |

We can notice from Table 2 that on average, our search space $S_t$ contains only
7.31% of $S_p$, but it covers the correct answers for more than 95% of test cases.
In other words, our method successfully includes the correct answers despite
the small search space. Besides, in most cases, correct objects lie in $S_{t,2}$ and
$S_{t,3}$. This phenomenon emphasizes the effectiveness of our clue texts and the
strong relationships between the correct objects and the incorrect objects. However, $S_{t,2}$ could not cover many correct entities in the locationCountry dataset
as the obtained keywords are names of small cities or locations that are not
important enough to appear in the abstract of the corresponding countries.
Meanwhile, $S_{t,3}$ is not so powerful when working with the birthPlace dataset
because sometimes there is not a direct link between $o$ and $o'$. For example,
an incorrect object is dbr:Barley (cereal grain), while its correct object is
dbr:Barley,_Hertfordshire (village). They share the name "Barley" but do
not immediately connect to each other. So, $S_{t,3}$ fails to include the correct object,
whereas $S_{t,2}$ relying on keywords is still effective in this dataset. Note that though
$S_{t,2}$ and $S_{t,3}$ work pretty well, $S_{t,1}$ is not useless due to two reasons: (1) there are
some correct objects which are covered only by $S_{t,1}$ and (2) it helps us prioritize
candidate objects in case the computed scores are equal.

## 5.3 Experiment 2: Evaluating the Whole Approach

The second experiment aims to assess the overall performance of our approach
in finding the correct objects or at least promoting them to top ranks as many as
possible. For each erroneous triple $t$, we ran all of the three scoring methods on
the search space $S_t$, sorted the candidates, and recorded the rank of the correct

object. We also compared our approach to several baseline methods used for entity search and knowledge graph completion.

**Entity Search.** As the incorrect object $o$ and the correct object $o'$ are usually related, one way to find $o'$ is searching for the object, with the required type $r_p$, that is most similar to $o$. So, we transform the problem of repairing range violation errors into an entity search problem. Various baseline methods for entity search in DBpedia were compared to our approach.

- *String similarity* ranks all entities in $S_p$ using similarity scores between the label of each entity and the label of $o$. For comparison, we selected four similarity measures, from different categories, which performed best for the title matching task [9]. These measures are Jaccard Index on bigrams [11], Damerau-Levenshtein [5], Dice coefficient on tokens [1], and Soft-TFIDF [4].
- *DBpedia lookup*[6] is a service to find DBpedia URIs from keywords, i.e., the label of $o$ in our task. An entity is matched based on not only its label but also anchor texts frequently used to represent that entity in Wikipedia. The search results are sorted by the number of inlinks pointing from other Wikipedia pages to an entity's Wikipedia page, and we also used this number as a score for ranking the matched entities.
- *wikiPageDisambiguates* is a DBpedia object property that connects a set of ambiguous terms to a corresponding disambiguation entity. For example, dbr:Acadian_(disambiguation) has dbo:wikiPageDisambiguates links to entities whose label is similar to "Acadian" such as dbr:Acadians (Ethnic group), dbr:Acadiana (Place), dbr:Acadian_French (Language), etc. So, if the incorrect object $o$ has a disambiguation entity $d$ pointing to it, we can pick the object(s) which have the type $r_p$ and also connect to $d$ as an answer.

**Knowledge Graph Completion.** As discussed in Sect. 3, we can use only $s$ and $p$ of the erroneous triple $t$ to find $o'$ using knowledge graph completion techniques. Thus, we applied two well-known completion techniques for comparison.

- *TransE* [3] considers a triple as a relation-specific translation of the subject to the object in a vector space. Let $V_p$ be a translation vector of property $p$, and $V_s$ and $V_e$ be latent features of entity $s$ and $e$, respectively. TransE assigns a distance between $V_e$ and $V_s + V_p$ as a score of an object $e \in S_p$. The closer, the better. In this experiment, we used the "bern" setting [21] and trained the whole DBpedia data using 1,000 epoch to obtain these vectors of length 100. The distance between vectors was calculated using L1 norm.
- *AMIE+* [8] is a rule mining system for knowledge bases with the open world assumption. To find the correct object, it selects the rules of property $p$ obtained from association rule mining, sets the rules' subject to $s$, and collects all objects $e \in S_p$ which satisfy the other side of one of the rules. The score of $e$ is the maximum of PCA Confidence values of the rules $e$ satisfies. Please note that we used only 40% of DBpedia data for rule training because the whole DBpedia takes too long time to do so.

---

[6] http://wiki.dbpedia.org/projects/dbpedia-lookup.

**Evaluation.** Our approach and some baseline methods were implemented in Python by us except Damerau-Levenshtein from pyxdameraulevenshtein[7], Soft-TFIDF from py_stringmatching[8], TransE [13], and AMIE+ [8]. In addition, we used three measures as our evaluation metric: *(i)* Mean reciprocal ranking (MRR) averaging out the multiplicative inverse of the ranks of correct entities. *(ii)* HITS@1 calculating the proportion of correct entities ranked as the first object in the sorted candidate list and *(iii)* HITS@10 calculates the proportion of correct entities staying in the top-10 of the sorted candidate list. We chose HITS@1 and HITS@10 as the metric because they are widely used to evaluate techniques for knowledge graph completion, a similar task of our problem. However, we did not use the simple average rank because it may mislead when a large number of entities get equal scores. For example, if there is no entity satisfying the AMIE+ rules of the property, AMIE+ will give a zero score to all entities and the rank of the correct object is just a random number between 1 and $|S_p|$, but it greatly affects the average rank value. So, we used the MRR instead because it is more robust to random ranks and prioritizes only top ranks.

**Results.** The results of this experiment are presented in Table 3. M, @1, and @10 stand for Mean reciprocal ranking (MRR), HITS@1, and HITS@10, respectively. The letter $T$ indicates that we could not get the results although we had executed the method on that dataset for more than two weeks. This is an unacceptable situation for fixing only one hundred triples. The mark - for some datasets of AMIE+ means that there is not any association rule of that property, and hence AMIE+ cannot find the correct object for any triple.

It is noticeable that, for all datasets, our methods outperform all baseline methods in finding the correct objects. The combined method performs best – gaining the maximum MRR scores for five datasets with an average MRR 0.66. Following the champion, the graph and the keyword methods win two datasets each. Their average MRRs are 0.63 and 0.60 respectively.

**Discussion.** All baseline methods from entity search relies on the similarity between the labels of $o$ and $o'$. This feature can be noticed in college, language, and birthPlace datasets, and also some triples in targetAirport dataset. However, these baseline methods have a different degree of robustness which directly affects their performance. DBpedia lookup is too strict about spelling and does not allow any keyword to miss when it searches for an entity. Using the property wikiPageDisambiguates is functional only if the labels are so similar that Wikipedia creates a disambiguation page to prevent users from confusion. Even though string similarity methods work without this limitation, the correct object might not get the first rank if there is another entity which is more similar to the incorrect object. In any case, entity search methods cannot understand associations between entities behind their labels. They do not know that dbr:Athens and dbr:Greece are closely related. This is the main reason why the baseline methods from entity search are not practical for repairing range violation triples.

---

[7] https://github.com/gfairchild/pyxDamerauLevenshtein.
[8] https://pypi.python.org/pypi/py_stringmatching.

**Table 3.** The performance of our approach compared to other baseline methods for each dataset

| Method | locationCountry | | | language | | | targetAirport | | | routeEnd | | |
|---|---|---|---|---|---|---|---|---|---|---|---|---|
| | M | @1 | @10 | M | @1 | @10 | M | @1 | @10 | M | @1 | @10 |
| Bi-Jaccard | 0.04 | 3 | 8 | 0.29 | 19 | 55 | 0.31 | 26 | 41 | 0.02 | 0 | 4 |
| Damerau-Leven | 0.03 | 2 | 5 | 0.22 | 10 | 50 | 0.14 | 11 | 17 | 0.02 | 0 | 6 |
| Dice | 0.02 | 2 | 3 | 0.24 | 21 | 37 | 0.37 | 33 | 42 | 0.01 | 0 | 4 |
| Soft-TFIDF | 0.03 | 3 | 3 | 0.35 | 34 | 38 | 0.42 | 38 | 47 | 0.02 | 0 | 3 |
| DBpedia lookup | 0.02 | 1 | 3 | 0.11 | 11 | 11 | 0.44 | 41 | 48 | 0.02 | 1 | 3 |
| wikiDisambiguates | 0.03 | 3 | 3 | 0.40 | 38 | 42 | 0.15 | 14 | 15 | 0.00 | 0 | 0 |
| TransE | 0.19 | 11 | 37 | 0.07 | 2 | 21 | 0.02 | 0 | 3 | 0.00 | 0 | 1 |
| AMIE+ | 0.43 | 42 | 44 | 0.10 | 9 | 13 | – | – | – | 0.00 | 0 | 0 |
| Graph Method | **0.95** | **93** | **99** | 0.81 | 70 | 93 | 0.78 | 71 | **91** | **0.50** | **33** | **89** |
| Keyword Method | 0.61 | 51 | 84 | 0.63 | 47 | 85 | 0.77 | 68 | **91** | 0.42 | 26 | 75 |
| Combined Method | **0.95** | **93** | **99** | **0.85** | **76** | **95** | **0.83** | **78** | **91** | 0.45 | 25 | **89** |

| Method | formerTeam | | | college | | | employer | | | birthPlace | | |
|---|---|---|---|---|---|---|---|---|---|---|---|---|
| | M | @1 | @10 | M | @1 | @10 | M | @1 | @10 | M | @1 | @10 |
| Bi-Jaccard | 0.17 | 16 | 19 | 0.53 | 42 | 72 | 0.07 | 5 | 10 | 0.24 | 17 | 36 |
| Damerau-Leven | 0.16 | 15 | 18 | 0.40 | 33 | 50 | 0.05 | 3 | 7 | 0.11 | 5 | 19 |
| Dice | 0.17 | 16 | 20 | 0.65 | 52 | 88 | 0.10 | 8 | 13 | 0.35 | 22 | 61 |
| Soft-TFIDF | 0.18 | 17 | 21 | 0.52 | 41 | 70 | $T$ | $T$ | $T$ | $T$ | $T$ | $T$ |
| DBpedia lookup | 0.06 | 6 | 6 | 0.27 | 27 | 27 | 0.07 | 6 | 10 | 0.30 | 23 | 41 |
| wikiDisambiguates | 0.02 | 2 | 3 | 0.17 | 14 | 21 | 0.06 | 4 | 12 | 0.33 | 21 | 58 |
| TransE | 0.02 | 1 | 1 | 0.02 | 1 | 2 | 0.01 | 0 | 4 | 0.09 | 6 | 15 |
| AMIE+ | 0.04 | 4 | 4 | – | – | – | 0.00 | 0 | 0 | 0.15 | 12 | 19 |
| Graph Method | 0.41 | 30 | 64 | 0.96 | 95 | 97 | 0.25 | 15 | 51 | 0.35 | 20 | 69 |
| Keyword Method | **0.68** | **60** | **84** | 0.91 | 88 | 97 | 0.24 | **16** | 44 | **0.50** | **41** | 70 |
| Combined Method | 0.45 | 33 | 67 | **1.00** | **99** | **100** | **0.27** | 15 | **55** | 0.46 | 25 | **83** |

The performance of methods from knowledge graph completion is worse than the ones from entity search for most datasets. AMIE+ works quite well only on the locationCountry dataset. One important factor defining the performance of AMIE+ is the amount and quality of rules it obtains from rule mining. If a property does not have any rule (e.g., targetAirport, college), or the rules are not applicable to the characteristics of erroneous triples (e.g., routeEnd, employer), we could not expect good results from AMIE+. Regarding TransE, its model performs vaguely well for locationCountry which is mostly a unique property for an entity. By contrast, it is not effective for 1-to-many relationships like formerTeam and college. This is due to the limitation of TransE that has too few variables compared to the number of equations it has to fulfill [23]. Another weakness of knowledge graph completion is that a correct entity, based on real-world facts, may be returned as an answer but it is not the right one to replace the incorrect object in our task. This is because knowledge graph completion uses only the subject and the predicate without exploiting the incorrect object.

Though our combined method performs best in this experiment, there is no universal method which works successfully for all of the datasets due to their differences in error characteristics. For example, the graph method is very effective when an incorrect object corresponds to only one (and correct) entity in $S_p$ (as in locationCountry and college datasets), because there are relatively many objects connecting this entity pair in DBpedia. However, if an incorrect object is not strongly related to its correct object via graph structure, the graph method may give us a wrong answer. This can be seen in the case of birthPlace dataset. For example, the correct object of the triple <dbr:Buddie_Petit, dbo:birthPlace, dbr:White_Castle_(restaurant)> is dbr:White_Castle,_Louisiana, but the graph method returns dbr:United_States, where the White Castle hamburger restaurant chain was founded, as the answer while putting the real correct object to rank 51$^{st}$. In contrast, the keyword method successfully returns the correct object because of the keywords "White" and "Castle" from the incorrect object's label. As a result, the combined method can push the correct object into the third rank though its score from the graph method is not high. Overall, the collaboration mechanism like this helps the combined method achieve a higher HITS@10 compared to both methods in the birth-Place dataset. However, this is not always potent for all cases especially when one of the methods is far worse than the other one. Therefore, selecting the most appropriate method to repair each erroneous triple is a key factor of this task. Also, combining scores from both methods using weighted average is an interesting idea when we have enough training data to define weights. We plan to study both choices in the near future.

All in all, one major concept of our approach which makes it outperform other baseline methods is making capital out of incorrect objects. It is effective because the incorrect objects are not random objects, but almost correct objects. So, starting from almost correct objects as we do is definitely better than starting from scratch as knowledge graph completion methods do. Furthermore, our approach also utilizes the information from the subjects and the properties (via related properties and clue texts), since the correct objects or their traces may appear somewhere in the profile of the subjects. This helps our approach beat baseline methods from entity search using only the incorrect objects' labels.

## 6 Conclusion

The aim of this paper is to fix range violation errors in DBpedia automatically by finding correct objects to replace the incorrect objects. We developed an algorithm to construct a small search space of candidate objects and three scoring methods to evaluate the candidates. Owing to our novel idea to exploit meaningful information from all components of the erroneous triples including the incorrect objects, our proposed approach is effective to find the correct objects as demonstrated in the experiments. For future work, we plan to devise a heuristic to select the most suitable method for repairing each erroneous triple and examine the feasibility of using supervised learning to address this problem.

# References

1. Adamson, G.W., Boreham, J.: The use of an association measure based on character structure to identify semantically related pairs of words and document titles. Inf. Storage Retr. **10**(7–8), 253–260 (1974)
2. Auer, S., Bizer, C., Kobilarov, G., Lehmann, J., Cyganiak, R., Ives, Z.: DBpedia: a nucleus for a web of open data. In: Aberer, K., et al. (eds.) ASWC/ISWC -2007. LNCS, vol. 4825, pp. 722–735. Springer, Heidelberg (2007). https://doi.org/10.1007/978-3-540-76298-0_52
3. Bordes, A., Usunier, N., Garcia-Durán, A., Weston, J., Yakhnenko, O.: Translating embeddings for modeling multi-relational data. In: 26th NIPS, pp. 2787–2795. Curran Associates Inc., USA (2013)
4. Cohen, W.W., Ravikumar, P., Fienberg, S.E.: A comparison of string distance metrics for name-matching tasks. In: The Workshop on IIWeb, pp. 73–78 (2003)
5. Damerau, F.J.: A technique for computer detection and correction of spelling errors. Commun. ACM **7**(3), 171–176 (1964)
6. DBpedia: Dbpedia 2016–04 statistics. http://wiki.dbpedia.org/dbpedia-2016-04-statistics. Accessed 13 May 2017
7. Dimou, A., Kontokostas, D., Freudenberg, M., Verborgh, R., Lehmann, J., Mannens, E., Hellmann, S., Van de Walle, R.: Assessing and refining mappings to RDF to improve dataset quality. In: 14th ISWC, pp. 133–149 (2015)
8. Galárraga, L., Teflioudi, C., Hose, K., Suchanek, F.M.: Fast rule mining in ontological knowledge bases with amie+. VLDB J. **24**(6), 707–730 (2015)
9. Gali, N., Mariescu-Istodor, R., Frnti, P.: Similarity measures for title matching. In: 23rd ICPR, pp. 1548–1553 (2016)
10. Gangemi, A., Nuzzolese, A.G., Presutti, V., Draicchio, F., Musetti, A., Ciancarini, P.: Automatic typing of DBpedia entities. In: Cudré-Mauroux, P., et al. (eds.) ISWC 2012. LNCS, vol. 7649, pp. 65–81. Springer, Heidelberg (2012). https://doi.org/10.1007/978-3-642-35176-1_5
11. Jaccard, P.: Étude comparative de la distribution florale dans une portion des Alpes et des Jura. Bulletin del la Société Vaudoise des Sciences Naturelles **37**, 547–579 (1901)
12. Lehmann, J., Isele, R., Jakob, M., Jentzsch, A., Kontokostas, D., Mendes, P.N., Hellmann, S., Morsey, M., Van Kleef, P., Auer, S., et al.: Dbpedia-a large-scale, multilingual knowledge base extracted from Wikipedia. Semant. Web **6**(2), 167–195 (2015)
13. Lin, Y., Liu, Z., Sun, M., Liu, Y., Zhu, X.: Learning entity and relation embeddings for knowledge graph completion. In: 29th AAAI, pp. 2181–2187. AAAI Press (2015)
14. Liu, S., d'Aquin, M., Motta, E.: Towards linked data fact validation through measuring consensus. In: 2nd Workshop on LDQ (2015)
15. Ma, Y., Gao, H., Wu, T., Qi, G.: Learning disjointness axioms with association rule mining and its application to inconsistency detection of linked data. In: Zhao, D., Du, J., Wang, H., Wang, P., Ji, D., Pan, J.Z. (eds.) CSWS 2014. CCIS, vol. 480, pp. 29–41. Springer, Heidelberg (2014). https://doi.org/10.1007/978-3-662-45495-4_3

16. Paulheim, H.: Data-driven joint debugging of the dbpedia mappings and ontology. In: Blomqvist, E., Maynard, D., Gangemi, A., Hoekstra, R., Hitzler, P., Hartig, O. (eds.) ESWC 2017. LNCS, vol. 10249, pp. 404–418. Springer, Cham (2017). https://doi.org/10.1007/978-3-319-58068-5_25

17. Paulheim, H.: Knowledge graph refinement: a survey of approaches and evaluation methods. Semant. Web **8**(3), 489–508 (2017)

18. Paulheim, H., Bizer, C.: Improving the quality of linked data using statistical distributions. IJSWIS **10**(2), 63–86 (2014)

19. Paulheim, H., Gangemi, A.: Serving DBpedia with DOLCE – more than just adding a cherry on top. In: Arenas, M., et al. (eds.) ISWC 2015. LNCS, vol. 9366, pp. 180–196. Springer, Cham (2015). https://doi.org/10.1007/978-3-319-25007-6_11

20. Tonon, A., Catasta, M., Demartini, G., Cudré-Mauroux, P.: Fixing the domain and range of properties in linked data by context disambiguation. In: 8th LDOW (2015)

21. Wang, Z., Zhang, J., Feng, J., Chen, Z.: Knowledge graph embedding by translating on hyperplanes. In: 28th AAAI, pp. 1112–1119. AAAI Press (2014)

22. Wienand, D., Paulheim, H.: Detecting incorrect numerical data in DBpedia. In: Presutti, V., d'Amato, C., Gandon, F., d'Aquin, M., Staab, S., Tordai, A. (eds.) ESWC 2014. LNCS, vol. 8465, pp. 504–518. Springer, Cham (2014). https://doi.org/10.1007/978-3-319-07443-6_34

23. Xiao, H., Huang, M., Zhu, X.: From one point to a manifold: knowledge graph embedding for precise link prediction. In: 25th IJCAI, pp. 1315–1321 (2016)

# Entity Linking in Queries Using Word, Mention and Entity Joint Embedding

Zhichun Wang$^{(\boxtimes)}$, Rongyu Wang, Danlu Wen, Yong Huang, and Chu Li

College of Information Science and Technology, Beijing Normal University,
Beijing 100875, People's Republic of China
zcwang@bnu.edu.cn, {wangrongyu,dlwen,yhuang,chuli}@mail.bnu.edu.cn

**Abstract.** Entity linking in queries is an important task for connecting search engines and knowledge bases. This task is very challenging because queries are usually very short and there is very limited context information for entity disambiguation. This paper proposes a new accurate and efficient entity linking approach for search queries. The proposed approach first jointly learns word, mention and entity embeddings in a unified space, and then computes a set of features for entity disambiguation based on the learned embeddings. The entity linking problem is solved as a ranking problem in our approach, a ranking SVM is trained to accurately predict entity links. Experiments on real data show that our proposed approach achieves better performance than comparison approaches.

**Keywords:** Entity linking · Query · Embedding

## 1 Introduction

Recently, large-scale Knowledge Bases have been successfully used to enhance web search engines' search result with semantic information. For example, Google uses its knowledge graph to provide structured and detailed information about the search topic in addition to a list of links to websites. To combine search engines and knowledge bases, linking entities in search queries to knowledge bases is a very important task.

Identifying entities in text and linking them to a given knowledge base is usually called entity linking. The task of entity linking is challenging because of name variations and entity ambiguity. On one hand, one entity can be mentioned in text by different names; for example, both "Beijing" and "Peking" can refer to the same entity "Beijing City". On the other hand, the same mention can refer to multiple different entities; for example "Apple" may refer to "Apple Inc" and the fruit "Apple", etc. Lots of work has been done on the problem of entity linking, [20] gives detailed review of all kinds of entity linking approaches.

Entity linking in queries of search engines is more challenging than traditional entity linking tasks. First, queries are usually very short and therefore the context information for entity linking is very limited. Second, entity linking in

© Springer International Publishing AG 2017
Z. Wang et al. (Eds.): JIST 2017, LNCS 10675, pp. 138–150, 2017.
https://doi.org/10.1007/978-3-319-70682-5_9

queries needs to be done very efficiently because the responding time is crucial for web search engines. Little work has been done on query entity linking till now. The most related work on this problem was done by Blanco et al. [2]; they proposed a probabilistic model for entity linking in queries; their model ignores any dependency between the different entity candidates in order to run fast; hashing and compression techniques are used to reduce memory usage. Some other work focuses on linking entities in short texts like tweets or web tables, but more context information can be used and the linking efficiency is not the primary consideration.

This paper proposes a new accurate and efficient entity linking approach for search queries. Comparing with Blanco's work, our approach uses a more efficient method to compute relatedness between entities and their contextual information; our approach also explores information in the same search session to disambiguate entity links, which was not used in Blanco's work; our approach uses machine learning algorithm to optimize the scoring function for entity linking, which leads to more accurate linking results. Specifically, main contributions of our work include

- We propose a method to embed Words, Mentions, and Entities to the same low-dimensional space; the learned embeddings enable our approach to compute different kinds of relatedness efficiently, which is very important for entity disambiguation.
- We propose to use previous queries in the same search session to expand the context information for entity linking, which is proved to be helpful for improving the accuracy of entity linking.
- We define a set of local and global features to capture mention-entity relatedness and query level coherence, which are fed to a learning-to-rank algorithm to get the optimized scoring function for entity linking.

We evaluate our approach on Yahoo's Webscope query log data, the experimental results show that our approach gets more accurate linking results than the compared approaches.

The rest of this paper is organized as follows: Sect. 2 presents the method of jointly learning word, mention and entity; Sect. 3 describes the proposed entity linking approach; Sect. 4 presents the evaluation results; Sect. 5 discusses some related work and finally Sect. 6 concludes this work.

## 2 Word, Mention and Entity Joint Embedding

In the process of entity linking, computing semantic relatedness between entities and contextual context is very important for entity disambiguation. In this paper, we propose to use Skip-gram model [15, 16] to jointly map words, mentions and entities to one unified low-dimensional vector space. By using the jointly learned vectors, various relatednesses, such as *entity-word* relatedness, *mention-word* relatedness and *entity-entity* relatedness, can be efficiently computed.

## 2.1  The Skip-Gram Model

The skip-gram model is a recently published learning framework to learn continuous word vectors from text corpora. Each word in the text corpora is mapped to a continuous embedding space. The model is trained to find word representations that are good at predicting the surrounding words in a sentence or a document. Given a sequence of training words $w_1, w_2, ..., w_T$, the objective of the model is to maximize the average log probability

$$O = \frac{1}{T} \sum_{t=1}^{T} \sum_{-c \leq j \leq c, j \neq 0} \log p(w_{t+j} | w_t) \tag{1}$$

where $c$ is the size of training context, $p(w_{t+j} | w_t)$ is defined as

$$p(w_O | w_I) = \frac{\exp(v'_{w_O}{}^T v_{w_I})}{\sum_{w=1}^{W} \exp(v'_{w_O}{}^T v_{w_I})} \tag{2}$$

where $v_w$ and $v'_w$ are the input and output vector representations of $w$, and $W$ is the number of words in the vocabulary. The learned vectors of words can capture the semantic similarity of words; similar words are mapped to nearby places in a low-dimensional vector space.

## 2.2  Joint Embedding by Skip-Gram Model

The Skip-gram model is initially designed to learn embeddings of words. In order to extend the word model to a joint model of word, entity and mention, we add mentions and entities in the training corpus which only contains words before. Let the original corpus be $\mathcal{C} = \{w_1, w_2, ..., w_N\}$, if a certain word sequence $s = \{w_i, ..., w_{i+k}\}$ in $\mathcal{C}$ is a mention to an entity $e$ in the knowledge base, we replace $s$ with two tokens $\{MEN(w_i w_{i+1}...w_{i+k}), ENT(e)\}$; after that, the original word sequence containing $s$ in $\mathcal{C}$ becomes $\{w_{i-1}, MEN(w_i w_{i+1}...w_{i+k}), ENT(e), w_{i+k+1}\}$. After annotating all the mentions and their corresponding entities, $\mathcal{C}$ converts to a hybrid corpus $\mathcal{C}'$ that contains words, mentions, and entities. $\mathcal{C}'$ is then used to train the Skip-gram model, which will generate representations in the same vector space for words, mentions, and entities. Figure 1 shows an example of using the Skip-gram model to predict the surrounding tokens of the word *capital* in the example sentence.

## 2.3  Using Wikipedia as Training Corpus

Annotating mentions and entities in a corpus is a time-consuming task. Fortunately, since Wikipedia contains all the annotations we need, we can use it as the knowledge base to improve working efficiency. Figure 2 shows part of the page of Beijing in Wikipedia and its source text in editing model. In Wikipedia, an internal hyperlink is annotated by [[**entity** | **mention**]]; it also could be [[**entity**]]

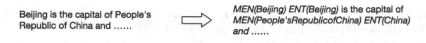

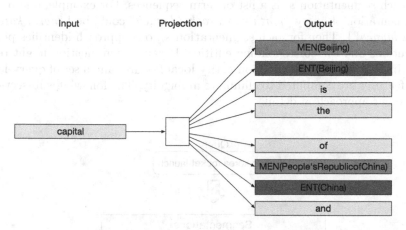

**Fig. 1.** An example of using the Skip-gram model to predict surrounding tokens of a specific token; here a token can be a word, a mention or an entity.

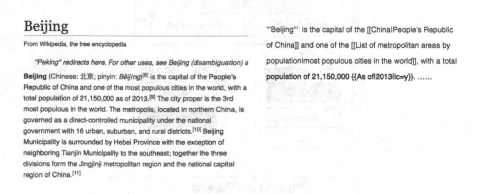

**Fig. 2.** Part of the Wiki page of Beijing and its source text

when the entity is mentioned by the exact name. Processing these inner links, we can generate the corpus which contains words, mentions and entities together. So in this paper, Wikipedia is used as the target knowledge base that entities link to, and its articles are processed to train the Skip-gram model.

## 3  Query Entity Linking Approach

Given a query $q$, our proposed approach aims to identify all the entities in Wikipedia that are mentioned in the query. The result of entity linking can be represented as $r = \{\langle m_1, e_1 \rangle, ..., \langle m_h, e_h \rangle\}$, where $m_j$ denotes a mention, and $e_j$ denotes the corresponding entity that $m_j$ refers to.

Figure 3 shows the general framework of our approach. For a query, our approach first generates all the possible segmentations $S = \{s_1, ..., s_k\}$ of query $q$, in which each segmentation $s_i$ is a list of term sequences. For example, one possible segmentation of query "north korea rocket launch" could be { "north korea", "rocket launch"}. Then for each segmentation $s_i$, our approach identifies possible mentions and the corresponding entities. Because each mention might refer to multiple entities, a set of mention-entity local features and a set of query-level global features are computed to eliminate ambiguity. The following sub-sections describe our approach in detail.

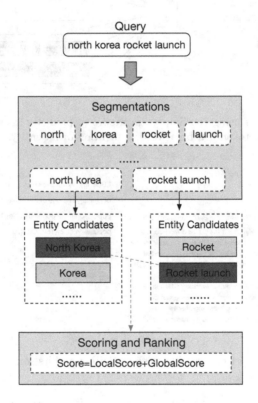

**Fig. 3.** Framework of proposed approach

## 3.1   Identifying Mentions and Its Candidate Entities

In order to identify whether a term sequence in a segmentation is a mention or not, and what entity a mention might link to, we build a Mention-to-Entity map $D = \{(m_i, E_{m_i})\}_{i=1}^{n}$, where $m_i$ is a possible mention, and $E_{m_i} = \{e_{m_i}^1, e_{m_i}^2, ...\}$ is the set of entities that $m_i$. Like some other Wikipedia-based entity linking approaches, the entity pages, redirect pages, disambiguation pages and the hyperlinks in Wikipedia are used to build this Mention-to-Entity map.

If a term sequence in a segmentation exactly matches a mention in the Mention-to-Entity map, it will be identified as a possible mention. And the candidate entities of the identified mention can be directly obtained from the Mention-to-Entity map.

## 3.2   Features for Entity Disambiguation

After identifying possible mentions and their candidate entities, our approach computes a set of features to predict the entity that each mention links to. All the features are divided into two groups: Mention-Entity level features (local features) and Query level features (global features).

**Mention-Entity Level Features.** Given a query $q$ and a identified mention $m$ in it, the likelihood of $m$ referring to an entity $e$ is estimated from different aspects, by computing the following features of the mention-entity pair $(m, e)$ for query $q$. These features are computed independently from other mentions in the current query $q$, so they are local features of mention-entity pairs.

**Link Probability.** This feature approximates the prior probability that mention $m$ links to entity $e$:

$$LF_1(m, e) = \frac{count(m, e)}{count(m)} \tag{3}$$

where $count(m, e)$ denotes the frequency that $m$ links to $e$ in the whole Wikipedia, and the $count(m)$ denotes the frequency that $m$ appears in Wikipedia.

**Entity-Query Relatedness.** This feature represents how the candidate entity $e$ relates to the current query $q$, which is computed as the cosine similarity between vectors of entity and the query. Because we don't directly have the vector representation of query $q$, the centroid of the vectors of words in $q$ is computed as the vector of $q$.

$$LF_2(m, e, q) = cos(\boldsymbol{v}_e, \boldsymbol{v}_q) \tag{4}$$

where $cos(\boldsymbol{v}_1, \boldsymbol{v}_2)$ is the cosine similarity between vectors:

$$cos(\boldsymbol{v}_1, \boldsymbol{v}_2) = \frac{\boldsymbol{v}_1 \cdot \boldsymbol{v}_2}{\| \boldsymbol{v}_1 \| \| \boldsymbol{v}_2 \|} \tag{5}$$

**Session-based Entity-Entity Relatedness.** Usually, users may search for several times in a single searching session to get their desired information. If there are previous queries in the current session, these queries and the identified entities in them can be used as important context information for entity linking in the following queries. This feature is defined to capture the relatedness between the candidate entity $e$ and the identified entities in the previous queries. Let $E$

denote the set of identified entities, and the centroid of these entities' vectors $\boldsymbol{v}_E$ is firstly computed; the session-based entity-entity relatedness is computed as:

$$LF_3(m, e, E) = cos(\boldsymbol{v}_e, \boldsymbol{v}_E) \tag{6}$$

**Session-based Entity-Query Relatedness.** This feature represents the relatedness between the candidate entity $e$ and the words in previous search queries in the current session. Let $Q$ denote the set of all the words in the previous queries, $\boldsymbol{v}_Q$ denote the centroid of all the words' vectors, the session-based entity-query relatedness is computed as:

$$LF_4(m, e, Q) = cos(\boldsymbol{v}_e, \boldsymbol{v}_Q) \tag{7}$$

**Query Level Features.** To get more precise entity linking results, query level features capturing relations among different mention-entity pairs are defined in this section. Let $q$ be the query, and $s$ be a segmentation of $q$; $M$ is the set of identified mentions, and $E$ is the set of entities that the mentions link to; the query level features are defined as follows.

**Granularity of Segmentation.** This feature favors long mentions in a query, and it is defined as:

$$GF_1(q, s) = 1 - \frac{\#Term\_Sequences(s)}{\#Words(q)} \tag{8}$$

where $\#Term\_Sequences(s)$ and $\#Words(q)$ denote the number of term sequences in the current segmentation and the number of words in the query, respectively.

**Mention-Query Relatedness.** This feature computes the average relatedness between mentions and the query:

$$GF_2(q, M) = \frac{1}{|M|} \sum_{m \in M} cos(\boldsymbol{v}_m, \boldsymbol{v}_{q/m}) \tag{9}$$

where $q/m$ represents the words in $q$ but not in $m$, $\boldsymbol{v}_{q/m}$ is the centroid of words' vectors in $q/m$; $\boldsymbol{v}_m$ is the vector of mention $m$.

**Mention-Mention Relatedness.** This feature computes the average relatedness between mentions:

$$GF_3(M) = \frac{1}{|M|} \sum_{m \in M} cos(\boldsymbol{v}_m, \boldsymbol{v}_{M/m}) \tag{10}$$

where $M/m$ represents the set of mentions without $m$, $\boldsymbol{v}_{M/m}$ is the centroid of mentions' vectors in $M/m$.

**Entity-Entity Relatedness.** This feature computes the average relatedness between entities that might be referred by different mentions:

$$GF_4(E) = \frac{1}{|E|} \sum_{e \in E} cos(\boldsymbol{v}_e, \boldsymbol{v}_{E/e}) \tag{11}$$

where $E/e$ represents the set of candidate entities without $e$, $\boldsymbol{v}_{E/e}$ is the centroid of entities' vectors in $E/e$.

**Session-based Entity-Entity Relatedness**

$$GF_5(E, E_s) = \frac{1}{|E|} \sum_{e \in E} cos(\boldsymbol{v}_e, \boldsymbol{v}_{E_s}) \tag{12}$$

where $E_s$ represents the set of entities that are identified in the previous queries of the same session.

## 3.3   Entity Linking as Ranking

Let $R_q = \{r_1, r_2, ..., r_n\}$ be all the possible results of entity linking for query $q$, where $r_i = \{\langle m_{i1}, e_{i1} \rangle, ..., \langle m_{ih}, e_{ih} \rangle\}$, $m_{ij}$ and $e_{ij}$ denote mention and entity respectively. A scoring function (Eq. 13) is defined to rank these results; if a link result $r^*$ has the maximum score and $Score(r^*) > \delta$, then $r^*$ is taken as the linking result of query $q$; Here, $\delta$ is a predefined threshold which filters out possible inaccurate linking results.

$$Score(r) = \frac{1}{|r|} \sum_{i=1}^{|r|} \sum_{j=1}^{4} \omega_j \cdot LF_j + \sum_{k=1}^{5} \mu_k \cdot GF_k \tag{13}$$

$\omega_j$ and $\mu_k$ are weights of the corresponding features, and their satisfy $\sum_{j=1}^{4} \omega_j = 1$ and $\sum_{k=1}^{5} \mu_k = 1$.

In order to get accurate linking results, setting appropriate weights in the scoring function is very important. Here we use machine learning algorithm to learn the optimal weights from training data. Specifically, $SVM^{Rank}$ [9,10] is used in our approach. Given pairs of ranked objects $\{\langle o_{i1}, o_{i2} \rangle\}_i$, $SVM^{Rank}$ learns a weight vector $\boldsymbol{\omega}$ that ensures $\boldsymbol{\omega}F(o_{i1}) > \boldsymbol{\omega}F(o_{i2})$, where $F(o)$ denotes the feature vector of a object $o$. Here, we use $SVM^{Rank}$ to separately learn the weights $\omega_j$ and $\mu_k$ in Eq. 13.

**Weight Learning for Local Features.** As described in Sect. 3.1, an identified mention $m$ might have multiple candidate entities $E_m = \{e_m^1, e_m^2, ...\}$. According to the known entity links in the training data, if the correct entity $e_m^* \in E_m$, pairs of entities $\{\langle e_m^*, e_m^k \rangle | e_m^k \in E_m \wedge e_m^k \neq e_m^* \}$ will be generated as training object pairs for $SVM^{Rank}$. Here the feature vector $F(e_m) = \langle LF_1, LF_2, LF_3, LF_4 \rangle$.

**Weight Learning for Global Features.** For a query $q$, $R_q = \{r_1, r_2, ..., r_n\}$ is all the candidate linking results. If $r^* \in R_q$ is the desired result according to the training data, pairs of linking results $\{\langle r^*, r_k \rangle | r_k \in R_q \wedge r_k \neq r^* \}$ are generated to fed $SVM^{Rank}$. Here the feature vector $F(r) = \langle GF_1, GF_2, GF_3, GF_4, GF_5 \rangle$.

When the weights for local and global features are learned by $SVM^{Rank}$, they are normalized to ensure $\sum_{j=1}^{4} \omega_j = 1$ and $\sum_{k=1}^{5} \mu_k = 1$; then normalized weights are used in Eq. 13 to predict entity links in queries.

## 4    Evaluation

This section presents the evaluation results of our proposed approach.

### 4.1    Dataset

English Wikipedia is used as the target Knowledge Base for entity linking. The dataset of *Yahoo Search Query Log To Entities*[1] is used for the evaluation. This dataset contains manually identified links to entities in Wikipedia. In total, there are 2,635 queries in 980 search sessions, 4,691 mentions are annotated which link to 4,725 entities in Wikipedia.

### 4.2    Comparison Systems

We compared our approach with two entity linking systems, Illinois Wikifier [6, 19] and DBpedia Spotlight [7].

- Illinois Wikifier is an entity linking system that was developed by University of Illinois at Urbana-Champaign. The system was built based on the work [6, 19].
- DBpedia Spotlight is a system for automatically annotating text documents with DBpedia URIs. Because DBpedia [1] is built from Wikipedia and each DBpedia URI corresponds to a Wikipedia entity, the results of DBpedia Spotlight can be easily converted to entity links of Wikipedia.

### 4.3    Evaluation Results

We evaluate our approach and the comparison systems on the Yahoo dataset, using the Precision (Pre.), Recall (Rec.) and F1-Measure (F1.) as the evaluation metrics. Let $L$ be the set of manually annotated entity links in the Yahoo dataset, $A$ be the set of returned links by a system, the above metrics are computed as follows.

$$Pre = \frac{|A \cap L|}{|A|} \tag{14}$$

$$Rec = \frac{|A \cap L|}{|L|} \tag{15}$$

$$F1 = \frac{2Pre \cdot Rec}{Pre + Rec} \tag{16}$$

Figure 4 shows the evaluation results of three different approaches. The precision and recall of each approach are evaluated. According to the results, our approach achieves the best precision and recall. It shows that the joint embedding model and learning to rank model work effectively in entity linking problem.

---

[1] http://webscope.sandbox.yahoo.com/catalog.php?datatype=l&did=66.

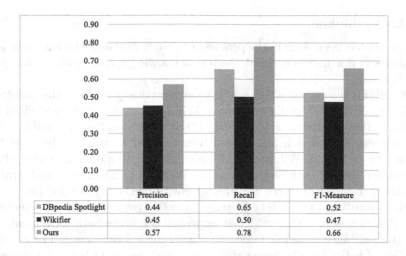

| | Precision | Recall | F1-Measure |
|---|---|---|---|
| ▪ DBpedia Spotlight | 0.44 | 0.65 | 0.52 |
| ▪ Wikifier | 0.45 | 0.50 | 0.47 |
| ▪ Ours | 0.57 | 0.78 | 0.66 |

**Fig. 4.** Evaluation results

## 5   Related Work

### 5.1   Entity Embedding

Recently, a number of approaches have been proposed to embed entities and relations from knowledge bases into a continuous vector space while preserving the original knowledge. The embeddings are usually learned by minimizing a global loss function of all the entities and relations in the knowledge base, which can be used for relation prediction, information extraction and some other tasks. Structured Embedding (SE) model [5] embeds entities into $\mathbb{R}^k$, and relations into two matrices $L_1 \in \mathbb{R}^{k \times k}$ and $L_2 \in \mathbb{R}^{k \times k}$; the embeddings are learned to ensure that $d(L_1 h, L_2 t)$ is large for triples $(h, l, t)$ not in the knowledge base. Semantic Matching Energy (SME) model [3] embeds both entities and relations into vectors, and all the relations share the same parameters. TransE [4] represents both entities and relations as vectors in $\mathbb{R}^k$; if a triple $(h, r, t)$ holds, TransE wants that $h + r \approx t$. The embeddings are learned by minimizing a margin-based ranking criterion over the training set. Socher et al. [24] proposed to use Neural Tensor Network (NTN) for reasoning over relationships between entities in KBs; entities are represented as vectors, and each relation is defined by the parameters of a neural tensor network which can explicitly relate two entity vectors. TransH [25] was proposed for solving the problems of TransE in modeling $1:n/n:1/n:n$ relations. For a triple $(h, r, t)$ in the knowledge base, the embedding $h$ and $t$ are first projected to the hyperplane $w_r$, resulting in $h_\perp$ and $t_\perp$, respectively. Then $h_\perp + d_r \approx t_\perp$ is expected to hold, where $d_r$ is a translation vector corresponding to $r$. TransR [12] is another model that handles $1:n/n:1/n:n$ relations. It embeds entities and relations in distinct spaces, entity space and multiple relation spaces. For each triple $(h, r, t)$, entity vectors are first projected into $r$-relation space as $h_r$ and $t_r$; then $h_r + d_r \approx t_r$ is expected to hold.

## 5.2   Entity Linking

Lots of work has been done in the problem of *Entity Linking*, which aims to identify entities in documents and link them to a knowledge base, such as Wikipedia and DBpedia.

Wikify! [14] is a system which is able to automatically perform the annotation task following the Wikipedia guidelines. Wikify! first uses a unsupervised keyword extraction algorithm to identify and rank mentions; then it combines both knowledge-based approach and data-driven method to predict the links from mentions to entities in Wikipedia. Milne et al. [17] proposed a learning-based approach for linking entities from text to Wikipedia. Their approach trains a C4.5 classifier based on three features of entity-mention pairs for link disambiguation. Kaulkarni et al. [11] proposed a collective approach for annotating Wikipedia entities in Web text. Their approach combines both local mention to entity compatibility and global document level topical coherence. The collective prediction of entity links improves the accuracy of results. Other collective entity linking approaches include [8, 21, 22].

The above entity linking approaches mainly handle long documents, there are also some work on linking entities in tweets to knowledge graphs [13, 23]. To perform entity linking in short tweets, these approaches usually use users' other information to help disambiguate entities in tweets, such as current user's other tweets or current user's social network information. There are several approaches for entity linking in English queries. For example, Radhakrishnan et al. [18] proposed an approach for entity linking for English queries by utilizing Wikipedia inlinks; Blance et al. [2] proposed an approach for fast and space-efficient entity linking for English queries.

## 6   Conclusion

This paper proposes a new entity linking approach for search queries. In order to solve the problem of limited context information, we propose to jointly learn word, mention and entity embeddings by Skip-gram model, which are used to compute different features for entity disambiguation. Experiments show that our approach works effectively, and outperforms the compared approaches.

**Acknowledgement.** The work is supported by NSFC (No. 61772079) and the project of Beijing Advanced Innovation Center for Future Education (BJAICFE2016IR-002).

## References

1. Bizer, C., Lehmann, J., Kobilarov, G., Auer, S., Becker, C., Cyganiak, R., Hellmann, S.: DBpedia - a crystallization point for the web of data. Web Semant. Sci. Serv. Agents World Wide Web **7**(3), 154–165 (2009)
2. Blanco, R., Ottaviano, G., Meij, E.: Fast and space-efficient entity linking for queries. In: Proceedings of the Eighth ACM International Conference on Web Search and Data Mining (WSDM 2015), pp. 179–188, New York, NY, USA. ACM (2015)

3. Bordes, A., Glorot, X., Weston, J., Bengio, Y.: A semantic matching energy function for learning with multi-relational data. Mach. Learn. **94**(2), 233–259 (2014)
4. Bordes, A., Usunier, N., Garcia-Duran, A., Weston, J., Yakhnenko, O.: Translating embeddings for modeling multi-relational data. In: Advances in Neural Information Processing Systems, pp. 2787–2795 (2013)
5. Bordes, A., Weston, J., Collobert, R., Bengio, Y., et al.: Learning structured embeddings of knowledge bases. In: AAAI (2011)
6. Cheng, X., Roth, D.: Relational inference for wikification. In: Proceedings of the 2013 Conference on Empirical Methods in Natural Language Processing (2013)
7. Daiber, J., Jakob, M., Hokamp, C., Mendes, P.N.: Improving efficiency and accuracy in multilingual entity extraction. In: Proceedings of the 9th International Conference on Semantic Systems (I-Semantics) (2013)
8. Han, X., Sun, L., Zhao, J.: Collective entity linking in web text: a graph-based method. In: Proceedings of the 34th International ACM SIGIR Conference on Research and Development in Information Retrieval, pp. 765–774 (2011)
9. Joachims, T.: Optimizing search engines using click through data. In: Proceedings of the Eighth ACM SIGKDD International Conference on Knowledge Discovery and Data Mining (KDD 2002), pp. 133–142, New York, NY, USA. ACM (2002)
10. Joachims, T.: Training linear SVMS in linear time. In: Proceedings of the 12th ACM SIGKDD International Conference on Knowledge Discovery and Data Mining (KDD 2006), pp. 217–226, New York, NY, USA. ACM (2006)
11. Kulkarni, S., Singh, A., Ramakrishnan, G., Chakrabarti, S.: Collective annotation of wikipedia entities in web text. In: Proceedings of the 15th ACM SIGKDD International Conference on Knowledge Discovery and Data Mining, pp. 457–466 (2009)
12. Lin, Y., Liu, Z., Sun, M., Liu, Y., Zhu, X.: Learning entity and relation embeddings for knowledge graph completion. In: Proceedings of the Twenty-Nineth AAAI Conference on Artificial Intelligence (2015)
13. Liu, X., Li, Y., Wu, H., Zhou, M., Wei, F., Lu, Y.: Entity linking for tweets. In: Proceedings of the 51st Annual Meeting of the Association for Computational Linguistics (ACL 2013) (2013)
14. Mihalcea, R., Csomai, A.: Wikify! Linking documents to encyclopedic knowledge. In: Proceedings of the Sixteenth ACM Conference on Information and Knowledge Management, pp. 233–242 (2007)
15. Mikolov, T., Chen, K., Corrado, G., Dean, J.: Efficient estimation of word representations in vector space. arXiv preprint (2013). arXiv:1301.3781
16. Mikolov, T., Sutskever, I., Chen, K., Corrado, G.S., Dean, J.: Distributed representations of words and phrases and their compositionality. In: Burges, C.J.C., Bottou, L., Welling, M., Ghahramani, Z., Weinberger, K.Q. (eds.) Advances in Neural Information Processing Systems, vol. 26, pp. 3111–3119. Curran Associates Inc. (2013)
17. Milne, D., Witten, I.H.: Learning to link with wikipedia. In: Proceedings of the 17th ACM Conference on Information and Knowledge Management, pp. 509–518 (2008)
18. Radhakrishnan, P., Bansal, R., Gupta, M., Varma, V.: Exploiting wikipedia inlinks for linking entities in queries. In: Proceedings of the First International Workshop on Entity Recognition & #38; Disambiguation (ERD 2014), pp. 101–104, New York, NY, USA. ACM (2014)

19. Ratinov, L., Roth, D., Downey, D., Anderson, M.: Local and global algorithms for disambiguation to wikipedia. In: Proceedings of the 49th Annual Meeting of the Association for Computational Linguistics: Human Language Technologies (HLT 2011), vol. 1, pp. 1375–1384, Stroudsburg, PA, USA. Association for Computational Linguistics (2011)
20. Shen, W., Wang, J., Han, J.: Entity linking with a knowledge base: issues, techniques, and solutions. IEEE Trans. Knowl. Data Eng. **27**(2), 443–460 (2015)
21. Shen, W., Wang, J., Luo, P., Wang, M.: LIEGE: link entities in web lists with knowledge base. In: Proceedings of the 18th ACM SIGKDD International Conference on Knowledge Discovery and Data Mining, pp. 1424–1432 (2012)
22. Shen, W., Wang, J., Luo, P., Wang, M.: LINDEN: linking named entities with knowledge base via semantic knowledge. In: Proceedings of the 21st International Conference on World Wide Web, pp. 449–458 (2012)
23. Shen, W., Wang, J., Luo, P., Wang, M.: Linking named entities in tweets with knowledge base via user interest modeling. In: Proceedings of the 19th ACM SIGKDD International Conference on Knowledge Discovery and Data Mining (KDD 2013), pp. 68–76, New York, NY, USA. ACM (2013)
24. Socher, R., Chen, D., Manning, C.D., Ng, A.: Reasoning with neural tensor networks for knowledge base completion. In: Advances in Neural Information Processing Systems, pp. 926–934 (2013)
25. Wang, Z., Zhang, J., Feng, J., Chen, Z.: Knowledge graph embedding by translating on hyperplanes. In: Proceedings of the Twenty-Eighth AAAI Conference on Artificial Intelligence, pp. 1112–1119 (2014)

# Publishing E-RDF Linked Data for Many Agents by Single Third-Party Server

Dongsheng Wang[1], Yongyuan Zhang[2,3], Zhengjun Wang[4], and Tao Chen[3(✉)]

[1] Department of Computer Science, University of Copenhagen, Copenhagen, Denmark
dswang2011@gmail.com
[2] Department of Library, Information and Archives, Shanghai University, Shanghai, China
zhangyj@sibs.ac.cn
[3] Shanghai Information Center for Life Sciences, Chinese Academy of Sciences, Beijing, China
chentao01@sibs.ac.cn
[4] Department of Chemical Physics, Lund University, Lund, Sweden
zhengjun.wang@chemphys.lu.se

**Abstract.** Linked data is one of the most successful practices in semantic web, which has led to the opening and interlinking of data. Though many agents (mostly academic organizations and government) have published a large amount of linked data, numerous agents such as private companies and industries either do not have the ability or do not want to make an additional effort to publish linked data. Thus, for agents who are willing to open part of their data but do not want to make an effort, the task can be undertaken by a professional third-party server (together with professional experts) that publishes linked data for these agents. Consequently, when a single third-party server is on behalf of multiple agents, it is also responsible to organize these multiple-source URIs (data) in a systematic way to make them referable, satisfying the 4-star data principles, as well as protect the confidential data of these agents. In this paper, we propose a framework to leverage these challenges and design a URI standard based on our proposed E-RDF, which extends and optimizes the existing 5-star linked data principles. Also, we introduce a customized data filtering mechanism to protect the confidential data. For validation, we implement a prototype system as a third-party server that publishes linked data for a number of agents. It demonstrates well-organized 5-star linked data plus E-RDF and shows the additional advantages of data integration and interlinking among agents.

**Keywords:** Semantic web · E-RDF · Web service · Linked data · Knowledge representation · Data integration

## 1 Introduction

Linked data has become a significant product split from the development of semantic web. Many agents ranging from academic organizations to government publish a large

---

D. Wang and Y. Zhang contributed equally to the work and serve as co-first authors.

© Springer International Publishing AG 2017
Z. Wang et al. (Eds.): JIST 2017, LNCS 10675, pp. 151–163, 2017.
https://doi.org/10.1007/978-3-319-70682-5_10

amount of linked data depicted in [2]. However, the more numerous agents like private and small companies either do not have the capacity or are not well motivated to make an effort to publish linked data. Even assuming they do publish their linked data, the data quality is often not guaranteed (many are only 2-star level), and is produced with a casual effort.

Therefore, instead of encouraging agents to publish linked data on their own, we assume the task can be undertaken by a professional third-party organization (maintained by professional experts) for these multiple agents. The scenario, hence, is different from publishing data by agents themselves since third-party servers face the problems of agent privacy (classified data) and multiple data source management, while attempting to meet with the 4-star or 5-star data principles.

In this paper, we explore the methodology for such third-party servers when publishing linked data for multiple agents. The contribution of our work is:

(1) We illustrate the difference between publishing linked data by general agents themselves as opposed to by a single third-party server, and accordingly designed a systematic standard to organize URIs in a third-party server for multiple agents while satisfying the 4-star principles;
(2) We propose a notion called E-RDF, which adopts the combination of ER (Entity Relationship Diagram) and RDF, by representing the schemes in systematic RDFS/OWL as like the scheme graph in relational databases, in contrast with many RDF resources that do not have explicit schemes;
(3) We introduce the framework and mechanisms to ensure data privacy of agents;
(4) We demonstrate the ease of improving data into 5-star in such centralized third-party servers by introducing an adapted algorithm to conduct data inter-linking.

According to our best knowledge, we are the first study to exploit the methodology of such third-party servers, publishing linked data for multiple agents.

In order to validate our framework and methodology, we implement a prototype system of one third-party server that publishes linked data for a great number of agents. It demonstrates clean 5+E star linked data and shows the ease of conceptual ontology integration and interlinking among agents.

## 2    Related Works

The concept of Semantic web [12] was initially raised in 1998 and has been developed for decades with a variety of semantic technologies generated during the progress. The technology such as ontologies (OWL and RDF) and linked data are adopted by many applications. One significant application is that many research organizations released large-scale knowledge bases, including YAGO, DBPedia [13], WordNet, etc., using ontology.

Academic organizations and research communities have had a great progress on ontologies creation and generation, either manually or automatically to contribute to the quantity of ontology such as the form-based ontology creation in [7] and non-experts community-based way to participate in ontology generation in [6]. Some knowledge base (ontology) is generated based on structured and unstructured web text mining, such as the semantic knowledge base construction in [10].

However, these kinds of approach could not result in accurate and large amount ontology even if the recall and accuracy could be highly improved, due to the approach itself and the second-hand data source they employ. In addition, many of the published data does not ensure the quality, reusability and reproducibility [3].

As a consequence, another more straightforward way is to encourage organizations to release linked data directly from their databases. There are some existing technologies and tools available for us to publish linked data directly based on local databases. As the Relational Database is still dominant among the majority of the database systems, one of the mostly employed tools is the D2R [8], which have implemented the relevant standard by W3C - R2RML (Relational database to RDF Mapping Language) [9]. However, many organizations such as small and private companies are not well motivated to make an effort to do this, since it is a professional and self-employed approach.

Moreover, though publishing linked data is widely adopted by many organizations and agents, data quality is not satisfied and data formats are diverse, due to the effort of publishing. Hence, Tim Berners-Lee has specified a star evaluation system for these published data. The following Fig. 1 demonstrates the principles that evaluate the published data as from one star to five stars.

- One star: Available on the Web (whatever format) under an open license
- Two Star: Available as structured data (e.g., Excel instead of image scan of a table)
- Three Star: As non-proprietary open format (e.g., CSV as well as of Excel)
- Four Star: Use URIs to denote things, so that people can point at your stuff (e.g. RDF and SPARQL)
- Five Star: Link your data to other data to provide context

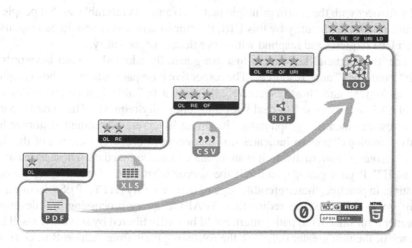

**Fig. 1.** Evaluation of five star data [1]

Some researchers follow these principles and attempt to satisfy the four principles of linked data such as the work in [4]. However, when we took a closer observation, it is not explicitly required that the scheme should be included in published linked data, which lead to diverse understanding and publishing of the Data.

# 3   Systematic URI Standard of Third-Party Server

We define agent as all kinds of organizations, companies, schools, governments, etc. that generally have their own database systems. URIs in linked data is significantly important because it not only makes the specific data unique in our global context, but also is referable and tractable, in other words, they are supposed to give response when we visit the URI with HTTP request. Thus, when agents' data are published as linked data by a third-party server, the very third-party server becomes responsible for such HTTP request for these multiple URIs, instead of agents themselves.

In this section, we discuss in detail the design and URI standard of such third-party servers in order to solve the problem of multiple data source management and of multiple HTTP requests. For professional third-party experts, it is easy to satisfy the first three principles, 3-star data; hence, we focus on satisfying the fourth and fifth principles in this paper. In Sect. 3.1, we illustrate the most crucial three keywords and their corresponding URI patterns. In Sect. 3.2, we discuss the pattern of "graph" for multiple data management, followed by "ontology" pattern for concept and properties in Sect. 3.3 while illustrating the difference of publishing linked data by agents themselves as opposed to by third-party server; and finally the pattern of "resource" for instances in Sect. 3.4.

## 3.1   URI Definition Standard

In order to meet with the fourth principle that each entity is referable so that people can track and identify each entity by this URI, the third-party server should be responsible for such URI requests and respond with more details of the entity.

Hence, three crucial key words that are generally adopted as path keywords are "graph", "ontology" and "resource". The respective base path pattern is "host/graph/*", "host/ontology/*" and "host/resource/*". As shown in Table 1, it demonstrates what kinds of URIs are supposed to deal with different entity requests. The "graph" pattern could be used to identify "graph name" in semantic server, indicating the storage location; the "ontology" pattern indicates it is a concept or property (schema of the data); and the "resource" pattern means it is an instance or individual data. These base patterns work as HTTP path gates, which tell the server what type of entities the users are requesting. In practice, these referable page of certain entity (HTTP REST Request) can be implemented by uniform redirection, SPARQL query processing and description page generation since fixed path pattern could be easily filtered by server. We will illustrate each of them in greater details in the following parts from section B to D. (For all the remaining examples concerned, we will all use the host URL "http://www.usources.cn" as a sample.)

**Table 1.** Three keywords adopted in path

| Keyword | URL patterns | Example |
| --- | --- | --- |
| Graph | host/graph/* | http://www.usources.cn/graph/* |
| Ontology | host/ontology/* | http://www.usources.cn/ontology/* |
| Resource | host/resource/* | http://www.usources.cn/resource/* |

### 3.2  Graph: Multiple Agents Management

As discussed before, one third-party server is concerned with a number of data sources; thus, we store each of them into different named graphs like distinct storage spaces. This is a virtual and clean way to separate their models in one semantic database server. On the contrary, when agents publish their linked data on their own, they do not necessarily store the data into different graphs. Even some of them do, the meaning of using the named graphs is generally different and varied in customized ways.

As shown in Table 2, the "agentName" in the pattern of the third-party server is supposed to be unique for the graph name, followed after the "graph" keyword. For "agentName", we propose a format that takes advantage of the agent's website URLs, that is, the "agentName" can be the URL transformation in which all the dots are replaced with "_" and "www" being removed. For instance, for the website "http://baike.baidu.com", we can transform it into "baike_baidu_com" as the corresponding "agentName". In this way, the graph name will be (1) globally unique and (2) also fit into the principle of URI itself without odd characters (e.g. #, %, ^, 提, 안, etc.). As shown in Table 2, it shows two examples of how named graph of "http://baike.baidu.com" and "http://www.facebook.com" can be expressed in such format.

**Table 2.**  Graph URI pattern

| | Agent | Third-party server |
| --- | --- | --- |
| Graph Name | Null | host/graph/agentName |
| Example | Null | http://www.usources.cn/graph/baike_baidu_com |
| | Null | http://www.usources.cn/graph/facebook_com |
| | ...... | ...... |

### 3.3  Ontology: Concept and Property

Different from publishing linked data by agents themselves (such as the examples listed in Table 3), it is critically crucial for concepts and properties to locate their named graph in third-party server since they are from different agents and stored in distinct named graphs. In other words, many entities might share the same names since there are large amounts of multi-source data and overlapping entity names. So, we put the agent name (marked as "agentName") in URI pattern, which can in turn be used to trace the graph name. The format of a concept or property URI is defined in Table 4, where the "ontology" keyword means the URI pattern is a concept or property, and then the following keyword "agentName" can be employed to obtain the named graph the entity belongs to. For example, the concept "http://www.usources.cn/ontology/

baike_baidu_com/Person" could locate its graph name as "http://www.usources.cn/graph/baike_baidu_com". Thus, we can retrieve the corresponding information of this entity from this specific named graph to improve the efficiency.

**Table 3.** Concepts and property URI pattern in agent itself

|                  | Agents                                    |
| ---------------- | ----------------------------------------- |
| Concept/Property | host/ontology/ConceptOrProperty           |
| Example          | http://baike.baidu.com/ontology/Person    |
|                  | http://www.police_sh.cn/ontology/name     |

**Table 4.** Concept and property URI pattern in third-party server

|                  | Third-party server                                        |
| ---------------- | --------------------------------------------------------- |
| Concept/Property | host/ontology/agentName/ConceptOrProperty                 |
| Example          | http://www.usources.cn/ontology/baike_baidu_com/Person    |
|                  | http://www.usources.cn/ontology/plice_sh_cn/name          |

By comparing Table 3 with Table 4, we can see the third-party server is relatively consistent in URI format while the URIs in agents is varying from one to another.

### 3.4 Resource: Instance or Individual

The case of Resource pattern is largely close to that of Ontology (concepts and properties), where the only difference is the path keyword - "resource". As illustrated in Table 5, "resource" in the pattern identifies the URI as instance, followed by "agentName" and finally the "instance ID". We can detect the graph name in the same way as in section C and obtain more information about certain instance.

**Table 5.** Instance URI pattern

|          | Third-party server                                    |
| -------- | ----------------------------------------------------- |
| Instance | host/resource/agentName/instanceID                    |
| Example  | http://www.usources.cn/resource/baike_baidu_com/lina  |
|          | http://www.usources.cn/resource/plice_sh_cn/id2       |

## 4    E-RDF: An Extension Notion of RDF

E-RDF is an extension notion of RDF, which explicitly requires that the scheme of RDF can be extracted and interpreted as like ER (Entity Relationship) Diagrams, which in terns, called the ER-(R)DF or E-RDF. In contrast with the traditional emphasis of relation or predicate search in RDF triple, our proposed E-RDF advocates that the object-oriented or ER-like diagram is the best way for people and machine to understand the structure data. For instance, when we retrieve information from RDF dataset or linked data from the perspective of triple or relation, the overwhelming could be reached easily given multiple RDF sources; but if we refer to the concepts, and their data properties or object properties, the SPARQL query is easier for us to organize.

Thus, we prefer to consider RDF as a condensed format utilized for data sharing and distributing rather than a format for data understanding and interpreting. The proposed notion of E-RDF could take advantage of both the well-known object-oriented understanding and the advantage of distributing and sharing.

E-RDF is practically an evaluation principle that evaluate whether the instances of a RDF set could be mostly mapped to their Concepts or Classes in their context. Subsequently, we propose the combination of 5-star data principle and E-RDF notation as 5+E star data evaluating principles. To be specific, if the data is RDF format and the scheme for instances could be extracted and expressed in ER-like graph, we call it as 4+E Star; if the data is linked to other sources as well, we call it 5+E star data, which is the highest evaluation. In contrast, if a RDF dataset is linked to other sources but without scheme or concepts, the data is hard for human to understand though it is also evaluated as 5 star data. In short, the E-RDF is an extension principle in addition to 5-star principles.

Given the ER scheme, it is significant to visualize it for human, which is also the best practice so far in most database systems. Concept level visualization is more straightforward and intuitive than intense instance level visualization. As shown in the following Fig. 2, with the tool of VISOWL, it enables us to organize a SPARQL query more easily and effectively according to the concept relationship, as well as their properties.

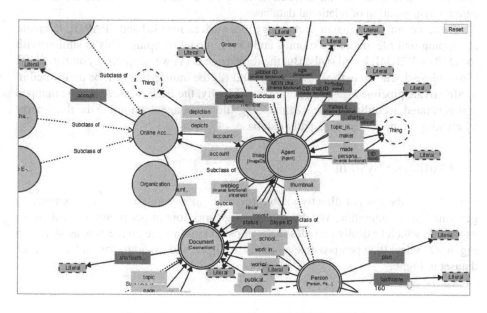

**Fig. 2.** Concept visualization of E-RDF for BIBO

## 5  Confidential Data Protection

For most agents, privacy and confidential data is always a sensitive and significant issue. We propose a lightweight framework that enables agents to select what kind of data they are intending to open and which they wish to keep private. The overall workflow is depicted in Fig. 3 from agent to third-party server.

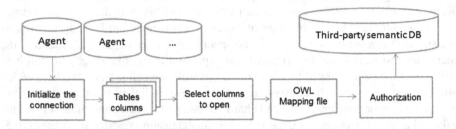

**Fig. 3.** The overall architecture of confidential data protection, mapping, and transforming

First of all, a connection is initialized to the agent database in a safe environment (like an intranet or local server of the agent). Subsequently, it will demonstrate all the tables and their attributes or schemas of their databases. It should be conducted by the agent's side to select the table and attributes they want to open and sometimes even specify conditions for certain attribute to filter out certain values. These are close to general manipulation of relational databases.

Then, we employ semantic technology tool such as Jena [5] and SPARQL to create a mapping owl file, defining columns and OWL model mapping. This is similar with the D2R or R2RML based tools. But the difference is (1) we adopt the systematic URI standard we discussed in the previous part and (2) the transferring will be performed in a safe private pipeline instead of public one. Finally, the RDF instance will be mapped and generated according to the owl mapping file when agent authorize the third-party by clicking submit to the third-party database.

## 6  Prototype System

The URI standard is not directly visible for users in system since it is automatically generated in configuration. We demonstrate the workflow of our prototype system for agent side which basically includes three steps: (1) initiate the connection as shown in Fig. 4; (2) select the open-possible columns in Fig. 5; and (3) authorize and submit the dataset in Fig. 6.

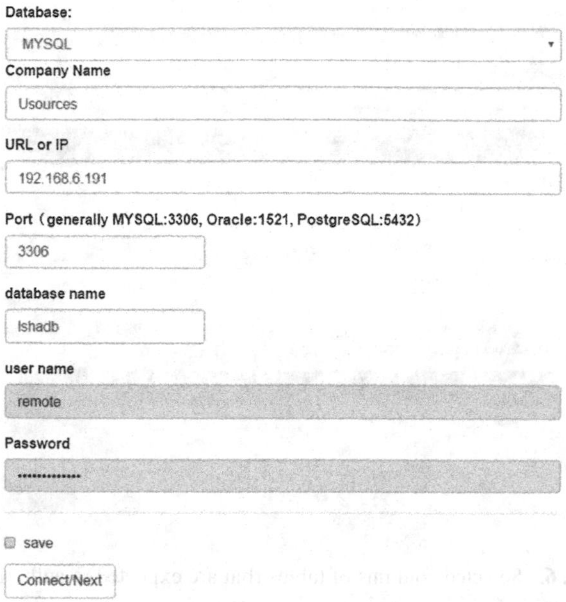

**Database:**

MYSQL ▾

**Company Name**

Usources

**URL or IP**

192.168.6.191

**Port（generally MYSQL:3306, Oracle:1521, PostgreSQL:5432)**

3306

**database name**

lshadb

**user name**

remote

**Password**

••••••••••••

☐ save

Connect/Next

**Fig. 4.** Initiate the connection

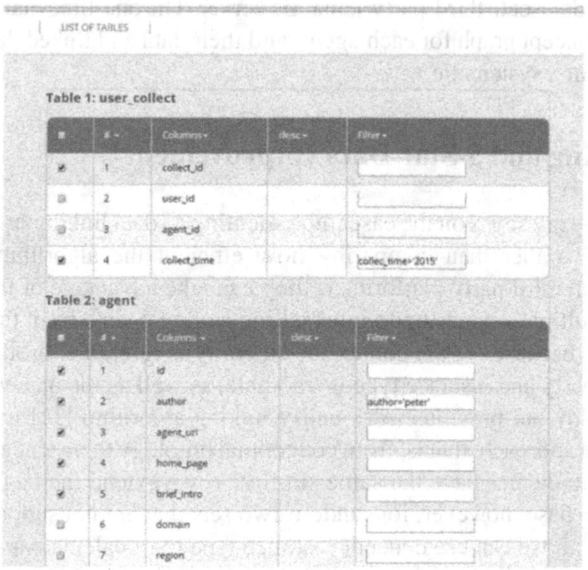

LIST OF TABLES

**Table 1: user_collect**

| ≡ | # ▾ | Columns ▾ | desc ▾ | Filter ▾ |
|---|---|---|---|---|
| ☑ | 1 | collect_id | | |
| ☐ | 2 | user_id | | |
| ☐ | 3 | agent_id | | |
| ☑ | 4 | collect_time | | collec_time>'2015' |

**Table 2: agent**

| ≡ | # ▾ | Columns ▾ | desc ▾ | Filter ▾ |
|---|---|---|---|---|
| ☑ | 1 | id | | |
| ☑ | 2 | author | | author='peter' |
| ☑ | 3 | agent_url | | |
| ☑ | 4 | home_page | | |
| ☑ | 5 | brief_intro | | |
| ☐ | 6 | domain | | |
| ☐ | 7 | region | | |

**Fig. 5.** Database connection and customized columns selection

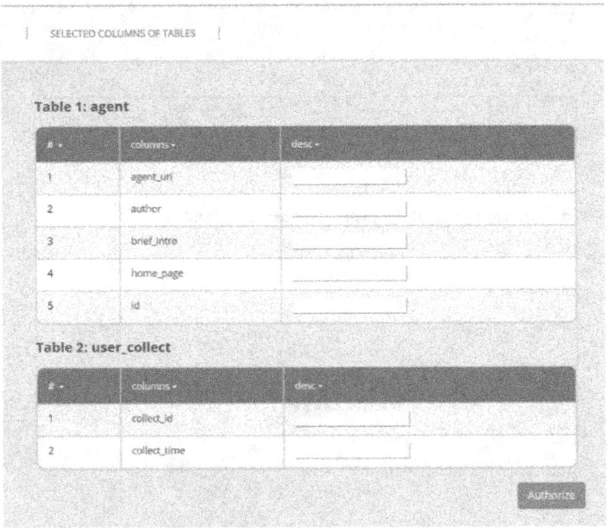

**Fig. 6.** Selected columns of tables that are expected to authorize

The mapping OWL file is generated after the three steps according to the URI standard we discussed previously. Finally, the graph name and password will be filled out to submit the dataset to third-party semantic server. The one-time manipulation result in a visualized concept graph for each agent, and their data and linked data are stored in different graphs in a systematic way.

## 7    Interlinking and 5-Star Data Improvement

In this section, we lay stress on the ease implementing of data linking or data improving into 5-star level, rather than comparing how efficient the algorithm itself is. The advantage of such third-party platforms is they can take advantage of the E-RDF, URI standards and multiple sources to do interlinking among agents from the same domain or cross domain, because (1) the context of an entity is referable through URI and (2) the concept/property and instance type is tractable, as well as the named graph.

We can employ our previous work entity linking algorithm [11] to integrate them using a top-down approach, that is, from conceptual ontology to instances. In [11], when more than two candidates have the same similarity, we extend the bag using encyclopedia knowledge base; however, for random two sets that are not supported by knowledge base material, we can use concept-instance type to re-calculate and rank instance candidate again. For instance, when we refer to "pianist", it could be a person title, but there is a movie also called "Pianist". So we could take advantage of concept information to filter it out and re-calculate. Actually, most concepts are obtainable in most original databases but often missed by human when publishing linked data. Thus, this is the reason E-RDF is explicitly required in our proposed 5+E evaluation framework.

The algorithm of Stepwise Bag-of-Words (S-BOW) [11] can be employed and adapted in the following one:

---

**Algorithm: Adapted Stepwise-BOW**

**Input: Two entity set E1 and E2**

**Output: Set P<ei,ej>**

s

1  **Begin**

2  Separate the E1 into concept set C1 and instance set I1; and the E2 into concept set C2 and instance set I2, via identifying URI

3  Obtain context ctxi of ci via URI request as target and ctxj of cj via URI request as knowledge base

4    Run the S-BOW for them (without extending bags).

5  Get context ctxi of ii through URI request as target and ctxj of ij as knowledge base,

6    Run the S-BOW for them without extending bags

7    If more than one sim(ctxi,ctxj) share the same value, use "rdf:type" attribute to filter out non-relevant concept instance and rank again

8  **End and Output**

---

As shown in the adapted algorithm, the context ctxi can be obtained through entity URI since the entity is referable and could be retrieved for more information such as "rdfs:label" and "rdfs:comment". Also, instead of extending, we separate the concepts and instances and use their relationship "rdf:type" to re-rank the candidates in the algorithm. We select 100 ambiguous entities from Hudong encyclopedia dataset together with corresponding entities from Baidu encyclopedia. Using the adapted algorithm, the accuracy is improved by 11.59% from 69 to 77 out of 100. It is lower than the extending methodology in [11] which is 14.9%; however, in the absence of knowledge base material (which is more common in real world), interlinking between randomly two data sets still can be improved than traditional BOW method and be handled in such a convenient third-party server. We can handle the interlinking on a large number of datasets. In this way, the data can meet with 5-star principles vastly and widely.

Finally, we could visualize E-RDF scheme graph as part of the linked data services.

# 8    Conclusion

In this paper, we assume that for agents who are willing to open part of their data but do not want to put in the effort to publish high quality data, the task can be uniformly undertaken by a professional third-party server which publishes linked data for these multiple agents. Consequently, we illustrated how the standard of URI could be designed for a single third-party server to be capable of managing multiple agents. The proposed URI standard not only provides systematic ways to organize multiple URIs but satisfies the 4-star data principles.

Moreover, we proposed the notion of E-RDF, an extension of RDF, which explicitly requires the scheme of RDF could be extracted and interpreted as ER (Entity Relationship) Diagrams, which in terns, called the ER-(R)DF or E-RDF. The proposed E-RDF claims that the object-oriented or ER diagram is the best way for people and machine to understand the structure data, as opposed to the emphasis of relation or predicate retrieving from RDF triple.

In addition, we proposed a lightweight framework that utilizes customized data filtering mechanisms to protect the classified data. For validation, we implemented a prototype system as a third-party server that publishes linked data for several agents. It demonstrated a well-organized 4-star linked data, and we showed the additional ease of data improvement to 5-star data, depicting the interlinking among multiple agents on one platform taking advantages of these standards. Moreover, the proposed method still showed an improvement of 11.59% between two randomly chosen data sets than traditional BOW method, even in the absence of knowledge base support.

**Acknowledgments.** This work was supported by Shanghai Institute of Life Sciences Information Center Foundation: Research Output Evaluation ($\pi$ Index) Based on Linked Data and Knowledge Graph Analysis (2016–2020).

# References

1. Berners-Lee, T.: 5 Star Open Data. http://5stardata.info/en/. Accessed 26 Sept 2017
2. Christain, B., Tom, H., Kingsley, I., Berners-Lee, T.: Linked data on the Web (LDOW 2008). In: 17th International Conference on World Wide Web (WWW 2008), pp. 1265–1266. ACM, USA. C. Bizer, New York (2008)
3. Sean, B., Iain, B., David, D.R., Paolo, M., John, A., Jiten, B., Philip, C., Don, C., Mark, D., Ian, D., Matthew, G., Danius, M., Stuart, O., David, N., Shoaib, S., Carole, G.: Why linked data is not enough for scientists. J. Future Gener. Comput. Syst. **29**, 599–611 (2013)
4. Edward, C., James, O.D., Edward, C., Souleiman, H., Marcus, K., Sean, O.: Linking building data in the cloud: integrating cross-domain building data using linked data. Adv. Eng. Inform. **27**, 206–219 (2013)
5. LNCS Homepage. https://jena.apache.org/. Accessed 26 Sept 2017
6. Hansaem, P., Jeungmin, L., Kyunglag, K., Jongsoo, S., Yunwan, J., Sungwoo, J., In-Jeong, C.: Collaborative ontology generation method using an ant colony optimization model. In: Park, J., Jin, H., Jeong, Y.S., Khan, M. (eds.) Advanced Multimedia and Ubiquitous Engineering. LNEE, vol. 393, pp. 541–549. Springer, Singapore (2016). https://doi.org/10.1007/978-981-10-1536-6_71

7. David, W., Embley, W., Tao C.: Form-based ontology creation and information harvesting. US Patent 8,103,962 [P] (2012)
8. Christian, B., Richard, C.: D2r server-publishing relational databases on the semantic web. In: The 5th International Semantic Web Conference, pp. 1–3. C. Bizer (2006)
9. Souripriya, D., Seema, S., Cyganiak, R.: R2RML: RDB to RDF mapping language. 3, 1–15 (2012)
10. Zeng, Y., Wang, D., Zhang, T., Wang, H., Hao, H., Xu, B.: CASIA-KB: a multi-source Chinese semantic knowledge base built from structured and unstructured web data. In: Kim, W., Ding, Y., Kim, H.-G. (eds.) JIST 2013. LNCS, vol. 8388, pp. 75–88. Springer, Cham (2014). https://doi.org/10.1007/978-3-319-06826-8_7
11. Zeng, Y., Wang, D., Zhang, T., Wang, H., Hao, H.: Linking entities in short texts based on a Chinese semantic knowledge base. In: Zhou, G., Li, J., Zhao, D., Feng, Y. (eds.) NLPCC 2013. CCIS, vol. 400, pp. 266–276. Springer, Heidelberg (2013). https://doi.org/10.1007/978-3-642-41644-6_25
12. Berners-Lee, T., Hendler, J., Lassila, O.: The semantic web. Sci. Am. 284(5), 28–37 (2001)
13. Mohanmed, M., Jens, L., Soren, A., Claus, S., Sebastian, H.: DBpedia and the live extraction of structured data from Wikipedia. Program 46(2), 157–181 (2012). Electronic Library and Information Systems

# Missing RDF Triples Detection and Correction in Knowledge Graphs

Lihua Zhao[1], Rumana Ferdous Munne[2,3]([✉]), Natthawut Kertkeidkachorn[2,3], and Ryutaro Ichise[1,2,3]

[1] National Institute of Advanced Industrial Science and Technology, Tokyo, Japan
lihua.zhao@aist.go.jp, ichise@nii.ac.jp
[2] SOKENDAI (The Graduate University for Advanced Studies), Hayama, Japan
{rfmunne,ichise}@nii.ac.jp
[3] National Institute of Informatics, Tokyo, Japan

**Abstract.** Knowledge graphs (KGs) have become a powerful asset in information science and technology. To foster enhancing search, information retrieval and question answering domains KGs offer effective structured information. KGs represent real-world entities and their relationships in Resource Description Framework (RDF) triples format. Despite the large amount of knowledge, there are still missing and incorrect knowledge in the KGs. We study the graph patterns of interlinked entities to discover missing and incorrect RDF triples in two KGs - DBpedia and YAGO. We apply graph-based approach to map similar object properties and apply similarity based approach to map similar datatype properties. Our propose methods can utilize those similar ontology properties and efficiently discover missing and incorrect RDF triples in DBpedia and YAGO.

**Keywords:** RDF triple · Knowledge graph · Word embedding · Ontology matching

## 1 Introduction

Knowledge Graphs (KGs) are structured knowledge bases that can represent real-world entities and their relationships. An entity is represented as a collection of Resource Description Framework (RDF) triples in the form of <subject, predicate, object>, where the former two are Uniform Resource Identifiers (URIs) of entities [10], and the latter is either an URI or a value [17]. There are two principal types of RDF triples: Object Property Triples (OPTs) and Datatype Property Triples (DPTs). OPTs describe the relationship between two entities and the predicates (or properties) are described using owl:Objectproperty [2]. DPTs use datatype properties (owl:DatatypeProperty) to link entities with data values, with the type of string, number, or date. Since these two types of RDF triples have different characteristics, we need to treat them differently.

© Springer International Publishing AG 2017
Z. Wang et al. (Eds.): JIST 2017, LNCS 10675, pp. 164–180, 2017.
https://doi.org/10.1007/978-3-319-70682-5_11

Although many large KGs have been published with millions of entities, there are still missing or incorrect knowledge in the KGs. DBpedia [11] and YAGO [13] are two large KGs derived from multilingual Wikipedia articles and there are millions of same entities interlinked between these two KGs. These interlinked same entities are essential resources to discover similar properties between two KGs.

However, it is challenging to discover similar properties because each KG is constructed with different ontologies, which causes the ontology heterogeneity problem. Ontology integration is one of the most popular methods to solve ontology heterogeneity problem by mapping similar properties of ontologies among different data sets. Ontologies are used in the KGs to describe entities, classes (concepts), attributes, and relations. With an integrated ontology, we can effectively combine data or information from multiple heterogeneous sources [22]. Another rising approach for matching concepts is using representation learning models that can represent entities with a low-dimensional embedding vector [3]. A tensor representation method was introduced to model the relations between entities in KGs [16].

In order to discover missing RDF triples in KGs, we apply both ontology integration and word embedding method. We will introduce how to discover different types of RDF triples between two KGs - DBpedia and YAGO. We apply graph-based ontology integration method for mapping similar object properties that link entities in the KGs. In order to map similar datatype properties, we apply typical string-based similarity measures and Recurrent Neural Network (RNN), specifically the GRU based word embedding method. Then we query on the KGs to discover missing RDF triples with the similar object properties and datatype properties.

The remainder of this paper is organized as follows. We introduce related works and discuss their limitations in Sect. 2. We describe how to discover similar object properties and datatype properties using ontology integration and word embedding approach, respectively, in Sect. 3. We show experiments for discovering similar properties and missing RDF triples in Sect. 4. We conclude our research and propose future work in Sect. 5.

## 2   Related Work

A semi-automatic ontology integration system, called the Framework for InTegrating ONtologies (FITON) was introduced to integrate heterogeneous ontologies among various Linked Data [24]. This system applied graph-based ontology integration method to map ontologies at both class and property level. The FITON framework constructs a high-quality integrated ontology, which is effective in knowledge acquisition from various data sets by using simple SPARQL queries. However, the authors simply applied string-based and WordNet-based similarity matching approaches for both object properties and datatype properties, which have different characteristics.

Also, the string-based approaches are not effective when textual strings are significantly different, while the WordNet-based similarity approaches require the lexical resource. Recently, embedding words into a continuous vector space (word2vec) [14] has gained more attentions. It tends to reveal the semantic relations between words better than the WordNet-based similarity approach [7]. However, the term of a property is usually a compound word, which cannot be observed during the training process of word embedding. To estimate the vector representation of unobserved compound words, many studies use the average vector representation to represent the compositional words [6,20,21]. In the study [9], the authors showed that estimating vector representation of compound words by using the RNN-based approach outperformed the average vector representation.

Many knowledge graph embedding methods have been introduced, which project RDF triples to vectors. A graph aware knowledge embedding (GAKE) method was proposed, which learns representations of vertices or edges of a directed knowledge graph [5]. This method used three different types of graph context for embedding: neighbor context, path context, and edge context. GAKE outperforms several state-of-art knowledge embedding methods such as TransE [4] in link classification tasks, which predicts the missing properties or missing object entities. However, they only focused on the relations between entities and did not consider datatype properties.

Similar research was introduced to learn entity and predicate embeddings for knowledge graph completion [15]. The authors treated an RDF graph as a labeled directed multigraph and focused on reducing learning time in embedding models. While most of these knowledge graph embedding research focused on entity and relation embeddings, Lin et al. considered attribute (datatype property) embeddings [12]. The experiments were conducted for predicting entities, relations, and attributes for knowledge graph completion.

Although above knowledge graph embedding methods can predict entities and relations for KGs, the ranking scores in prediction still need improvements for real-world applications. Therefore, we take advantages of both technologies used in Semantic Web and Natural Language Processing, specifically, ontology integration and knowledge embedding methods to discover missing RDF triples in KGs.

## 3    Knowledge Graph Analysis for Discovering RDF Triples

Knowledge Graphs (KsG) consist of RDF triples that describe relations between entities and attributes of entities. Here, the relations between entities are defined as object properties (owl:ObjectProperty) and the attributes are defined as datatype properties (owl:DatatypeProperty) in Web Ontology Language (OWL), which is a vocabulary extension of RDF [2]. In this paper, we use OP to represent object properties (excluding *owl:sameAs*) and use DP to represent datatype properties. Therefore, a Knowledge Graph ($KG$) is defined as follows:

**Definition 1.** *A Knowledge Graph $KG = (V, E)$, where $V$ is a set of vertices and $E$ is a set of labeled edges.*

Here, $V$ is either an entity, a literal value, or a Class definition, and $E$ is either an object property (OP), a datatype property (DP), or rdf:type[1]. Therefore, we divide the RDF triples in a $KG$ into three groups: OP triples (OPT), DP triples (DPT), and Type triples (TT).

**Definition 2.** *An RDF triple is either an Object Property Triple (OPT), a Datatype Property Triple (DPT), or a Type Triple (TT):*

*OPT = <Ent1, OP, Ent2>, where OP is the relation between two entities Ent1 and Ent2.*
*DPT = <Ent, DP, Val>, where DP is the attribute that describes the value Val of an entity Ent.*
*TT = <Ent, rdf:type, Type>, where Type is the class information of an entity Ent.*

In this research, we mainly focus on OPTs and DPTs. The same entities in different KGs are interlinked by *owl:sameAs*, which are important resources to explore matching of OPs and DPs. Figure 1 shows interlinked entities of "Aaron_Burr" in YAGO and DBpedia. The circle vertices represent entities, triangle vertices represent literal values, and rectangular vertices represent class information. From the graph of two entities, we could observe matching OPTs and DPTs according to the linked same entities and same values. For example, we could observe that *yago:wasBornIn*[2] matches to *db-onto:birthPlace*[3] and *db-prop:placeOfBirth*[4] according to the interlinked instances of *yago:NJ* and *dbpedia:NJ*[5]. Furthermore, we could expect there should be a "child" or "children" OP link between *dbpedia:Aaron_Burr* and *dbpedia:Phillip_Schuyler*, which is missing in DBpedia.

In the following, we will introduce our approach to discover OPTs and DPTs based on the interlinked same entities. Considering the characteristics of RDF triples, we apply different approaches to discover similar OPs and DPs in the KGs. These matching similar OPs and DPs can be used to discover missing RDF triples in the KGs.

## 3.1   Graph-Based Approach for Object Property Triples

Object Properties (OPs) are used to link entities, and the same entities in different KGs are interlinked by owl:sameAs. Therefore, we can analyze the graph patterns of interlinked entities in various KGs to identify similar OPs to reduce ontology heterogeneity. Furthermore, with the similar OPs, we can also discover missing information in each KG. In this section, we will introduce a graph-based approach to discover similar OPs from interlinked entities and then describe how to discover missing OPTs using the similar OPs.

---

[1] http://www.w3.org/1999/02/22-rdf-syntax-ns#type.
[2] yago: http://yago-knowledge.org/resource/.
[3] db-onto: http://dbpedia.org/ontology/.
[4] db-prop: http://dbpedia.org/property/.
[5] dbpedia: http://dbpedia.org/resource/.

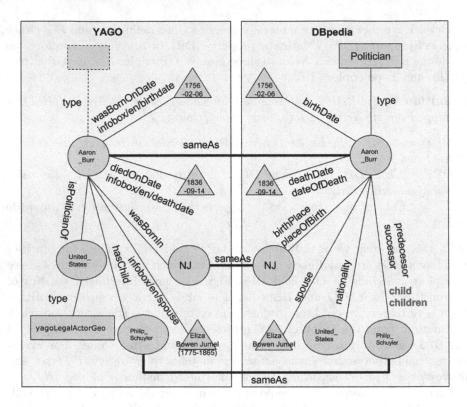

**Fig. 1.** Interlinked entities.

**Discover Similar Object Properties.** The main steps of discovering similar OPs are as follows:

Step 1: Randomly retrieve some interlinked entities.
Step 2: Retrieve preliminary similar OP pairs by checking the graph patterns of interlinked entities and memorize the co-occurrence of OP pairs.
Step 3: Retrieve frequent pairs that co-occur more than a predefined threshold.
Step 4: Based on the filtered frequent OP pairs, we group similar OPs.

In Step 1, we randomly chose some of the interlinked entities (e.g. 10% of the total number of interlinked same entities) from KGs to reduce computation time. In Step 2, we retrieve preliminary similar OP pairs by analyzing graph patterns. If there are four RDF triples $<Ent_1, owl:sameAs, Ent_2>$, $<Ent_1, OP_1, Ent_3>$, $<Ent_2, OP_2, Ent_4>$, and $<Ent_3, owl:sameAs, Ent_4>$, we can infer that $OP_1$ and $OP_2$ are similar. For example, in Fig. 1, we can infer that *yago:wasBornIn* matches to *db-onto:birthPlace* and *db-prop:placeOfBirth*. We also memorize the co-occurrence of these preliminary similar OP pairs. In Step 3, we need to filter out OP pairs that occur occasionally and then group similar OPs in the last step. For example, in Fig. 1, we group three OPs *yago:wasBornIn*, *db-onto:birthPlace*, and *db-prop:placeOfBirth*, which are used for describing interlinked birth place entities *yago:NJ* and *dbpedia:NJ*. These three OPs co-occur frequently in inter-linked entities of people and therefore we keep this group of OPs as similar properties.

**Discover Missing Object Property Triples.** Although KGs contain large amount of entities and RDF triples that describe the entities, there are still complemental knowledge in another KG. With the groups of similar OPs, we can easily query on the KGs to discover missing knowledge. With the help of owl:sameAs links and groups of similar OPs that co-occur frequently, we can discover missing Object Property Triples (OPTs) in each KG. For each pair of interlinked entities, we construct entity-only directed graphs that only contain OPTs.

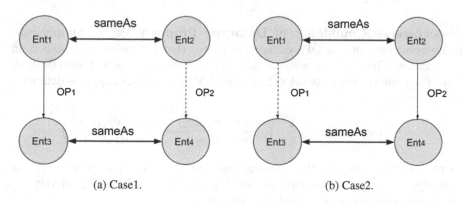

(a) Case1.                              (b) Case2.

**Fig. 2.** Missing OPT discovery.

As shown in Fig. 2, there are several possible cases of directed graphs of interlinked entities. Here, $\{Ent_1, Ent_3 \in KG_1\}$, $\{Ent_2, Ent_4 \in KG_2\}$, and $\{OP_1, OP_2 \in Group_i\}$, where $Group_i$ contains similar OPs such as $OP_1$ and $OP_2$.

In Fig. 2a, if there is an *owl:sameAs link* between $<Ent_1, Ent_2>$ and $<Ent_3, Ent_4>$ in any direction, we can infer that OPT $<Ent_2, OP_2, Ent_4>$ is missing in $KG_2$. In the same way, we can discover missing OPT $<Ent_1, OP_1, Ent_3>$ in $KG_1$ as shown in Fig. 2b because $OP_1$ and $OP_2$ both belong to the group $Group_i$.

### 3.2 Word Embedding Based Approach for Datatype Property Triples

Unlike Object Property Triples, Datatype Property Triples (DPTs) do not have *owl:sameAs* links that link the same values. Therefore, we need a different app-roach to discover similar DPTs from interlinked entities. We consider three aspects to discover similar DP pairs:

1. co-occurrence of two DP pairs $dp_1$ and $dp_2$.
2. similarity of values in DPTs.
3. similarity of the terms in DP pairs $dp_1$ and $dp_2$.

Therefore, the similarity between two DP pairs can be calculated as follows:

$$Sim(dp_1, dp_2) = conf\_co(dp_1, dp_2) *$$
$$\Big( \alpha * strSim(val_1, val_2) + \beta * weSim(dp_1, dp_2) \Big) \quad (1)$$

where the weights $\alpha + \beta = 1$, $conf\_co(dp_1, dp_2)$ is the co-occurrence confidence of DP pairs $dp_1$ and $dp_2$, $strSim(val_1, val_2)$ is the string-based similarity of $val_1$ and $val_2$ that correspond to $dp_1$ and $dp_2$, and $weSim(dp_1, dp_2)$ is the word embedding based similarity for the terms of DP pairs $dp_1$ and $dp_2$.

**Co-occurrence Confidence of Datatype Property Pairs.** Simply using the co-occurrence of two DPs cannot guarantee that DP pairs have high confidence values. Therefore, we also consider the occurrence of each single DP in the KGs. The co-occurrence confidence of two DPs ($conf\_co(dp_1, dp_2)$) is defined as follows:

$$conf\_co(dp_1, dp_2) = \frac{1}{2} * \Big( \frac{occur(dp_1, dp_2)}{occur(dp_1)} + \frac{occur(dp_1, dp_2)}{occur(dp_2)} \Big)$$

where $occur(dp_1, dp_2)$ is the co-occurrence of $dp_1$ and $dp_2$, $occur(dp_1)$ and $occur(dp_2)$ represent the occurrence of $dp_1$ and $dp_2$ in KGs, respectively. We assign the same weight $\frac{1}{2}$ to both $dp_1$ and $dp_2$.

**String-Based Similarity of DP Values.** The values in DPTs are mainly categorized into three types: number, date, and string. Here, we don't consider DPTs with the values in URLs because most of the DPs with values in type URL are used for any URL that are related to the entity. For the DP pairs with the values of number and date, we use exact matching method, which returns 1 if they are the same and returns 0 otherwise. For the string values, we use three string-based similarity measures, namely, JaroWinkler distance [23], EditDistance (Levenshtein), and n-gram [8] to calculate the similarity of values as follows:

$$strSim(val_1, val_2) = \lambda * JaroWinker(val_1, val_2) +$$
$$\mu * EditDistance(val_1, val_2) + \nu * ngram(val_1, val_2) \quad (2)$$

where $\lambda + \mu + \nu = 1$.

**Word Embedding Based Similarity for DP Terms.** Well designed ontologies are used to construct Knowledge Graphs and the terms of ontology properties are mostly meaningful individual words. Continuous Bag-of-Words (CBOW) model is one of the most popular neural network models for learning distributed representations of words, in other words, embedding words in a vector space (word2vec) [14]. In this paper we use CBOW to learn the representation of words. Given a sequence of training words $w_1, w_2, ..., w_n$, and a context window $c$, the learning function of CBOW model is to maximize the following average log probability:

$$\frac{1}{n} \sum_{t=1}^{n} log\ p(w_t|w_{t-c}, .., w_{t-1}, w_{t+1}, ..., w_{t+c})$$

where $p(w_t|w_{t-c}, .., w_{t-1}, w_{t+1}, ..., w_{t+c}) = \frac{exp(\bar{v} \cdot v_{w_t})}{\sum_{w=1}^{|V|} exp(\bar{v} \cdot v_w)}$.

Here, $v_w$ is the distributed vector representation of word $w$, $\bar{v}$ is an average of the distributed representation of words in the context $c$, and $V$ is the set of vocabularies.

Based on our observation, we found that the terms of the ontology properties are mostly compound words, e.g., "birthDate" or "dateOfBirth". The estimation of the representation of compound words using representations of individual words is required. Since distributed representation of a compound word cannot be directly represented as a sum of individual vectors, we apply Recurrent Neural Networks based approach, specifically, GRU-based approach to represent compound words as introduced in the study [9].

## 4    Experiment

In this section, we will discuss experiments in discovering knowledge from KGs. First, we will describe the KGs and tools we used for implementation. Then we will discuss experiments in discovering similar object properties, missing object property triples, similar datatype properties, and missing datatype property triples in the KGs.

### 4.1    Knowledge Graphs

We used DBpedia (2015-10) and YAGO 3 dump datasets, which are cross-domain data derived from Wikipedia articles. We have 721,992,850 RDF triples of DBpedia and 673,822,741 RDF triples of YAGO 3. There are 1,099 object properties and 1,734 datatype properties of DBpedia in our RDF triple store (Virtuoso[6]). YAGO 3 contains 75 properties that have confidence higher than 0.9, among which 31 are object properties and 44 are datatype properties.

In our RDF triple store, there are 2,637,892 bidirectional *owl:sameAs* links between two KGs, 248,514 *owl:sameAs* links from DBpedia to YAGO, and 1,058,659 *owl:sameAs* links from YAGO to DBpedia. These interlinked entites are the main resources for mapping similar OP and DP pairs.

---

[6] https://virtuoso.openlinksw.com/.

## 4.2   Implementation

SPARQL Protocol and RDF Query Language (SPARQL) [18] is used to access the KGs to retrieve graph patterns of interlinked entities. We also used discovered similar OPs and DPs in SPARQL queries to discover missing RDF triples in the KGs.

Wikipedia articles are used to train the distributed representation of individual words and compound words using GRU-based word embedding. The average length of compound words is 3.06 tokens and 16,369,076 compound words were used for the GRU-based approach. We used word2vec provided by gensim [19] to learn the distributed representation of words and used tensorflow [1] to implement RNN model using GRU unit. We used CBOW with dimension size as 200 and window size as 5. We discarded words that appear less than 25 times and used the token "UNK" to replace out of vocabulary words.

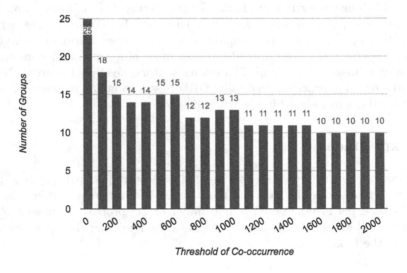

**Fig. 3.** The number of similar OP groups for varied co-occurrence thresholds.

## 4.3   Exp 1: Discover Similar Object Properties

In this experiment, we show similar object properties discovered using graph-based approach. By analyzing graph patterns of interlinked entities, we group similar object properties in the KGs. We set different thresholds of co-occurrence of OP pairs to observe the groups of similar object properties. Figure 3 shows the number of similar OP groups we discovered by setting different co-occurrence thresholds from 0 to 2000. Obviously, when we set a larger threshold, non-frequent pairs are removed and the quality of groups increase.

However, we may miss some interesting groups that have low frequency due to less available data in the KGs. For example, the group with *yago:hasAcademicAdvisor*, *db-onto:doctoralAdvisor*, and *db-prop:doctoralAdvisor* disappears when the threshold is higher than 600. On the other hand, when the threshold is low, there might be large groups that contain noisy OPs because of some coincident cases. For instance, if the birth place and death place are the same, they might be grouped together. Therefore, we manually checked the groups with different thresholds and constructed 18 high-quality groups of similar OPs.

Table 1 shows 15 groups (among 18 discovered groups) that have small number of similar object properties. The other three groups are *yago:hasChild*, *yago:playsFor*, and *yago:isLocatedIn*, which contain many similar DBpedia object properties.

**Table 1.** Groups of similar object properties.

| | | |
|---|---|---|
| yago:isMarriedTo<br>db-onto:spouse<br>db-prop:spouse | yago:influences<br>db-onto:influenced<br>db-prop:influenced | yago:isInterestedIn<br>db-onto:mainInterest |
| yago:hasCapital<br>db-onto:capital<br>db-prop:capital | yago:hasCurrency<br>db-onto:currency<br>db-prop:currency | yago:hasAcademicAdvisor<br>db-onto:doctoralAdvisor<br>db-prop:doctoralAdvisor |
| yago:happenedIn<br>db-onto:place<br>db-prop:place | yago:isCitizenOf<br>db-onto:nationality<br>db-prop:nationality | yago:graduatedFrom<br>db-onto:almaMaster<br>db-prop:almaMaster |
| yago:livesIn<br>db-onto:residence<br>db-prop:residence | yago:created<br>db-onto:knownFor<br>db-onto:notableWork | yago:worksAt<br>db-prop:workplaces<br>db-prop:workInstitutions |
| yago:wasBornIn<br>db-onto:birthPlace<br>db-prop:birthPlace<br>db-prop:placeOfBirth | yago:diedIn<br>db-onto:deathPlace<br>db-prop:deathPlace<br>db-prop:placeOfDeath | yago:hasWonPrize<br>db-onto:award<br>db-prop:awards<br>db-prop:prizes |

## 4.4   Exp 2: Discover Object Property Triples

This experiment shows the number of Object Property Triples (OPTs) we discovered using the groups of similar OPs in DBpedia and YAGO. We assume that existing OPTs in the KGs are accurate. We used similar object properties in each group as listed in Table 1 to construct SPARQL queries for retrieving missing OPTs. Table 2 shows the number of missing OPTs we retrieved for each group of similar object properties in DBpedia and YAGO, respectively. We pick the term of YAGO properties as the representative of each group because most of the groups contain one YAGO object property with several DBpedia object properties, except the *playsFor* group, which also contains *yago:isAffiliatedTo*.

The quality of discovered OPTs depends on the accuracy of existing knowledge in the KGs and the confidence of discovered similar OPs.

**Table 2.** The number of discovered missing OPTs.

| Property group | DBpedia | YAGO |
|---|---|---|
| isMarriedTo | 25,799 | 6,568 |
| influences | 31,188 | 2,451 |
| isInterestedIn | 316 | 732 |
| hasCapital | 6,538 | 2,739 |
| hasCurrency | 675 | 996 |
| hasAcademicAdvisor | 3,671 | 1,276 |
| happnedIn | 44,404 | 17,677 |
| isCitizenOf | 33,196 | 77,178 |
| graduatedFrom | 18,884 | 60,356 |
| livesIn | 16,355 | 36,542 |
| created | 458,386 | 23,496 |
| worksAt | 12,020 | 9,271 |
| wasBornIn | 254,574 | 841,270 |
| diedIn | 94,246 | 231,353 |
| hasWonPrize | 251,431 | 2,739 |
| hasChild | 178,724 | 186,753 |
| playsFor | 6,791,509 | 1,101,334 |
| isLocatedIn | 26,951,362 | 1,212,848 |

To deeply investigate our quality of results in Table 3, we sampled 1,000 data triples for each property group and then manually analyzed the characteristic of the missing triples as shown in Table 3. Here, we presented the statistical analysis of the correct OPTs, Missing OPT, other OPT. Correct OPT represents the number triples which has same object property group in both KGs. Missing OPT indicates the number of missing object property and other OPT defined as the number of triples that are connected with different types object property that are not in the same group.

Based on these result, we found that KGs have some missing OPTs which can be automatically added by our proposed method. As an example, we can consider the object property group "worksAt". In Dbpedia, the entity "Hideo Hosono" and the entity "Nagoya Institute of Technology" are connected by the object property "workInstitution", while there is no relation found correspond to such entities in YAGO. On the other hand, for same object property group "worksAt", YAGO has a "worksAt" relation connectivity with entity "Simon Hayhoe" and "Higher Colleges of Technology" but no relation exists between "Simon Hayhoe" and "Higher Colleges of Technology" in Dbpedia.

Moreover, we also discovered our method can suggest appropriate relations between entities where the relation representations are not clear enough.

**Table 3.** The statistical analysis of discovered OPTs in DBpedia and YAGO

| Property group | DBpedia | | | | YAGO | | | |
|---|---|---|---|---|---|---|---|---|
| | Correct OPT | Missing OPT | Other OPT | | Correct OPT | Missing OPT | Other OPT | |
| | | | WikiPage WikiLink | Other | | | LinkTo | Other |
| wasBornIn | 830 | 143 | 27 | 0 | 123 | 481 | 390 | 6 |
| LocatedIn | 38 | 277 | 675 | 10 | 834 | 155 | 11 | 0 |
| playsFor | 699 | 269 | 27 | 5 | 735 | 154 | 110 | 1 |
| livesIn | 635 | 235 | 105 | 25 | 285 | 544 | 168 | 6 |
| isMarriedTo | 540 | 211 | 248 | 1 | 813 | 121 | 66 | 0 |
| InterestedIn | 568 | 256 | 175 | 1 | 443 | 342 | 215 | 0 |
| isCitizenOf | 571 | 269 | 154 | 6 | 206 | 606 | 188 | 0 |
| influences | 403 | 363 | 234 | 0 | 744 | 160 | 96 | 0 |
| hasWonPrize | 124 | 104 | 770 | 2 | 383 | 389 | 228 | 0 |
| hasCurrency | 551 | 386 | 63 | 0 | 488 | 417 | 95 | 0 |
| hasChild | 141 | 180 | 678 | 1 | 183 | 165 | 648 | 4 |
| hasCapital | 487 | 248 | 253 | 12 | 485 | 181 | 332 | 2 |
| hasAcademicAdvisor | 586 | 215 | 199 | 0 | 756 | 215 | 29 | 0 |
| happenedIn | 311 | 302 | 387 | 0 | 480 | 373 | 139 | 8 |
| graduatedFrom | 846 | 84 | 70 | 0 | 231 | 477 | 292 | 0 |
| diedIn | 951 | 28 | 15 | 6 | 354 | 160 | 485 | 1 |
| created | 101 | 288 | 599 | 12 | 894 | 17 | 87 | 2 |
| worksAT | 741 | 28 | 131 | 0 | 28 | 109 | 860 | 3 |

In YAGO many entities are connected to each other with Linkto (similar to *Dbpedia:wikiPageWikiLink*) which typically means those entities have some relations. These types of linking do not state the exact relation between entities. Our method can separate such phenomenon distinguishably and can suggest modification with proper linking. As an example, entity "K.C. Sreedharan Pillai" and "United Nations" are connected with *DBpedia:workInstitution* property but in YAGO these two entities are connected as "Linkto", where we cannot understand their relationship explicitly. In Table 3, Dbpedia has 131 such triples where YAGO has 860 triples for "worksAt. We have found very few other type of object property which is mostly OP group specific such as, for "wasBornIn" group we found some triples are connected with "isCitizenOf" or "isLeaderOf" or "diedIn" which we mentioned as "other" in Table 3.

In this experiment, we discover the missing OPTs and also evaluate the accuracy of them. From our method, we can successfully recommend the KGs to add the missing triples and also to add the proper object property where it is not explicitly defined.

## 4.5    Exp 3: Discover Similar Datatype Properties

In this experiment, we show the advantage of representing compound words using GRU-based approach by comparing with simple word2vec based approach. The similarity of datatype properties are calculated according to the co-occurrence of two DP pairs, string-based similarity of values in DPTs, and word embedding based similarity of the terms in DP pairs. We filtered out DP pairs that have co-occurrence less than 100 times to avoid assigning high confidence values for the infrequent DP pairs. In Eq. 1, we set both $\alpha$ and $\beta$ as 0.5, and set $\lambda$, $\mu$, and $\nu$ as $\frac{1}{3}$ in Eq. 2. We used GRU-based RNN approach to model the terms of datatype properties and calculated word-embedding based cosine similarity between the terms.

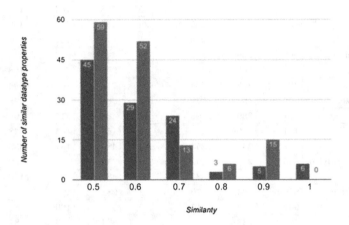

**Fig. 4.** Comparison of word2vec-based and GRU-based approach in discovering similar datatype properties.

Figure 4 shows the comparison of using word2vec and GRU-based approach in discovering similar datatype properties. We applied word2vec-based cosine similarity on the segmented terms of datatype properties, and calculate the average of maximal similarity. The left columns represent the number of DP pairs discovered using word2vec and the right columns represent the results of using GRU-based approach. The horizontal axis is the range of different similarity, e.g. [0.5, 0.6), [0.6, 0.7), etc. As a result, GRU-based approach can find more similar datatype properties that cannot be found with word2vec-based approach.

**Table 4.** Examples of cosine similarity for DP pairs.

| Terms of DP pairs | Word2Vec-based | GRU-based |
|---|---|---|
| lat & hasLatitude | 0.1565 | 0.6645 |
| dateOfBirth & wasBornOnDate | 0.5090 | 0.8406 |
| meaning & significance | 0.2105 | 0.7059 |
| cuisine & foodtype | 0.4497 | 0.7664 |
| startDate & happenedOnDate | 0.6015 | 0.8153 |

In total, we discovered 112 DP pairs with word2vec and 145 DP pairs with GRU-based approach. In most of the cases, GRU-based approach assigns higher cosine similarity to the semantically similar DP pairs than word2vec as shown in Table 4. For example, for the pair of *geo:lat*[7] and *yago:hasLatitude*, GRU-based similarity is 0.6645 while the word2vec-based similarity is 0.1565.

## 4.6    Exp 4: Discover Datatype Property Triples

This experiment shows discovered Datatype Property Triples (DPTs) by querying on the KGs with similar datatype properties. Since there are too many discovered similar DP pairs including infobox properties, we mainly focus on non-infobox DP pairs with similarity higher than 0.5 to discover missing DPTs in the KGs. First two column of Table 6 shows non-infobox datatype properties and the number of discovered DPTs from DBpedia and YAGO. Some of the YAGO datatype properties match to several DBpedia DPs because of the ontology heterogeneity problem.

**Table 5.** The statistical analysis of discovered DPTs in DBpedia and YAGO

| Property group | DBpedia | | | | YAGO | | | |
|---|---|---|---|---|---|---|---|---|
| | Match DPT | Missing DPT | Format mismatch DPT | Value mismatch DPT | Match DPT | Missing DPT | Format mismatch DPT | Value mismatch DPT |
| db-onto:birthDate yago:wasBornOnDate | 86 | 121 | 776 | 17 | 343 | 225 | 406 | 26 |
| db-onto:revenue yago:hasRevenue | 461 | 215 | 32 | 298 | 562 | 246 | 16 | 176 |
| db-onto:weight yago:hasWeight | 638 | 130 | 0 | 251 | 611 | 340 | 0 | 49 |
| geo:lat yago:hasLatitude | 745 | 106 | 0 | 149 | 692 | 237 | 0 | 71 |

Since single YAGO datatype properties may map to several DBpedia datatype properties, the discovered DPTs in DBpedia may have several triples that represent the same values. For example, either DPTs with *db-onto:startDate* or *db-onto:date* can represent the date of an event. Therefore, it is difficult to decide which RDF triple should be added to the event entity in DBpedia that is interlinked with YAGO entity.

Table 5, shows the pattern of the datatype property triples in DBpedia and YAGO for the selected groups (1000 sampled data). We have found four patterns. First group is "Match" in which both KGs have exact same value. Group "Missing" shows the number of DPT where the datatype property of triples exists in one KG but missing in another. As example, for any specific "Person" type entity might have date of birth information in DBpedia, but this information doesn't exist in YAGO. "Format mismatch" indicates the different representation of values, such as for Date type data format some property uses dd-mm-yyyy while

---

[7] http://www.w3.org/2003/01/geo/wgs84_pos#lat.

**Table 6.** The number discovered and mismatched DPT in DBpedia and YAGO.

| DBpedia | | YAGO | | Num of mismatch entities |
|---|---|---|---|---|
| db-onto:height | 11,947 | yago:hasHeight | 117,239 | 173 |
| db-onto:weight | 7,906 | yago:hasWeight | 51,420 | 1,488 |
| db-onto:length | 641 | yago:hasLength | 45,589 | 166 |
| db-onto:wikiPageLength | 10261 | yago:hasWikipediaArticleLength | 1,696,578 | 1,674,231 |
| db-onto:revenue | 698 | yago:hasRevenue | 5,337 | 1,103 |
| geo:long | 18,293 | yago:hasLongitude | 419,971 | 3,051 |
| geo:lat | 18,293 | yago:hasLatitude | 419,971 | 2,754 |
| db-onto:completionDate | 57,538 | yago:wasDestroyedOnDate | 4,278 | 468 |
| db-onto:deathDate | 31,024 | yago:diedOnDate | 18,940 | 8,819 |
| db-onto:deathYear | 30,530 | yago:diedOnDate | | 4,297 |
| db-prop:dateOfDeath | 34,443 | yago:diedOnDate | | 10,762 |
| db-prop:deathDate | 251,682 | yago:diedOnDate | | 9,614 |
| db-onto:runtime | 567 | yago:hasDuration | 184,772 | 413 |
| db-prop:runtime | 339 | yago:hasDuration | | 579 |
| db-onto:startDate | 69,664 | yago:happenedOnDate | 4,037 | 72 |
| db-onto:date | 67,014 | yago:happenedOnDate | | 763 |
| db-prop:dateOfBirth | 78,298 | yago:wasBornOnDate | 26,956 | 28,387 |
| db-onto:birthDate | 58,555 | yago:wasBornOnDate | | 29,194 |
| db-onto:birthYear | 70,833 | yago:wasBornOnDate | | 12,581 |
| db-onto:releaseDate | 825,830 | yago:wasCreatedOnDate | 11,713 | 36,978 |
| db-onto:foundingYear | 891,587 | yago:wasCreatedOnDate | | 3,738 |
| db-prop:pages | 594 | yago:hasPages | 7,277 | 852 |
| db-onto:numberOfPages | 799 | yago:hasPages | | 417 |

other use only yyyy representation. Final group is to define the mismatching values of the same data in the interlinked entities.

For example, the height of "Alec_Mazo" is 1.524 in DBpedia and 1.80 in YAGO. As shown in third column of table, we found 173 entities have different values of height, 1,488 entities have different values of weight, and 29,194 entities have different birthday with the pair of *db-onto:birthDate* and *yago:wasBornOnDate*, etc. The values are mostly dates and numbers, but in different format. For example, some date values use "1719-##-##" to represent the year "1719" while others use the number "1719", or the decimal numbers have different length. There are still some special cases that we did not deal with, such as "−212" and "c. 212 BC@en", "1985-4" and "Disappeared c. April 1978; Declared legally dead in 1985@en", etc.

# 5    Conclusion and Future Work

In this paper, we proposed a graph-based approach to discover similar object property pairs and a word embedding based approach, specifically GRU-based approach to discover similar datatype properties from two Knowledge Graphs - DBpedia and YAGO. With the similar property pairs, we discovered missing RDF triples in the KGs and also discovered mismatching values in the KGs. By using both Semantic Web technology and NLP methods, we could successfully discover missing RDF triples from the interlinked KGs that can be used for KG completion and validation.

Although, many RDF triples and mismatching values are discovered, it is still a challenging problem to evaluate the quality of discovered knowledge and there are some special cases that we have not solved to prove they are semantically the same. In future work, we will work on discovering *owl:sameAs* links, which needs consideration of both object property triples and datatype property triples. We will expand our RDF triple store with more KGs to evaluate our approach.

**Acknowledgements.** This work was partially supported by NEDO (New Energy and Industrial Technology Development Organization).

# References

1. Abadi, M., Agarwal, A., Barham, P., Brevdo, E., Chen, Z., Citro, C., Corrado, G.S., Davis, A., Dean, J., Devin, M., Ghemawat, S., Goodfellow, I.J., Harp, A., Irving, G., Isard, M., Jia, Y., Józefowicz, R., Kaiser, L., Kudlur, M., Levenberg, J., Mané, D., Monga, R., Moore, S., Murray, D.G., Olah, C., Schuster, M., Shlens, J., Steiner, B., Sutskever, I., Talwar, K., Tucker, P.A., Vanhoucke, V., Vasudevan, V., Viégas, F.B., Vinyals, O., Warden, P., Wattenberg, M., Wicke, M., Yu, Y., Zheng, X.: Tensorflow: large-scale machine learning on heterogeneous distributed systems. Computing Research Repository (CoRR) abs/1603.04467 (2016)
2. Bechhofer, S., van Harmelen, F., Hendler, J., Horrocks, I., McGuinness, D.L., Patel-Schneider, P.F., Stein, L.A.: OWL Web Ontology Language Reference. W3C Recommendation (2004). http://www.w3.org/TR/owl-ref/
3. Bengio, Y., Courville, A., Vincent, P.: Representation learning: a review and new perspectives. IEEE Trans. Pattern Anal. Mach. Intell. **35**(8), 1798–1828 (2013)
4. Bordes, A., Usunier, N., Garcia-Duran, A., Weston, J., Yakhnenko, O.: Translating embeddings for modeling multi-relational data. In: 26th International Conference on Neural Information Processing Systems (NIPS), pp. 2787–2795. Curran Associates, Inc. (2013)
5. Feng, J., Huang, M., Yang, Y., Zhu, X.: GAKE: graph aware knowledge embedding. In: 26th International Conference on Computational Linguistics, pp. 641–651 (2016)
6. Garten, J., Sagae, K., Ustun, V., Dehghani, M.: Combining distributed vector representations for words. In: NAACL-HLT, pp. 95–101 (2015)
7. Handler, A.: An empirical study of semantic similarity in WordNet and Word2Vec. Ph.D. thesis, University of New Orleans (2014)
8. Ichise, R.: An analysis of multiple similarity measures for ontology mapping problem. Int. J. Semant. Comput. **4**(1), 103–122 (2010)

9. Kertkeidkachorn, N., Ichise, R.: Estimating distributed representations of compound words using recurrent neural networks. In: Frasincar, F., Ittoo, A., Nguyen, L.M., Métais, E. (eds.) NLDB 2017. LNCS, vol. 10260, pp. 235–246. Springer, Cham (2017). doi:10.1007/978-3-319-59569-6_28

10. Klyne, G., Carroll, J.J.: Resource Description Framework (RDF): Concepts and Abstract Syntax. W3C Recommendation (2004). http://www.w3.org/TR/rdf-concepts/

11. Lehmann, J., Isele, R., Jakob, M., Jentzsch, A., Kontokostas, D., Mendes, P., Hellmann, S., Morsey, M., van Kleef, P., Auer, S., Bizer, C.: DBpedia - a large-scale, multilingual knowledge base extracted from Wikipedia. Semant. Web J. **6**(2), 167–195 (2015)

12. Lin, Y., Liu, Z., Sun, M., Liu, Y., Zhu, X.: Learning entity and relation embeddings for knowledge graph completion. In: 29th AAAI Conference on Artificial Intelligence, pp. 2181–2187 (2015)

13. Mahdisoltani, F., Biega, J., Suchanek, F.M.: YAGO3: a knowledge base from multilingual Wikipedias. In: 7th Biennial Conference on Innovative Data Systems Research (CIDR) (2015)

14. Mikolov, T., Chen, K., Corrado, G., Dean, J.: Efficient estimation of word representations in vector space. Computing Research Repository (CoRR) abs/1301.3781 (2013)

15. Minervini, P., Fanizzi, N., d'Amato, C., Esposito, F.: Scalable learning of entity and predicate embeddings for knowledge graph completion. In: 14th International Conference on Machine Learning and Applications, pp. 162–167. IEEE (2015)

16. Nickel, M., Murphy, K., Tresp, V., Gabrilovich, E.: A review of relational machine learning for knowledge graphs. Proc. IEEE **104**(1), 11–33 (2016)

17. Pan, J.Z., Vetere, G., Gomez-Perez, J.M., Wu, H.: Exploiting Linked Data and Knowledge Graphs in Large Organisations. Springer, Cham (2016). doi:10.1007/978-3-319-45654-6

18. Prud'hommeaux, E., Seaborne, A.: SPARQL query language for RDF. W3C Recommendation (2008). http://www.w3.org/TR/rdf-sparql-query/

19. Řehůřek, R., Sojka, P.: Software framework for topic modelling with large corpora. In: LREC 2010 Workshop on New Challenges for NLP Frameworks, pp. 45–50. ELRA (2010)

20. Shimaoka, S., Stenetorp, P., Inui, K., Riedel, S.: An attentive neural architecture for fine-grained entity type classification. In: AKBC (2016)

21. Socher, R., Perelygin, A., Wu, J.Y., Chuang, J., Manning, C.D., Ng, A.Y., Potts, C., et al.: Recursive deep models for semantic compositionality over a sentiment treebank. In: EMNLP, vol. 1631, p. 1642 (2013)

22. Wache, H., Voegele, T., Visser, U., Stuckenschmidt, H., Schuster, G., Neumann, H., Hubner, S.: Ontology-based integration of information - a survey of existing approaches. In: IJCAI-2001 Workshop: Ontologies and Information, pp. 108–117 (2001)

23. Winkler, W.E.: Overview of record linkage and current research directions. Technical report, Statistical Research Division U.S. Bureau of the Census (2006)

24. Zhao, L., Ichise, R.: Ontology integration for linked data. J. Data Semant. **3**(4), 237–254 (2014)

# Information Retrieval and Knowledge Discovery

# A New Sentiment and Topic Model for Short Texts on Social Media

Kang Xu$^{(\boxtimes)}$, Junheng Huang, and Guilin Qi

Department of Computer Science, Southeast Unversity, Nanjing, China
{kxu,jhhuang,gqi}@seu.edu.cn

**Abstract.** Nowadays plenty of user-generated posts, e.g., tweets and sina weibos, are published on social media and the posts imply the public's opinions towards various topics. Joint sentiment/topic models are widely applied in detecting sentiment-aware topics on the lengthy documents. However, the characteristics of posts, i.e., short texts, on social media pose new challenges: (1) context sparsity problem of posts makes traditional sentiment-topic models inapplicable; (2) conventional sentiment-topic models are designed for flat documents without structure information, while publishing users, publishing timeslices and hashtags of posts provide rich structure information for these posts. In this paper, we firstly devise a method to mine potential hashtags, based on explicit hashtags, to further enrich structure information for posts, then we propose a novel Sentiment Topic Model for Posts (STMP) which aggregates posts with the structure information, i.e., timeslices, users and hashtags, to alleviate the context sparsity problem. Experiments on Sentiment140 and Twitter7 show STMP outperforms previous models both in sentiment classification and sentiment-aware topic extraction.

**Keywords:** Topic model · Sentiment analysis · Sentiment classification · Topic extraction · Short text

## 1 Introduction

With the rapid growth of Web 2.0, a mass of user-generated posts, e.g., tweets and sina weibos, which capture people's interests, thoughts, sentiments and actions. The posts have been accumulating on the social media with each passing day. Sentiment analysis attempts to find user preference, likes and dislikes from the posts on social media, such as reviews, blogs and microblogs [17] and topic modeling attempts to discover the topics or aspects from reviews, blogs and microblogs [3]. Topic modeling and sentiment analysis on the posts are two significant tasks which can benefit many people. For example, we can discover a topic about "Apple Inc." and the overall sentiment of the topic. The sentiment of the topic about "Apple Inc." is implicitly associated with the stock trading of "Apple Inc.", because negative sentiments towards the company on social media can fall sales and financial gains but positive sentiments can improve sales [2].

© Springer International Publishing AG 2017
Z. Wang et al. (Eds.): JIST 2017, LNCS 10675, pp. 183–198, 2017.
https://doi.org/10.1007/978-3-319-70682-5_12

Topic modeling and sentiment analysis on the social media are complementary where sentiments on the social media often change over different topics and topics on the social media are always related to public sentiments. So jointly modeling topics and sentiments on the social media is a significative task and it can reflect people's sentiments on different topics, e.g., a topic about "Apple Inc." ('ipad', 'iphone', 'itouch', 'imac', 'beautiful' and 'popular') with the overall sentiment polarity "positive". Conventional sentiment-aware topic models, like Joint Sentiment/Topic Model (JST) [12] and Aspect/Sentiment Unification Model (ASUM) [9], are utilized for uncovering the hidden topics and sentiments on lengthy documents without structure information. However, the characteristics of posts, i.e., short texts, on social media pose both a challenge and an opportunity: (1) context sparsity problem of posts makes traditional sentiment-topic models inapplicable; (2) conventional sentiment-topic models are designed for documents without structure information, while users and timeslices of posts and hashtags contained in posts provide structure information for posts.

One simple and effective way to alleviate the sparsity problem is to aggregate short posts into lengthy pseudo-documents [6,29]. Inspired by the observations mentioned above, Xu et al. proposes a Time-User Sentiment/Topic Latent Dirichlet Allocation (TUS-LDA) [26]. TUS-LDA is based on the assumption that the posts on the social media are a mixture of two kinds of topics: temporal topics which are related to current events and stable topics which are related to personal interests. TUS-LDA takes advantage of structure information of posts, i.e., users (who publish posts) and timeslices (when posts are published), to aggregate posts for alleviating context sparsity problems of posts.

Moreover, we observe that hashtags, prefixing one or more characters with a hash symbol as "#hashtag", provide another kind of structure information for posts. Semantic relations between posts are built with hashtags, where posts under the same hashtags always talk about the similar topics. Since hashtags are strong topic indicators labeled by users on social media. Hence, hashtags are more effective, than timeslices and users, to be utilized for aggregating short texts on social media into pseudo-documents. We also find that there exist two kinds of hashtags for posts: explicit hashtags which are explicitly contained in the posts and potential hashtags which are not explicitly contained in posts but have semantic relevance with explicit hashtags. [24,25] proposes a method to mine potential hashtags based on explicit hashtags which exploit the co-occurrences of hashtags. For example, {D1: Healthy lunch if egg and broccoli **#cooking#food**; D2: Testing **#recipes** in my kitchen all day. I hated **#cooking**; D3: Wonderful day. nice movie thanks **#tweet**; D4: Powers of sping vegetable with chicken enrich breakfast **#recipes#tweet**}, "D1" has "#cooking" and "#food", "D2" has explicit hashtags,,"#cooking" and "#recipes"; for the occurrences of ("#cooking", "#recipes"), ("#cooking", "#food"), "D1" has a potential hashtag "#recipes" and "D2" has a potential hashtag "#food". However, the method only exploit the direct co-occurrences of hashtags will also introduce noisy hashtags. For example, "D3" will introduce a wrong hashtag "#recipes". To optimize the introduction of potential hashtags, embedding representations of all the words (including hashtags) in the posts are learnt using vector arithmetic, such as Glove [19] and Word2Vec [20]. Fine-grained semantic and syntactic regularities of hashtags are captured, where semantically close hashtags are adjacent on embedding space. Hence, we use cosine similarities to compute the semantic distances of hashtags to mine potential hashtags for each hashtag. However, hashtags cannot cover all the posts on social media [21]. Timeslices and users are also important structure information for posts to alleviate the context sparsity [26].

Furthermore, based on the analysis of the characteristics of topics and sentiments, we exploit the important observation of topics: A single post always talks about a single topic [29]. Although a post usually talks about a single topic, a post may talk about multiple aspects of the topic with different sentiment polarities [10, 14]. For better modeling topics and sentiments respectively, we follow the assumption that words in the same post should belong to the same topic, but they can have different sentiments.

In this paper, we propose a novel Sentiment Topic Model for Posts (STMP) which aggregates posts with structure information, i.e., timeslices, users or hashtags, to model topics and sentiments for posts on social media.

There exist three main contributions of our work: (1) We design a simple and effective method to mine potential hashtags for all the explicit hashtags and a new model, STMP, which aggregates posts in the same timeslice, user or hashtag as a pseudo-document to alleviate the context sparsity problem. (2) We design approaches of parameter inference and incorporating prior sentiment knowledge for STMP. (3) We implement experiments on two datasets to evaluate the effectiveness of sentiment classification and topic extraction in STMP.

## 2   Related Work

### 2.1   Topic Models on Short Texts

LDA [1] and PLSA [8] originally focus on mining topics from lengthy documents. Recently topic modeling in the posts on social media is popular, however, it also suffers from the context sparsity problem of the posts. To overcome the sparsity problem of posts on the social media, there exist some work of aggregating posts into pseudo-documents. In [29], Twitter-LDA aggregated posts published by a user into one lengthy pseudo-document and made words in the same post belong to the same topic. In [6], posts in TimeUserLDA were aggregated by timeslices or users for finding bursty topics where posts belong to two kinds of topics: personal topics and temporal topics. Similar to TimeUserLDA, posts in TUK-TTM [27] were also aggregated by timeslices or users and TUK-TTM was utilized for time-aware personalized hashtag recommendation. Although these models can alleviate the problem of the context sparsity of posts on social media, they did not model an extra aspect of posts, i.e., sentiment.

## 2.2   Joint Sentiment/Topic Models

Recently, some topic models have been extended to model topics and sentiments jointly. The first work of topic and sentiment modeling is Topic-Sentiment Mixture model (TSM) [15]. In TSM, a sentiment is a special kind of topic and each word is generated from either a sentiment or a topic. The relation between sentiments and topics cannot be mined by TSM. At the same time, TSM is based on PLSA and suffers from the problems of inferencing on new documents and overfitting the data. To overcome these shortcomings, Joint Sentiment-Topic model (JST) [12] which is a two-level sentiment-topic model based on Latent Dirichlet Allocation (LDA) was proposed. In JST, sentiment labels are associated with documents, under which topics are associated with sentiment labels and words are associated with both sentiment labels and topics. Reverse-JST (RJST) [13] is a variant of JST where the position of sentiment and topic layer is swapped. In JST, topics were generated conditioned on a sentiment polarity, while in RJST sentiments were generated conditioned on a topic. Topic-Sentiment (TS) model [4] is similar to RJST. Aspect/Sentiment Unification Model (ASUM) [9] is similar to JST. In ASUM, words in the same sentence belong to the same sentiment and topic. Sentiment Topic Model with Decomposed Prior (STDP) [30] is another variant of JST. STDP first determined whether the word is used as a sentiment word or ordinary topic words and then chose the accurate sentiments for sentiment words. Time-aware Topic-Sentiment Model (TTS) [5] extracted the hidden topics from texts and modeled the association between topics and sentiments and tracked the strength of topic-sentiment association over time. In TTS, time is viewed as a special word to bias the topic-sentiment distributions. But in our model, we use time to aggregate short texts and generate pseudo documents for modeling topics and sentiments. JST, RJST, ASUM, STDP and TTS are designed for normal texts where each piece of text has rich context to infer topics and sentiments, but our work models posts (i.e., short and informal texts) on social media and all of these models lose efficacy in the short and informal texts. MaxEnt-LDA [28] jointly discovers both aspects and aspect-specific opinion words by integrating a supervised maximum entropy algorithm to separate opinion words from objective ones. However, it does not further discover aspect-aware sentiment polarities of opinion words, which are very useful for sentiment analysis.

In our model, we focus on short and informal texts on social media. There exists some work about LDA-based sentiment analysis on social media. Twitter Opinion Topic Model (TOTM) [11] aggregated or summarized opinions of a product from tweets, which can discover target specific opinion words and improve opinion prediction. Topic Sentiment Latent Dirichlet Allocation (TSLDA) [18] utilized sentiments on social media for predicting stock price movement. TSLDA distinguished topic words and opinion words where topic words were drawn from the topic-word distribution and opinion words were drawn from the sentiment-topic-word distribution. Although these two work focuses on posts on social media, they do not consider and solve the context sparsity problem of posts. Xu et al. [26] proposed a Time-User Sentiment/Topic Latent Dirichlet Allocation

(TUS-LDA), which only utilized user and timeslice information and ignored the important topic information, i.e., hashtag.

## 3    The Proposed Work

In this section, we firstly propose a generative model, Sentiment Topic Model for Posts (STMP), to mine sentiment-aware topics on posts. A collapsed Gibbs sampling algorithm is developed for parameter estimation of STMP. Finally, we introduce a way to incorporate sentiment knowledge into STMP. All the notations needed in TUS-LDA and STMP are listed in Table 1.

**Table 1.** Notation used in the TUS-LDA and STMP model

| Symbol | Description |
|---|---|
| D, K | Number of documents, topics |
| V, U, T, H | Number of vocabulary, users, timeslices, hashtags |
| **Z, W, Y, X** | All the topics, words, switches of user and timeslice, switches of potential-explicit hashtags |
| **T, U, H** | All the timeslices, users and hashtags |
| **L, $\bar{L}$** | All the sentiments of posts and words |
| **R** | The potential hashtags of all the explicit hashtags |
| $N_d$ | Number of word tokens in post $d$ |
| $u_d, t_d, h_d$ | User, timeslice, hashtag of $d$ |
| $y_d, l_d, x_d$ | User-timeslice switches, sentiment and potential-explicit hashtag switches of $d$ |
| $l_{d,i}$ | Sentiment of word $w_{d,i}$ |
| $H_d, R$ | Explicit hashtags, potential hashtags |
| $\varepsilon$ | Beta distribution of stable topics and temporal topics |
| $\pi_d$ | Document-sentiment distribution, $\Omega = \{\pi_d\}_{d=1}^{D}$ |
| $\theta_{t,s}$ | Timeslice-sentiment topic distribution, $\Theta = \{\theta_{t,s}\}_{t=1,s=1}^{T*S}$ |
| $\delta_{u,s}$ | User-sentiment topic distribution, $\Phi = \{\delta_{u,s}\}_{u=1,s=1}^{U*S}$ |
| $o_{h,s}$ | Hashtag-sentiment topic distribution, $O = \{o_{h,s}\}_{h=1,s=1}^{H*S}$ |
| $\varphi_{s,k}$ | Sentiment-topic word distribution, $\Psi = \{\varphi_{s,k}\}_{s=1,k=1}^{S*K}$ |
| $\alpha$ | Hyperparameters of $\theta_{t,s}$ and $\delta_{u,s}$ |
| $\beta, \lambda$ | Hyperparameters of $\varphi_{s,k}, \pi_d$ |
| $\gamma$ | Hyperparameters of $\varepsilon$ |
| $\tau$ | Hyperparameters of $x_d$ |
| $\omega_s$ | Prior knowledge of $\varphi_{s,k}$ |

### 3.1    Brief Review of TUS-LDA

Time-User Sentiment/Topic Latent Dirichlet Allocation (TUS-LDA) [26] is a probabilistic model for sentiment-aware topics for posts on social media, which only utilized user and timeslice information. It is assumed that there exists a stream of $D$ posts, denoted as $d_1, d_2, ..., d_D$. Each post $d$ is generated by a user $u_d$ within a timeslice $t_d$ and the post $d$ contains a bag of words, denoted as $\{w_{d,1}, w_{d,2}, ..., w_{d,N_d}\}$.

The generative process of all the posts in the stream is as follow: When a user $u_d$ publishes a post $d$ within a timeslice $t_d$, the user first utilizes the variable $y_d$, which is drawn from the global user-timeslice switch distribution $\varepsilon$, to decide whether the post talks about a stable topic or a temporal topic. Then the user chooses a sentiment label $l_d$ for the post from the document-sentiment $\pi_d$. If the user chooses a stable topic $u_d$ and a sentiment label $l_d$, the user then selects a topic $z_d$ from $\delta_{u_d, l_d}$; otherwise, the user selects a topic $z_d$ from $\theta_{t_d, l_d}$. For each word $w_{d,i}$ in the post $d$, the user first chooses a sentiment label $l_{d,i}$; with the chosen topic $z_d$ and sentiment label $l_{d,i}$, the word is drawn from the sentiment-topic word distribution $\varphi_{l_{d,i}, z_d}$.

## 3.2   Sentiment Topic Model for Posts

Based on TUS-LDA [26], we further design a new sentiment and topic model, Sentiment Topic Model for Posts (STMP). STMP is a probabilistic genetive model that describe a process of generating posts on social media with structure information from users, timeslices, explicit and potential hashtags. When a user $u_d$ publishes a post $p_d$ within a timeslice $t_d$ and explicit hashtags $H_d$, if the post contains at least one hashtag ($|H_d| > 0$), the user firstly chooses a key hashtag $h_d^1$ from explicit hashtags $H_d$, then utilize the variable $x_d$, which is drawn from

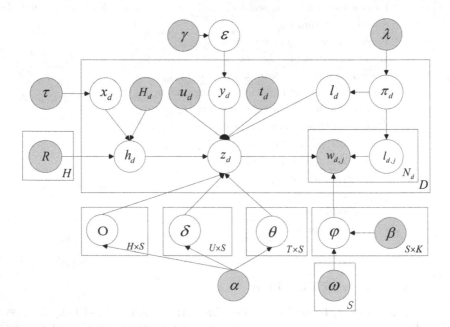

**Fig. 1.** Sentiment topic model for posts

$\tau$, to decide how to assign hashtags in post $d$; if $x_d = 0$, the representative $h_d$ is set as the key hashtag $h_d^1$; otherwise, the user chooses $h_d$ from $R(h_d^1)$, where $R(h_d^1)$ are the potential hashtags of $h_d^1$; if the post does not contain any hashtags ($|H_d| = 0$), the user first utilizes the variable $y_d$, which is drawn from the global user-timeslice switch distribution $\varepsilon$, to decide whether the post talks about a stable topic or a temporal topic. Then the user chooses a sentiment label $l_d$ for the post from the document-sentiment $\pi_d$. If the user chooses a hashtag topic $h_d$ and a sentiment label $l_d$, the user then selects a topic $z_d$ from $o_{h_d,l_d}$; if the user chooses a stable topic $u_d$ and a sentiment label $l_d$, the user then selects a topic $z_d$ from $\delta_{u_d,l_d}$; otherwise, the user selects a topic $z_d$ from $\theta_{t_d,l_d}$. For each word $w_{d,i}$ in the post $p_d$, the user first chooses a sentiment label $l_{d,i}$; with the chosen topic $z_d$ and sentiment label $l_{d,i}$, the word is drawn from the sentiment-topic word distribution $\varphi_{l_{d,i},z_d}$. Figure 1 shows the graphical representation of the generation process. Formally, the generative story for each post is as follows:

1. Draw $\varepsilon \sim Beta(\gamma)$
2. For each timeslice $t = 1, ..., T$
   i. For each sentiment label $s = 0, 1, 2$
      a. Draw $\theta_{t,s} \sim Dir(\alpha)$
3. For each user $u = 1, ..., U$
   i. For each sentiment label $s = 0, 1, 2$
      a. Draw $\delta_{u,s} \sim Dir(\alpha)$
4. For each hashtag $h = 1, ..., H$
   i. For each sentiment label $s = 0, 1, 2$
      a. Draw $o_{h,s} \sim Dir(\alpha)$
5. For each sentiment label $s = 0, 1, 2$
   i. For each topic $k = 1, ..., K$
      a. Draw $\varphi_{s,k} \sim Dir(\beta)$
6. For each post $p_d$, $d = 1, ..., D$
   i. Draw $\pi_d \sim Dir(\lambda)$
   ii. Draw $l_d \sim Multi(\pi_d)$
   iii. if $|H_d| > 0$, Draw $h_d^1 \sim Uniform(H_d)$; Draw $x_d \sim Bernoulli(\tau)$; if $x_d = 0$, $h_d = h_d^1$, else, Draw $h_d \sim Multi(R(h_d^1))$; Draw $z_d \sim Multi(o_{h_d,l_d})$
   iv. if $|H_d| = 0$, Draw $y_d \sim Bernoulli(\varepsilon)$; if $y_d = 0$, Draw $z_d \sim Multi(\theta_{u_d,l_d})$, else Draw $z_d \sim Multi(\delta_{t_d,l_d})$
   v. For each word $w$ $i = 1, ... N_d$
      a. Draw $l_{d,i} \sim Multi(\pi_d)$
      b. Draw $w_{d,i} \sim Multi(\varphi_{z_d,l_{d,i}})$

**Joint Distribution of STMP.** The joint probability of words, hashtags, explicit hashtags, potential hashtags, explicit hashtags and potential hashtags switches, users, timeslices, timeslices-user switches, topics and sentiments can be factored in Eq. 1, where $\varepsilon$, $\pi$, $\varphi$, $\delta$, $o$ and $\theta$ are integrated and $\overrightarrow{n}_d$ counts the number of three sentiment labels of a post and the words in the post.

$$P_{STMP}(\mathbf{Z}, \mathbf{W}, \mathbf{T}, \mathbf{U}, \mathbf{Y}, \mathbf{L}, \mathbf{L}, \bar{\mathbf{X}}, \mathbf{H}|\alpha, \gamma, \lambda, \beta, \omega, \tau) =$$

$$\tau^{n_x^1}(1-\tau)^{n_x^0} \prod_{d=1}^{D} \frac{1}{H_d} \sum_{j=1}^{H_d} p(h_j|H_d)p(h_d|h_j, R_{h_j}) \frac{\Delta(\overrightarrow{n_y} + \overrightarrow{\gamma})}{\Delta(\overrightarrow{\gamma})}$$

$$\times \prod_{d=1}^{D} \frac{\Delta(\overrightarrow{n}_d + \overrightarrow{\lambda})}{\Delta(\overrightarrow{\lambda})} \times \prod_{u=1}^{U} \prod_{s=1}^{S} \frac{\Delta(\overrightarrow{n}_{u,s} + \overrightarrow{\alpha})}{\Delta(\overrightarrow{\alpha})} \times \prod_{t=1}^{T} \prod_{s=1}^{S} \frac{\Delta(\overrightarrow{n}_{t,s} + \overrightarrow{\alpha})}{\Delta(\overrightarrow{\alpha})}$$

$$\times \prod_{h=1}^{H} \prod_{s=1}^{S} \frac{\Delta(\overrightarrow{n}_{h,s} + \overrightarrow{\alpha})}{\Delta(\overrightarrow{\alpha})} \times \prod_{s=1}^{S} \prod_{k=1}^{K} \frac{\Delta(\overrightarrow{n}_{s,k} + \overrightarrow{\beta})}{\Delta(\overrightarrow{\beta})};$$

$$\Delta = \frac{\prod_{k=1}^{dim\,\overrightarrow{x}} \Gamma(x_k)}{\Gamma(\prod_{k=1}^{dim\,\overrightarrow{x}} x_k)}, \overrightarrow{n}_y = \{n_y^0, n_y^1\}, \overrightarrow{n}_d = \{n_d^{pos}, n_d^{neg}\}$$

$$\overrightarrow{n}_{u,s} = \{n_{u,s}^k\}_{k=1}^{K}, \overrightarrow{n}_{t,s} = \{n_{t,s}^k\}_{k=1}^{K}, \overrightarrow{n}_{s,k} = \{n_{s,k}^v\}_{v=1}^{V}$$

(1)

### 3.3  Parameters Inference for STMP

Posterior distribution is estimated as follows: for the $i$-th post, the user $u_i$, timeslice $t_i$ and explicit hashtags $H_d$ are known. $x_i$, $h_i$, $y_i$, $z_i$ and $l_i$ can be jointly sampled given all other variables. Since the sampling formulas of users, timeslices ($y_i$) and sentiments of all the word in the posts $l_{i,w}$ are the same as TUS-LDA [26] (Eqs. 4 and 5). Here, we use $\mathbf{x}$ to denote all the hidden variables $x$ and $\mathbf{x}_{-i}$ to denote all the other $x$ except $x_i$. All the hyperparameters are omitted.

If $x_i = 0$, the $i$-th post choose a hashtag from explicit hashtags, the sampling formula is shown in Eq. 2. If $x_i = 1$, the $i$-th post chooses a hashtag from potential hashtags, the sampling formula is shown in Eq. 3.

$$P(x_i = 0, h_i = h, z_i = k, l_i = s|\mathbf{x}_{-i}, \mathbf{y}, \mathbf{h}_{-i}, \mathbf{z}_{-i}, \mathbf{l}_{-i}, \bar{\mathbf{l}}, \mathbf{w})$$

$$\propto \tau \frac{1}{|H_d|} \times \frac{\lambda_s + n_{d,-i}^s}{\sum_{s'=1}^{S} \lambda_{s'} + n_{d,-i}^s} \times \frac{\alpha_k + n_{h,s,-i}^k}{\sum_{k=1}^{K} \alpha_{k'} + n_{h,s,-i}^{k'}}$$

$$\times \frac{\prod_{v=1}^{V} \prod_{n_v=0}^{N_{(v)}-1}(\beta_{s,k}^v + n_{s,k,-i}^v + n_v)}{\prod_{n=0}^{N-1}(\sum_{v'=1}^{V}(\beta_{s,k}^{v'} + n_{s,k,-i}^{v'}) + n)}$$

(2)

$$P(x_i = 1, h_i = h, z_i = k, l_i = s|\mathbf{x}_{-i}, \mathbf{y}, \mathbf{h}_{-i}, \mathbf{z}_{-i}, \mathbf{l}_{-i}, \bar{\mathbf{l}}, \mathbf{w})$$

$$\propto (1-\tau) \sum_{j=1}^{|H_d|} p(h_j|H_d)p(h|h_j, R_{h_j}) \times \frac{\lambda_s + n_{d,-i}^s}{\sum_{s'=1}^{S} \lambda_{s'} + n_{d,-i}^s}$$

(3)

$$\times \frac{\alpha_k + n_{h,s,-i}^k}{\sum_{k=1}^{K} \alpha_{k'} + n_{h,s,-i}^{k'}} \times \frac{\prod_{v=1}^{V} \prod_{n_v=0}^{N_{(v)}-1}(\beta_{s,k}^v + n_{s,k,-i}^v + n_v)}{\prod_{n=0}^{N-1}(\sum_{v'=1}^{V}(\beta_{s,k}^{v'} + n_{s,k,-i}^{v'}) + n)}$$

$$P(y_i = 0, z_i = k, l_i = s | \mathbf{x}, \mathbf{y}_{-i}, \mathbf{z}_{-i}, \mathbf{L}_{-i}, \bar{\mathbf{l}}, \mathbf{w}) \propto \frac{\gamma_0 + n_{y,-i}^0}{\sum_{p=1}^2 \gamma_p + n_{y,-i}^p}$$

$$\times \frac{\lambda_s + n_{d,-i}^s}{\sum_{s'=1}^S \lambda_{s'} + n_{d,-i}^s} \times \frac{\alpha_k + n_{u,s,-i}^k}{\sum_{k=1}^K \alpha_{k'} + n_{u,s,-i}^{k'}} \tag{4}$$

$$\times \frac{\prod_{v=1}^V \prod_{n_v=0}^{N_{(v)}-1}(\beta_{s,k}^v + n_{s,k,-i}^v + n_v)}{\prod_{n=0}^{N-1}(\sum_{v'=1}^V(\beta_{s,k}^{v'} + n_{s,k,-i}^{v'}) + n)}$$

If $y_i = 0$, the $i$-th post talks about a stable topic, the sampling formula is shown in Eq. 4; otherwise, the $i$-th post talks about a temporal topic, the sampling formula is shown in Eq. 5.

$$P(y_i = 1, z_i = k, l_i = s | \mathbf{x}, \mathbf{y}_{-i}, \mathbf{z}_{-i}, \mathbf{L}_{-i}, \bar{\mathbf{l}}, \mathbf{w}) \propto \frac{\gamma_1 + n_{y,-i}^1}{\sum_{p=1}^2 \gamma_p + n_{y,-i}^p}$$

$$\times \frac{\lambda_s + n_{d,-i}^s}{\sum_{s'=1}^S \lambda_{s'} + n_{d,-i}^s} \times \frac{\alpha_k + n_{t,s,-i}^k}{\sum_{k'=1}^K \alpha_{k'} + n_{t,s,-i}^{k'}} \tag{5}$$

$$\times \frac{\prod_{v=1}^V \prod_{n_v=0}^{N_{(v)}-1}(\beta_{s,k}^v + n_{s,k,-i}^v + n_v)}{\prod_{n=0}^{N-1}(\sum_{v'=1}^V(\beta_{s,k}^{v'} + n_{s,k,-i}^{v'}) + n)}$$

For the $j$-th word in the $i$-th post, the sample formula of is shown in Eq. 6.

$$P(\bar{l}_{ij} = s | \mathbf{z}, \bar{\mathbf{l}}_{-ij}, \mathbf{w}, \mathbf{y}, \mathbf{l}) \propto \frac{\lambda_s + n_{m,-ij}^s}{\sum_{s'=1}^S (\lambda_{s'} + n_{m,-ij}^{s'})}$$

$$\times \frac{\beta_{s,k}^v + n_{s,k.-ij}^v}{\sum_{v'=1}^V (\beta_{s,k}^{v'} + n_{s,k.-ij}^{v'})} \tag{6}$$

Samples obtained from MCMC are then utilized for estimating the distributions $\pi$ (Eq. 7), $\delta$ and $\theta$ (Eq. 8), $\phi$ and $o$ (Eq. 9).

$$\pi_d^s = \frac{\lambda_s + n_d^s}{\sum_{s'=1}^S (\lambda_{s'}' + n_d^{s'})} \tag{7}$$

$$\delta_{u,s}^k = \frac{\alpha_k + n_{u,s}^k}{\sum_{k'=1}^K (\alpha_{k'} + n_{u,s}^{k'})}, \theta_{t,s}^k = \frac{\alpha_k + n_{t,s}^k}{\sum_{k'=1}^K (\alpha_{k'} + n_{t,s}^{k'})} \tag{8}$$

$$\varphi_{s,k}^v = \frac{\beta_{s,k}^v + n_{s,k}^v}{\sum_{v'=1}^V (\beta_{s,k}^{v'} + n_{s,k}^{v'})}, o_{h,s}^k = \frac{\alpha_k + n_{h,s}^k}{\sum_{k'=1}^K (\alpha_{k'} + n_{h,s}^{k'})} \tag{9}$$

## 3.4 Incorporating Prior Knowledge

Drawing on the experience of JST and RJST [13], we also add an additional dependency link of $\varphi$ on the matrix $\omega$ of size $S * V$, which is utilized for encoding word prior sentiment information into STMP. To incorporate prior knowledge into STMP, we first set all the values of $\omega$ as 1. Then the matrix $\omega$ is updated with a sentiment lexicon which contains words with the corresponding sentiment labels, i.e., positive and negative. For each term $w \in \{1, ..., V\}$ in the corpus, if $w$ is found in the sentiment lexicon with the sentiment label $l \in \{1, ..., S\}$, the element $\omega_{lw}$ is set as 1 and other elements of the word $w$ are set as 0. The element $\omega_{lw}$ is updated as follows:

$$\omega_{lw} = \begin{cases} 1 & \text{if } S(w)=l \\ 0 & \text{otherwise} \end{cases}$$

The Dirichlet prior $\beta$ of the size $S * K * V$ are multiplied by the matrix $\omega$ (a transformation matrix) to capture the word prior sentiment polarity.

## 4    Experiment

### 4.1    Dataset Description

For experiments, we performed sentiment-aware topic discovery and sentiment classification on tweets, which are characterized by their limited 140 characters text. We selected tweets, which are related to electronic products such as camera and mobile phones, from Tweet7[1]. These tweets contain the description and reviews of various electronic products and correspond to multiple sentiment-aware topics. Besides, each tweet contains the content, the release timeslice, the user information. Due to the lack of sentiment labels on the Tweet7, we utilized the Sentiment140[2] [7], which contains 1.6 million tweets, for sentiment classification evaluation. Each tweet in Sentiment140 has the content, a release timeslice, a user and the overall polarity label (positive or negative). The number of positive and negative tweets are nearly identical.

### 4.2    Sentiment Lexicon

In JST [12] and our model, each sentiment label is viewed as a special kind of topic that we have known in advance. To improve the accuracy of sentiment detection, we need to incorporate prior knowledge or subjectivity lexicon (i.e., words with positive or negative polarity). Here, we chose PARADIGM [22], which consists of a set of positive and negative words, e.g., happy and sad. It defines the positive and negative semantic orientation of words. Moreover, emoticons are also strong emotion indicators on social media. The entire list of emotions is taken from Wikipedia[3] (Table 2).

---

[1] https://snap.stanford.edu/data/twitter7.html.
[2] http://help.sentiment140.com/for-students/.
[3] https://en.wikipedia.org/wiki/List_of_emotions.

**Table 2.** Emoticons

| Positive | Negative |
|---|---|
| :-) :o) :] :3 :c | >:-( >:[ :-( :c |
| :> =] 8) : } :-D | :@ >:( ;( ;-( |
| ;-D :D 8-D \o/ ^̈ | :'-(:'( D; (T_T) (;_;) |
| :} (ö)/ ()/ | (;_:) T.T !_! |

## 4.3  Parameter Settings

To optimize the number of topics $K$, we empirically ran the models with four values of $K$: 10, 20, 50 and 100 in Sentiment140 and ran the models with three values of $K$: 10, 20, 50 in Twitter7 (In Twitter7, these tweets only contain a small number of electronic product-related topics). In our model, we simply selected symmetric Dirichlet prior vectors as is empirically done in JST and ASUM. For JST, ASUM and TS, $\alpha = \frac{50}{K}, \beta = 0.01$ and $\gamma = 0.01$. For STMP and TUS-LDA, we set $\alpha = 0.5, \gamma = 0.01, \tau = 0.01, \lambda = 0.01$ and $\beta = 0.01$. These LDA-based models are not sensitive to the hyperparameters [23]. In all the methods, Gibbs sampling was run for 1,000 iterations with 200 burn-in periods.

## 4.4  Sentiment Classification

In this section, we performed a sentiment classification task to predict the sentiment labels of the test data in Sentiment140. Note that the Sentiment140 tweets do not contain neutral tweets. We determined the polarity of a tweet $d$ by selecting the polarity $s$ that has a higher probability in $\pi_d^s$ ($\pi_d$ is the sentiment distribution of the $d$-th post).

$$polarity(d) = \underset{s=\{neg,pos\}}{argmax} \; \pi_d^s \tag{10}$$

We present the results of sentiment classification with *Accuracy*, which is the proportion of true results (both true positive results and true negative results) among the total number of cases examined in the binary classification.

Based on the results of sentiment classification, for *Accuracy* (Fig. 2), although TUS-LDA (aggregating tweets in timeslices or users) performed better than other models, STMP still performed better than TUS-LDA. It shows that hashtags are more important topic indicators than users and timeslices. Moreover, STMP performed best in Sentiment140 when $K = 20$.

**Topic Coherence.** Another goal of TUS-LDA and STMP is to extract coherent sentiment-aware topics from user-generated post collection and evaluate the effectiveness of topic and sentiment captured by our models. In order to conduct quantitative evaluation of topic coherence, we used an automated metric proposed in [16], $C(t; V^{(t)}) = \sum_{m=2}^{M} \sum_{l=1}^{m-1} log \frac{D(v_m^{(t)}, v_l^{(t)})+1}{D(v_l^{(t)})}$, where topic coherence, denoted as $D(v)$, is the document frequency of word $v$, $D(v, v')$ is the

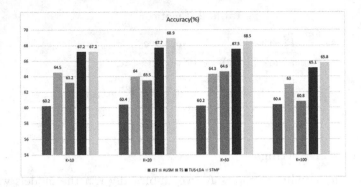

**Fig. 2.** Accuracy of sentiment classification on all the models with Sentiment140

**Table 3.** Average coherence score on the top $T$ words in the $K$ topics discovered on tweets of electronic products

| T | Top 5 | | | Top 10 | | | Top 20 | | |
|---|---|---|---|---|---|---|---|---|---|
| K | 10 | 20 | 50 | 10 | 20 | 50 | 10 | 20 | 50 |
| JST | −39.88 | −42.08 | −41.68 | −242.74 | −246.79 | −251.97 | −1139.43 | −1145.36 | −1142.01 |
| ASUM | −38.02 | −39.86 | −39.58 | −240.47 | −243.97 | −246.47 | −1135.44 | −1131.96 | −1135.27 |
| TS | −38.85 | −39.55 | −39.50 | −241.17 | −245.23 | −248.83 | −1139.42 | −1141.69 | −1142.21 |
| TUS-LDA | −33.91 | −35.7 | −35.61 | −233.08 | −234.72 | −241.78 | −1030.83 | −1127.64 | −1130.02 |
| STMP | −32.51 | −34.7 | −35.1 | −231.13 | −232.51 | −240.73 | −1007.13 | −1085.34 | −1122.31 |

co-document frequency of word $v$ and $v'$ and $V^{(k)} = (v_1^{(k)}, ..., v_T^{(k)})$ is a list of the $T$ most probable words in topic $k$. The key idea of the coherence score is that if a word pair is related to the same topic, they will co-occur frequently in the corpus. In order to quantify the overall coherence of the discovered topics, the average coherence score, $\frac{1}{K} \sum^k C(z_k; V^{(z_k)})$, was utilized. We conducted and evaluated the topic extraction experiments on the tweets of electronic products. Here we also compared STMP with four sentiment-topic models: JST, ASUM, TS and TUS-LDA. In this collection, we set the number of topics $K = 10, 20, 50$ for all the methods. The result is listed in Table 3. From the topic coherent results, it is clear that aggregating tweets in timeslices, users or hashtags (STMP) directly leads to significant improvement of topic coherent and STMP performed best in the topic coherent.

$$C(t; V^{(t)}) = \sum_{m=2}^{M} \sum_{l=1}^{m-1} log \frac{D(v_m^{(t)}, v_l^{(t)}) + 1}{D(v_l^{(t)})} \qquad (11)$$

**Human Evaluation.** As our objective is to discover more coherent sentiment-aware topics, so we chose to evaluate the topics manually which is based on human judgement. Without enough knowledge, the annotation will not be credible. Following [16], we asked two human judges, who are familiar with common

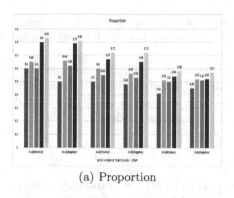
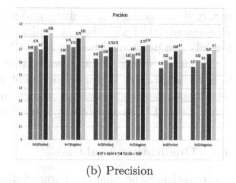

(a) Proportion                              (b) Precision

**Fig. 3.** (a) Proportion of *coherent* topics generated by each model in $K = 10, 20, 50$ (b) Average Precision @20 (p @20) of words in *coherent* topics generated by each model in $K = 10, 20, 50$

knowledge and skilled in looking up the test tweet dataset, to annotate the discovered sentiment-aware topics manually. To ensure the annotation reliable, we labeled the generated topics by all the baseline models and our proposed model at learning iteration 10.

**Topic Labeling:** Following [16], we asked the judges to label each sentiment-aware topic as *coherent* or *incoherent*. Each sentiment-aware topic is represented as a list of 20 most probable words in word distribution $\varphi$ of the topic. Here they annotated a sentiment-aware topic as *coherent* when at least half of top 20 words were related to the same semantic-coherent concept (e.g., an event, a hot topic) and the sentiment polarities of the words are accurate, others were *incoherent*.

**Word Labeling:** Then we chose *coherent* sentiment-aware topics which were judged before and asked judges to label each word of the top 20 words among these *coherent* sentiment-aware topics. When a word was in accordance with the main semantic-coherent concept that represents the topic, the word was annotated as *correct* and others were *incorrect*. After topic labeling, the judges had known the concept of each sentiment-aware topic and the overall sentiment of the topic, it is easy to label words of each sentiment-aware topic (Table 4).

**Table 4.** Cohen's Kappa for pairwise inter-rater agreements

|  | Topic labeling | Word labeling | | |
|---|---|---|---|---|
|  |  | p@5 | p@10 | p@20 |
| Kappa | 0.820 | 0.911 | 0.821 | 0.816 |

Figure 3(a) shows that STMP can discover more *coherent* topics than JST, ASUM, TS, TUS-LDA. Thereinto, TUS-LDA and STMP can also discover the nearly equal number of positive and negative topics. Figure 3 gives the average *Precision*@20 of all coherent topics. STMP performed better than other four models and performed best in $K = 10$.

From the above, we can observe that aggregating posts in the same timeslice, user or hashtag as a single document can indeed improve the performance in sentiment classification and sentiment-aware topic extraction in user-generated posts, i.e., STMP consistently outperformed the baseline models.

## 4.5   Qualitative Analysis

To investigate the quality of topics discovered by STMP, we randomly choose some topics for visualization. We randomly selected six topics, i.e., three positive topics and three negative topics. For each topic, we choose the top 10 words which can most represent the topic.

**Table 5.** Example of topics extracted by STMP

| Positive sentiment label | | | Negative sentiment label | | |
|---|---|---|---|---|---|
| Topic 1 | Topic 2 | Topic 3 | Topic 1 | Topic 2 | Topic 3 |
| camera | ipod | xbox | printer | window | phone |
| digit | song | game | ink | vista | **problem** |
| canon | phone | live | print | us | information |
| nikon | listen | sale | cartridge | microsoft | security |
| new | music | console | **low** | install | **strange** |
| len | love | microsoft | laser | download | **risk** |
| photograph | tone | play | color | software | finance |
| **sharp** | play | playstate | laserjet | file | mobile |
| panason | shuffle | ps3 | paper | **slow** | digit |
| slr | **good** | **new** | scanner | server | on-line |

Table 5 presents the top words of the selected topics. The three topics with a positive sentiment label respectively talk about "Camera", "apple music product" and "game" and these topics are listed in the left columns of Table 5; the three negative topics are related to "printer", "window product" and "phone" are listed in right columns of Table 5. As we can see clearly from Table 5, the six topics are quite explicit and coherent, where each of them tried to capture the topic of a kind of electronic product. In terms of topic sentiment, by checking each of the topics in Table 5, it is clear that all the 6 topics can indeed bear positive and negative sentiment labels respectively where all the sentiment words are written in bold. By manually examining the tweet data, we observe that the sentiment labels of these topics are accurate. The analysis of these topics shows that STMP can indeed discover coherent sentiment-aware topics.

# 5   Conclusion and Future Work

In this paper, we studied the problem of sentiment-aware topic detection from the user-generated posts on the social media. The existing work is not suitable for the short and informal posts, we proposed a new sentiment/topic model for posts on social media that considers the time, user and hashtag information of posts to jointly model topics and sentiments. Based on the different characteristics of sentiments and topics, we limited that words in the same post belong to the same topic, but they can belong to different sentiments. We compared our model with JST, ASUM, TS and TUS-LDA on two Twitter datasets. Our quantitative evaluation showed that our model outperformed other models both in sentiment classification and topic coherence. At the same time, we asked two judges to evaluate our models and baseline methods and the result also showed that our model STMP performed best in sentiment-aware topic extraction. Moreover, we also chose six examples to visualize some sentiment-aware topics. In the future work, we will incorporate word embedding into our model to improve the performance of modeling posts on social media.

# References

1. Blei, D.M., Ng, A.Y., Jordan, M.I.: Latent dirichlet allocation. J. Mach. Learn. Res. **3**, 993–1022 (2003)
2. Bollen, J., Mao, H., Zeng, X.: Twitter mood predicts the stock market. J. Comput. Sci. **2**(1), 1–8 (2011)
3. Chen, Z., Mukherjee, A., Liu, B., Hsu, M., Castellanos, M., Ghosh, R.: Leveraging multi-domain prior knowledge in topic models. In: Proceedings of IJCAI, pp. 2071–2077. AAAI (2013)
4. Dermouche, M., Khouas, L., Velcin, J., Loudcher, S.: A joint model for topic-sentiment modeling from text. In: Proceedings of SAC, pp. 819–824. ACM (2015)
5. Dermouche, M., Velcin, J., Khouas, L., Loudcher, S.: A joint model for topic-sentiment evolution over time. In: Proceedings of ICDM, pp. 773–778. IEEE (2014)
6. Diao, Q., Jiang, J., Zhu, F., Lim, E.-P.: Finding bursty topics from microblogs. In: Proceedings of ACL, pp. 536–544. ACL (2012)
7. Go, A., Bhayani, R., Huang, L.: Twitter sentiment classification using distant supervision. CS224N Project Report, Stanford, pp. 1–12 (2009)
8. Hofmann, T.: Probabilistic latent semantic indexing. In: Proceedings of SIGIR, pp. 50–57. ACM (1999)
9. Jo, Y., Oh, A.H.: Aspect and sentiment unification model for online review analysis. In: Proceedings of WSDM, pp. 815–824. ACM (2011)
10. Kiritchenko, S., Zhu, X., Cherry, C., Mohammad, S.: NRC-Canada-2014: detecting aspects and sentiment in customer reviews. In: SemEval, pp. 437–442. ACL (2014)
11. Lim, K.W., Buntine, W.: Twitter opinion topic model: extracting product opinions from tweets by leveraging hashtags and sentiment lexicon. In: Proceedings of CIKM, pp. 1319–1328. ACM (2014)
12. Lin, C., He, Y.: Joint sentiment/topic model for sentiment analysis. In: Proceedings of CIKM, pp. 375–384. ACM (2009)
13. Lin, C., He, Y., Everson, R., Rüger, S.: Weakly supervised joint sentiment-topic detection from text. IEEE Trans. Knowl. Data Eng. **24**(6), 1134–1145 (2012)

14. Lu, B., Ott, M., Cardie, C., Tsou, B.K.: Multi-aspect sentiment analysis with topic models. In: Proceedings of ICDMW, pp. 81–88. IEEE (2011)
15. Mei, Q., Ling, X., Wondra, M., Su, H., Zhai, C.: Topic sentiment mixture: modeling facets and opinions in weblogs. In: Proceedings of WWW, pp. 171–180. ACM (2007)
16. Mimno, D., Wallach, H.M., Talley, E., Leenders, M., McCallum, A.: Optimizing semantic coherence in topic models. In: Proceedings of EMNLP, pp. 262–272. ACL (2011)
17. Mukherjee, S., Basu, G., Joshi, S.: Joint author sentiment topic model. In: SDM, pp. 370–378. SIAM (2014)
18. Nguyen, T.H., Shirai, K.: Topic modeling based sentiment analysis on social media for stock market prediction. In: Proceedings of ACL, pp. 1354–1364. ACL (2015)
19. Pennington, v., Socher, R., Manning. C.D.: Glove: global vectors for word representation. In: Proceedings of EMNLP, pp. 1532–1543. ACL (2014)
20. Rong, X.: Word2Vec parameter learning explained. arXiv preprint arXiv:1411.2738 (2014)
21. Tsur, O., Rappoport, A.: What's in a hashtag? Content based prediction of the spread of ideas in microblogging communities. In: Proceedings of WSDM, pp. 643–652. ACM (2012)
22. Turney, P.D., Littman, M.L.: Measuring praise and criticism: inference of semantic orientation from association. ACM Trans. Inf. Syst. **21**(4), 315–346 (2003)
23. Wallach, H.M., Mimno, D.M., McCallum, A.: Rethinking LDA: why priors matter. In: NIPS, pp. 1973–1981 (2009)
24. Wang, Y., Liu, J., Huang, Y., Feng, X.: Using hashtag graph-based topic model to connect semantically-related words without co-occurrence in microblogs. IEEE Trans. Knowl. Data Eng. **28**(7), 1919–1933 (2016)
25. Wang, Y., Liu, J., Qu, J., Huang, Y., Chen, J., Feng, X.: Hashtag graph based topic model for tweet mining. In: Proceedings of ICDM, pp. 1025–1030. IEEE (2014)
26. Xu, K., Qi, G., Huang, J., Wu, T.: A joint model for sentiment-aware topic detection on social media. In: Procedings of ECAI, pp. 338–346. IOS Press (2016)
27. Zhang, Q., Gong, Y., Sun, X., Huang, X.: Time-aware personalized hashtag recommendation on social media. In: Proceedings of COLING, pp. 203–212. ACL (2014)
28. Zhao, W.X., Jiang, J., Yan, H., Li, X.: Jointly modeling aspects and opinions with a MaxEnt-LDA hybrid. In: Proceedings of EMNLP, pp. 56–65. ACL (2010)
29. Zhao, W.X., Jiang, J., Weng, J., He, J., Lim, E.-P., Yan, H., Li, X.: Comparing Twitter and traditional media using topic models. In: Clough, P., Foley, C., Gurrin, C., Jones, G.J.F., Kraaij, W., Lee, H., Mudoch, V. (eds.) ECIR 2011. LNCS, vol. 6611, pp. 338–349. Springer, Heidelberg (2011). https://doi.org/10.1007/978-3-642-20161-5_34
30. Zheng, C., Chengtao, L., Jian-Tao, S., Zhang, J.: Sentiment topic model with decomposed prior. In: Proceedings of SDM, pp. 767–775. SIAM (2013)

# Semi-supervised Stance-Topic Model for Stance Classification on Social Media

Kang Xu[✉], Sheng Bi, and Guilin Qi

Department of Computer Science, Southeast Unversity, Nanjing, China
{kxu,bisheng,gqi}@seu.edu.cn

**Abstract.** Stance detection aims to automatically determine from text whether the author of the text is in favor of, against, or neutral towards a issue. Social media, such as Sina Weibo, reflects the general public's stances towards different issues. Detecting and summarizing stances towards specific issues from social media is an important and challenging task. Although stance detection on social media has been studied before, previous work, most of which are based on supervised learning, may not work well because they suffer from its heavy dependence on training data. Other weakly supervised method also use some heuristic rules to select the posts with specific stances as training data, but these selected posts often concentrate on a few subtopics of the specific issue, these weakly supervised method can only train a biased stance classifier. To better detect stances toward specific issues, we consider to detect stances with a small number of labeled training data and a mass of unlabeled data. To integrate the supervised information into our model, we combine a discriminative maximum entropy (Max-Ent) component with the generative component. The Max-Ent component leverages hand-crafted features from labeled data to separate different stances. In this paper, we propose a semi-supervised topic model, Semi-Supervised Stance Topic Model (SSTM), that model stances and topics of the posts on social media. Since the posts on social media are short texts, we also incorporate the structural information of the posts, i.e., gender information, location information and time information, to aggregate posts for alleviating the context sparsity of the posts. The model has been evaluated on the selected posts on sina weibo, which talk about "the verbal battle of Han han and Fang zhouzi", to classify the stance of each posts. Preliminary experiments have shown promising results achieved by SSTM. Moreover, we also analyze the common difficulties in stance detection on social media. Finally, we also visualize the subtopics of the given issue generated by SSTM.

**Keywords:** Topic model · Stance detection · Sina weibo · Short text

## 1 Introduction

With the rapid growth of Web 2.0, a mass of user-generated posts, e.g., tweets and sina weibos, which capture the public's interests, thoughts, sentiments and

© Springer International Publishing AG 2017
Z. Wang et al. (Eds.): JIST 2017, LNCS 10675, pp. 199–214, 2017.
https://doi.org/10.1007/978-3-319-70682-5_13

stances. The posts have been accumulating on the social media with each passing day. Stance detection attempts to automatically determine the author of the post is in favor of, against, or neutral towards a specific issue (e.g., an event, or a controversial topic) or a specific entity (a person, a company) [22]. This is an interesting task to research on social media for there exist a mass of personalized and opinionated language [7].

Traditional supervised learning-based work [12,22,28,32] firstly choose hand-crafted features, then use common classifiers, such as SVM, Naive Bayes (NB) and Max-Ent. Another representation learning-based approaches, such as CNN-based approach [10], utilize word embeddings as input and train classifier with neural networks. These supervised learning approaches suffer from its heavily dependence on training data. Since manually labeled data is labour-consuming, it's difficult to get a mass of labeled data. Moreover, emerging topics appear continually in the posts on social media, the labeled data cannot cover the emergent topics. Weakly supervised approaches, such as [5], combines rules and supervised learning for detecting stance, where rules can automatically label data with regard to stance (i.e., favor, against or none). However, the weakness of weakly supervised methods are topic bias, i.e., the posts towards the same target can cover many topics with specific stances, but the rule-based labeled method can only cover a minority of topics. Our goal is to learn a model to automatically label new posts. In our scenario, there exist two kinds of data we can utilize, labeled data and unlabeled data. Compared to unlabeled data, which are usually easily obtained in volume, collecting a number of manually labeled data is expensive and, in many cases, impractical to build a model with higher labeling accuracy with a limited number of labeled data [24]. Hence, we attempt to train on both labeled data and unlabeled data in a semi-supervised way [20]. Our model aims to discover latent structure (sub-topics) in the data based on labeled data and unlabeled data, and based on the limited number of labeled data, learn the association between sub-topics and stances. We want to build our model by extending the ideas of Joint Sentiment and Topic Model (JST) [16], where our model can jointly discovers a set of sub-topics exhibited by a corpus, learns associations between each sub-topic and stances, labels each post with a stance. To integrate the supervised information into our model, we combine a discriminative maximum entropy (Max-Ent) component with the generative component. The Max-Ent component leverages hand-crafted features to separate different stances.

For the short and informal characteristic of the posts, applying the models to the short posts on the social media directly always suffers from the context sparsity problem. One simple and effective way to alleviate the sparsity problem is to aggregate short posts into lengthy pseudo-documents [4,30,35]. Social media platforms, such as Twitter and Sina weibo, provides a wealth of information: the time of published posts, users who publish the posts and their profile (such as gender, location). Xu et al. propose Time-User Sentiment/Topic Latent Dirichlet Allocation (TUS-LDA), where TUS-LDA utilize structural information, user and time, to alleviate the context sparsity. In our scenario, most of individuals only publish one or a few posts to express their stances towards specific issues. Hence, in our work, we ignore the user information. However, the user's profile information, i.e., gender and location, can be utilized to model stances and topics. In social media, different genders often express different stances towards the same issues. On the contrary, people with the same gender often the same stances towards specific issues. For example, If "Han han" publishes posts about women discrimination, most of women will oppose him. Hence, the structural information, gender, is a good way to aggregate posts. Similarly, location is another way to model stances and topics, because people living in a close location will express the same stance towards many issues. For example, the government release a policy which is beneficial for the citizens in a particular area, the people there will be in favor of the policy. In our work, we adopt the structural information, time, gender and location, to model stances and topics. We give preference to the three kinds of structural information to model. If the three ways are not suitable for modeling, we choose the corpus-level way to model stances and topics, i.e., we can get global stance-topic distributions.

In this paper, a semi-supervised topic model, Semi-Supervised Stance Topic Model (SSTM), that model stances and topics of the posts which may contain the stances toward the given issue, where SSTM combines a discriminative maximum entropy (Max-Ent) component with the generative component. Since the posts on social media are short texts, we also incorporate the structural information of the posts, i.e., gender information, location information and time information, to alleviate the context sparsity of the posts.

There exist four main contributions of SSTM: (1) We propose Semi-Supervised Stance Topic Model (SSTM), which can model stances and topics towards a specific issue, with labeled data and unlabeled data in a semi-supervised way. (2) In our model, we propose the structural information, i.e., time, gender and location, to alleviate the context sparsity problem. (3) We choose a dataset that contains a issue about "Han han" and "Fang zhouzi" for stance detection, and manually 3000 posts with three kinds of stances (FAVOR, AGAINST and OTHER). (4) We implement experiments on the dataset to evaluate the effectiveness of stance classification in SSTM and visualize stance-specific topics towards the given issue discovered by SSTM.

## 2 Related Work

Detecting stance in tweets is a new task proposed for SemEval-2016 [22]. The aim of the task is to determine user stance (FAVOR, AGAINST, or OTHER) in a dataset of posts on the given issues. In this section, we briefly summarize the related work of stance detection from the following perspectives: supervised stance classification, and weakly-supervised or semi-supervised stance classification.

### 2.1 Supervised Stance Classification

Most of work for the existing stance classification are fully based on labelled samples. Zarrella et al. employed a recurrent neural network, which was initialized from pre-trained features in similar tasks, to encode world knowledge via weak external supervision [32]. Wei et al. developed a convolutional neural network for stance detection in tweets, and designed a "vote scheme" using softmax results to predict the label of test data [28]. Tutek et al. designed a stance classification system that integrated multiple supervised classifier, trained with lexical and task-dependent features, then optimized with a genetic approach [25]. Zhang et al. decomposed stance detection into two steps: relevance detection (decide whether the post is relevant to the specific topic) and orientation detection (whether the post is in favor of the specific target). In both steps, they selected corresponding features for classification [33]. Liu et al. utilized a set of supervised classifiers (SVM, random forest, gradient boosting decision trees) and an ensemble classifier trained with a bag of words and word vectors as features [17]. Vijayaraghavan et al. combined the word-level and character-level CNN for stance classification [26]. Bøhler et al. proposed an approach which automatically detected stance in posts by building a supervised system combining shallow features and pre-trained word vector feature [2]. Igarashi et al. compared feature-based and representation learning-based approaches on the supervised stance classification for posts and the results of their approaches showed that CNN-based approach performed better than feature-based approach in the cross validation on the training data, while feature-based approach outperformed the CNN model on the test data [10]. Patra et al. [23] designed an approach that identifying the stance in posts using dependency and semantic features. Lai et al. [14] defined a set of features to incorporate the context surrounding a specific target aiming to predict the stance towards the target and trained the training data with Gaussian Naive Bayes classifier. Li et al. designed a deep memory network for stance detection towards a given set of targets from texts and they decomposed the tasks as: (1) target identification, (2) polarity classification [15]. Although these supervised methods achieved some effects, supervised learning suffered from its heavy dependence on training data where labelling samples are labour-expensive. Moreover, supervised methods explicitly modeled the label for the purpose of stance prediction, but they did not discover the latent patterns in the corpus (both labelled and unlabelled data) or align the patterns with stance labels.

## 2.2   Weakly-supervised or Semi-supervised Stance Classification

To reduce the human effort for labeling data, some researchers also proposed some semi-supervised or weakly-supervised methods for stance detection. Dias et al. [5] proposed a weakly supervised approach for stance classification which is solely based on textual content. In their approach, they firstly manually generate rules for automatically labeling instances with regard to stance (i.e., favor, against or none). There exist two kinds of functions of rules: (1) automatically generate training data; (2) complement the predictive model when determining the stance of new data. Then, they build a supervised classifier with automatically generated labelled data. Johnson et al. [11] generate a small set of stance-indicative patterns and label the tweets as positive and negative firstly; then, trained statistical relation learner with noisy-labeled posts to classify stance of other tweets and authors. The model aimed to constraint tweets with similar contexts (content and user) to be assigned to similar stance labels. Similarly, the supervision of [7] is only a small seed set of issue and frame indicators which characterize the stance of tweets (based on lexical heuristics), they designed a weakly supervised joint model through Probabilistic Soft Logic (PSL). Misra et al. [21] also explored a semi-supervised approach to stance classification utilizing stance-bearing hashtags. Our work also attempts to detect stance in a semi-supervised way and focuses on discovering a set of topics and learning associations between each topic and stance labels with labeled data and unlabeled data in a semi-supervised way. Based on the association between topics and stance labels, stance labels are assigned to all the unlabeled data.

## 3   Our Proposed Model

In this section, we firstly propose a generative model, Semi-Supervised Stance Topic Model (SSTM), to model stances and topics on posts. Then, we introduce a way to incorporate supervised information, with a maximum entropy model, into SSTM. Finally, a collapsed Gibbs sampling algorithm is developed for parameter estimation of SSTM. All the notations needed in SSTM are listed in Table 1.

**Table 1.** Notation used in the SSTM model

| Symbol | Description |
|---|---|
| D, K, S, V | Number of documents, topics, stances, vocabulary |
| P, T, G, L | Number of positions, timeslices, genders, class label |
| $\mathbf{Z}, \mathbf{W}, \mathbf{Y}$ | All the topics, words, switches |
| $\mathbf{T}, \mathbf{U}, \mathbf{H}$ | All the timeslices, users and hashtags |
| $\mathbf{L}$ | All the stances of posts |
| $N_d$ | Number of word tokens in post $d$ |
| $p_d, t_d, g_d$ | Position, timeslice, gender of $d$ |
| $y_d, l_d$ | Switch, stance of $d$ |
| $\varepsilon$ | Beta distribution of switches |
| $x_d, f_d$ | Document-stance distribution, document stance feature |
| $\tau_s$ | Global stance topic distribution |
| $\rho_{g,s}$ | Gender-stance topic distribution |
| $\delta_{p,s}$ | Position-stance topic distribution |
| $\theta_{t,s}$ | Timeslice-stance topic distribution |
| $\varphi_{s,k}$ | Sentiment-topic word distribution |
| $\alpha$ | Hyperparameters of $\tau$, $\rho$, $\theta$ and $\delta$ |
| $\beta$ | Hyperparameters of $\varphi_{s,k}$ |
| $\gamma$ | Hyperparameters of $\varepsilon$ |
| $\phi$ | Max-ent model |

### 3.1  Generative Process

We now describe the generative process of our model: When a user (gender: $g_d$) in the location of $p_d$ publishes the post $d$ at the time $t_d$, the user firstly choose the stance $l_d$ from the stance probability distribution $x_d$ (with the feature $f_d$). If $l_d = 0$ (i.e., there exist no specific stance in the post towards the specific issue, OTHER), we draw the topic $z_d$ from general topic distribution $\tau_0$, where $\tau_0$ corresponds to other topics; if $l_d! = 0$ (i.e., the post describes the stance towards the specific issue, FAVOR or AGAINST), the user utilizes the variable $y_d$, which is drawn from switch distribution $\varepsilon$ to decide whether the post describes a gender-related, location-related, time-related or global topic. If $y_d = 0$ (global topic), we draw the topic $z_d$ from general topic distribution $\tau_{l_d}$, if $y_d = 1$ (gender-related topic), we draw the topic $z_d$ from gender-related topic distribution $\rho_{l_d}$; if $y_d = 2$ (location-related topic), we draw the topic $z_d$ from location-related topic distribution $\delta_{l_d}$; if $y_d = 3$ (time-related topic), we draw the topic $z_d$ from time-related topic distribution $\theta_{l_d}$. For each word $w_{d,i}$ in the post $d$, based on the chosen $l_d$ and $z_d$ of the post $d$, the word is drawn from the sentiment-topic word distribution $\varphi_{l_{d,i},z_d}$. Figure 1 shows our model using the plate notation. The formal generative process is as follows:

1. Draw $\varepsilon \sim \mathrm{Dir}(\gamma)$
2. For each gender $g = 1, ..., G$
    i. For each stance $s = 1, 2$
        a. Draw $\rho_{g,s} \sim \mathrm{Dir}(\alpha)$
3. For each position $p = 1, ..., P$
    i. For each stance $s = 1, 2$
        a. Draw $\delta_{p,s} \sim \mathrm{Dir}(\alpha)$
4. For each timeslice $t = 1, ..., H$
    i. For each stance $s = 1, 2$
        a. Draw $\theta_{t,s} \sim \mathrm{Dir}(\alpha)$
5. For each stance $s = 0, 1, 2$
    a. Draw $\tau_s \sim \mathrm{Dir}(\alpha)$
6. For each stance $s = 0, 1, 2$
    i. For each topic $k = 1, ..., K$
        a. Draw $\varphi_{s,k} \sim \mathrm{Dir}(\beta)$
7. For each post $m_d$, $d = 1, ..., D$
    i. Draw $l_d \sim \mathrm{Multi}(x_d)$, the generation of $x_d$ will be discussed later
    ii. Draw $y_d \sim \mathrm{Multi}(\varepsilon_d)$
    iii. if $l_d = 0$, Draw $z_d \sim \tau_{l_d}$; else if $y_d = 0$, Draw $z_d \sim \rho_{l_d}$; if $y_d = 1$, Draw $z_d \sim \delta_{l_d}$; if $y_d = 2$, Draw $z_d \sim \theta_{l_d}$; if $y_d = 3$, Draw $z_d \sim \tau_{l_d}$
    iv. For each word $w$ $i = 1, ...N_d$
        a. Draw $w_{d,i} \sim \mathrm{Multi}(\varphi_{l_d,z_d})$

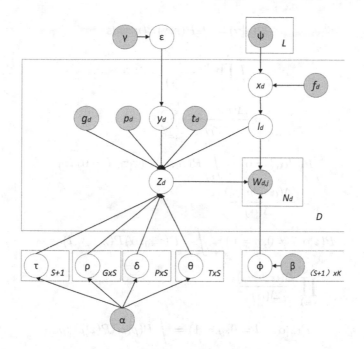

**Fig. 1.** The plate notation of our model

The full probability of our model is as Eq. 1.

$$
\begin{aligned}
&P(\varepsilon, y, g, p, t, \psi, x, f, l, z, w, \tau, \rho, \delta, \theta, \phi; \alpha, \beta, \gamma) = \\
&P(x; f, \psi)P(l|x)P(\tau; \alpha)P(\rho; \alpha)P(\delta; \alpha)P(\theta; \alpha)P(\varepsilon; \gamma) \\
&P(\phi; \beta)P(y|\varepsilon)P(z|l = 0, \tau)P(z|y = 0, l \neq 0, \tau)P(z|y = 1, l \neq 0, g, \rho) \quad (1)\\
&P(z|y = 2, l \neq 0, p, \delta)P(z|y = 3, l \neq 0, t, \theta)P(w|z, l, \phi) \\
&l \in \{0, 1, 2\}, y \in \{0, 1, 2, 3\}, z \in \{0, ..., K-1\}
\end{aligned}
$$

The first part in Eq. 1, we obtain Eq. 2. The second part, by integrating $\varepsilon$, we obtain Eq. 3. Thereinto, $\Delta(r) = \frac{\prod_{m=1}^{dim\,\vec{r}} \Gamma(r_m)}{\Gamma(\prod_{m=1}^{dim\,\vec{r}} r_m)}$, where $r$ is a vector, and $dim$ is the dimension of the vector. The third part, by integrating $\tau_0$, we obtain Eq. 4. The fourth part, by integrating $\tau_{l=0}$, we obtain Eq. 5. The fifth part, by integrating $\rho$, we obtain Eq. 6. The sixth part, by integrating $\delta$, we obtain Eq. 7. The seventh part, by integrating $\theta$, we obtain Eq. 8. The final part, by integrating $\phi$, we obtain Eq. 9.

$$
P(x; f, \psi)P(l|x) = \prod_{d=0}^{D-1} \prod_{s=0}^{S} x_{d,s}^{n_{d,s}}; \sum_{s=0}^{S} x_{d,s} = 1 \quad (2)
$$

$$P(y|\gamma) = \int_\varepsilon P(\varepsilon|\gamma)P(y|\varepsilon)d\varepsilon$$

$$= \int_\varepsilon \prod_{i=0}^Y \varepsilon_i^{n_y^i} \frac{1}{\Delta(\gamma)} \varepsilon_i^{\gamma_i-1} d\varepsilon \tag{3}$$

$$= \frac{\Delta(n_y+\gamma)}{\Delta(\gamma)}; \sum_{i=0}^{Y-1} \varepsilon_i = 1$$

$$P(z|\alpha, l=0) = \int_{\tau_0} P(\tau_0; \alpha)P(z|\tau_0, l=0)d\tau_0$$

$$= \frac{\Delta(\alpha + n_{l=0})}{\Delta(\alpha)} \tag{4}$$

$$P(z|\alpha, l \neq 0, y=0) = \int_{\tau_{\neq 0}} P(\tau_{\neq 0}; \alpha)P(z|\tau_{\neq 0}, l)d\tau_{\neq 0}$$

$$= \prod_{s=1}^S \frac{\Delta(\alpha + n_s)}{\Delta(\alpha)} \tag{5}$$

$$P(z|\alpha, g, l \neq 0, y=1) = \int_\rho P(\rho; \alpha)P(z|\rho)d\rho$$

$$= \prod_{g=0}^{G-1} \prod_{s=1}^S \frac{\Delta(n_{g,s}+\alpha)}{\Delta(\alpha)} \tag{6}$$

$$P(z|\alpha, p, l \neq 0, y=2) = \int_\delta P(\delta; \alpha)P(z|\delta)d\delta$$

$$= \prod_{p=0}^{P-1} \prod_{s=1}^S \frac{\Delta(n_{p,s}+\alpha)}{\Delta(\alpha)} \tag{7}$$

$$P(z|\alpha, t, l \neq 0, y=3) = \int_\theta P(\theta; \alpha)P(z|\theta)d\theta$$

$$= \prod_{t=0}^{T-1} \prod_{s=1}^S \frac{\Delta(n_{t,s}+\alpha)}{\Delta(\alpha)} \tag{8}$$

$$P(w|l, z; \beta) = \int_\phi P(w|\phi, l, z)P(\phi|\beta)d\phi$$

$$= \prod_{s=0}^S \prod_{k=0}^{K-1} \frac{\Delta(\beta + n_{s,k})}{\Delta(\beta)} \tag{9}$$

## 3.2   Setting $x$ with a Maximum Entropy Model

In the full generative probabilistic model, the prior distribution of $x$ is Dirichlet distribution. But it is impossible for a fully unsupervised topic model to discriminate the stances of posts as FAVOR, AGAINST or OTHER. Hence, we must incorporate supervised information, which can guide our model to choose the right stances, into our topic model. Here, supervised information is trained on a small number of posts with labeled stances, the stances of posts will be uncertain with different contexts on the new posts. Hence, it is inappropriate to identify stances by fully depending on supervised information. Consequently, we just give a stance probability, based on supervised model, for each post to guide stance classification.

Aiming to incorporate features such as to discriminate between FAVOR, AGAINST and OTHER posts, we design a maximum entropy model under feature vector $f_d$ (associated with post $d$) to set $x_d$, where $f_d$ can encode features which are discriminative for stance classification. Our MaxEnt-Topic Model is motivated by the previous model of opinion mining [18,22].

### 3.3 Inference

Like LDA, exact inference is intractable in our models. Hence approximate estimation approaches, such as Gibbs Sampling [9], are utilized to solve the problem. Gibbs Sampling, a special case of Markov Chain Monte Carlo (MCMC) [8], is a relatively simple algorithm of approximate inference for our models. The final formulas are given here.

If $l_d = 0$, the $d$-th post describes a topic without specific stance towards the specific issue, the sampling formula is shown in Eq. 10. If $l_d! = 0$, the $d$-th post describes the specific issue, there exist four ways, based on the switch $y_d$, to model topics with the chosen stance $l_d$.

If $y_d = 0$, the $d$-th post talks about a global topic, the sampling formula is shown in Eq. 11. If $y_d = 1$, the $d$-th post talks about a gender-related topic, the sampling formula is shown in Eq. 12. If $y_d = 2$, the $d$-th post talks about a location-related topic, the sampling formula is shown in Eq. 13. If $y_d = 3$, the $d$-th post talks about a time-related topic, the sampling formula is shown in Eq. 14.

$$P(l_d = 0, z_d = k|...) \propto x_{d,0} \cdot \frac{\alpha + n_{0,k}}{K \cdot \alpha + n_0} \cdot \frac{\prod_{v=0}^{V-1} \prod_{m=0}^{E_{(v)}-1}(n_{0,k,v} + m + \beta)}{\prod_{m=0}^{E_{(.)}-1}(n_{0,k} + m + V \cdot \beta)} \quad (10)$$

$$P(l_d \neq 0, y_d = 0, l_d = s, z_d = k|...) \propto x_{d,s} \cdot \frac{\gamma_0 + n_{y=0}}{Y * \gamma + n_y} \cdot \frac{\alpha_k + n_{s,k}}{K * \alpha + n_s}$$
$$\cdot \frac{\prod_{v=0}^{V-1} \prod_{m=0}^{E_{(v)}-1}(n_{s,k,v} + m + \beta)}{\prod_{m=0}^{E_{(.)}-1}(n_{s,k} + m + V \cdot \beta)} \quad (11)$$

$$P(l_d \neq 0, y_d = 1, l_d = s, z_d = k|...) \propto x_{d,s} \cdot \frac{\gamma_1 + n_{y=1}}{Y * \gamma + n_y} \cdot \frac{\alpha_k + n_{g,s,k}}{K * \alpha + n_{g,s}}$$
$$\cdot \frac{\prod_{v=0}^{V-1} \prod_{m=0}^{E_{(v)}-1}(n_{s,k,v} + m + \beta)}{\prod_{m=0}^{E_{(.)}-1}(n_{s,k} + m + V \cdot \beta)} \quad (12)$$

$$P(l_d \neq 0, y_d = 2, l_d = s, z_d = k|...) \propto x_{d,s} \cdot \frac{\gamma_0 + n_{y=2}}{Y * \gamma + n_y} \cdot \frac{\alpha_k + n_{p,s,k}}{K * \alpha + n_{p,s}}$$
$$\cdot \frac{\prod_{v=0}^{V-1} \prod_{m=0}^{E_{(v)}-1} (n_{s,k,v} + m + \beta)}{\prod_{m=0}^{E_{(.)}-1} (n_{s,k} + m + V \cdot \beta)} \tag{13}$$

$$P(l_d \neq 0, y_d = 3, l_d = s, z_d = k|...) \propto x_{d,s} \cdot \frac{\gamma_3 + n_{y=3}}{Y * \gamma + n_y} \cdot \frac{\alpha_k + n_{t,s,k}}{K * \alpha + n_{t,s}}$$
$$\cdot \frac{\prod_{v=0}^{V-1} \prod_{m=0}^{E_{(v)}-1} (n_{s,k,v} + m + \beta)}{\prod_{m=0}^{E_{(.)}-1} (n_{s,k} + m + V \cdot \beta)} \tag{14}$$

**Table 2.** Corpus statistics in the selected posts

| Types | Number |
|---|---|
| Number of posts | 27,584 |
| Number of vocabulary | 51,439 |
| Number of words | 626,198 |
| Number of locations | 37 |
| Number of timeslices | 353 |
| Types of genders | 2 |

# 4    Experiment Analysis

In this section, we conducted experiments on real-world posts in Chinese sina weibo[1] to demonstrate the effectiveness of our proposed approach. We take four baseline methods, i.e., Naive Bayes, SVM, Max-Ent [3], CNN [13] and RNN [29].

## 4.1    Dataset Description

To verify the effectiveness of SSTM on the posts, we carried experiments on a Chinese sina weibo dataset which contains the posts published in 2012. From sina weibo dataset, we queried the posts that contains keywords "韩寒" (Han Han) or "方舟子" (Fang Zhouzi), where we can detect stances and topics about "方舟子和韩寒之间的关于代笔的骂战" (the verbal battle between their camps about Han han's Ghostwriting) from these posts on sina weibo. For evaluating the quality of stance classification on our proposed model and other baseline methods, we asked two volunteers to annotate 3,000 posts as 1,922 FAVOR (64% support Han Han), 782 AGAINST (26% support Fang Zhouzi), 296 OTHER (10%). In the training set, 1,500 posts were divided into three parts: 961 FAVOR,

---

[1] http://weibo.com/.

391 AGAINST, 148 OTHER. The assignment of testing data was the same as the training set. Other unlabeled data was utilized for training our model.

We followed the preprocessing steps in BTM [31]. To improve the quality of our model, we added two extra steps: (1) Part-of-speech tagging of posts using the Chinese Part-of-speech tagger, JieBa[2], retaining the words tagged as nouns, verbs or adjectives; (2) Lemmatizing words tagged as noun, verb, which was used to reduce inflectional forms and sometimes related forms of a word to a common base form. After preprocessing, as is shown in Table 2, we left 27,584 valid posts, 51,439 distinct words, 353 timeslices (days) and 37 locations (provinces) in the selected posts.

During training our model, the stance labels of training data were given, but the labels of testing data and unlabeled data were unknown. Hence, we need to give prior probabilities for testing data and unlabeled data based on training data, and these probabilities can guide our model to discriminate the stances of the posts. Motivated by the previous work [34], we trained the training data with Max-Ent classifier which can generate the stance probabilities towards the given issue. For Max-Ent, NB and SVM, we also used the following features in $f$ based on the work of [14]: tfidf, lexical features (POS tag etc.), sentiment-based features (Chinese Sentiment Lexicon, HowNet [6]) and topic features (pre-trained by LDA [1]). For CNN and RNN, we used the pre-trained word embedding as inputs, where word embedding was training with word2vec [19] based on all the crawled sina weibo posts in 2012.

## 4.2    Parameter Settings

To optimize the number of topics $K$, we empirically ran the models with four values of $K$: 10, 20, 30, 40. In our model, we simply selected symmetric Dirichlet prior vectors as is empirically done in TUS-LDA [30]. For our model, we set $\alpha = 0.5, \beta = 0.01, \gamma = 0.01$ and $\beta = 0.01$. LDA-based models are not sensitive to the hyper-parameters [27]. In all the topic model-based methods with different topic numbers, Gibbs sampling was run for 1,000 iterations with 200 burn-in periods.

## 4.3    Stance Classification

In this section, we performed stance classification to predict the stance labels of the test data and unlabeled data in sina weibo. We determined the stance of a post $d$ by the value of the hidden variable $l_d$ in SSTM.

We presented the F1-score of stance classification with AVERAGE, FAVOR, AGAINST and OTHER (as is shown in Table 3).

Based on the results of stance classification, we can see that, when $K = 30, 40$, our proposed SSTM outperformed the supervised baseline methods in $F1$ (Table 3). As is shown in the result, we can observe that $F_{FAVOR}$ much better performed better than $F_{AGAINST}$ and $F_{OTHER}$. We analyzed the statistics

---

[2] https://github.com/fxsjy/jieba.

**Table 3.** F1-score Results for Stance Classification towards 'Han Han' and 'Fang Zhouzi'

| Method | $F_{Average}$ | $F_{FAVOR}$ | $F_{AGAINST}$ | $F_{OTHER}$ |
| --- | --- | --- | --- | --- |
| CNN | 0.43 | 0.58 | 0.15 | 0.38 |
| RNN | 0.45 | 0.61 | 0.13 | 0.26 |
| Max-Ent | 0.54 | 0.71 | 0.21 | 0.37 |
| NB | 0.58 | 0.7 | 0.35 | 0.4 |
| SVM | 0.59 | 0.73 | 0.36 | 0.38 |
| SSTM-10 | 0.55 | 0.72 | 0.23 | 0.3 |
| SSTM-20 | 0.59 | 0.75 | 0.31 | 0.29 |
| SSTM-30 | 0.63 | 0.79 | 0.34 | 0.3 |
| SSTM-40 | 0.6 | 0.77 | 0.3 | 0.28 |

of our dataset by randomly sampling posts and found that the number of the users published the posts that support "Han Han" is much more than others. In our experiment, CNN and RNN performed worst, we can speculate that the number of labeled data is too small, so that the implicit association between posts and stances cannot be mined with neural networks. Our model performed best in $F_{FAVOR}$ than SVM and NB. In $F_{AGAINST}$ and $F_{OTHER}$, the performance is close to them. Since topic models are based on high-frequency word co-occurrences, topic model can perform better in the topics (i.e., topics about Han Han's camp) with high frequencies.

### 4.4 Case Analysis for Stance Detection

In this section, we listed some cases to analyze the difficulties encountered in stance classification of posts on social media. Topic model-based methods are based on word co-occurrences, hence we gave some cases to interpret the difficulties of topic model-based in stance detection.

(1) There exist more and more emerging words or emoticons which never occur in training data, such as "倒韩" (Against Han han), "方肘子" (Fang zhouzi, a nickname of "方舟子" (Fang zhouzi)), most of the existing methods are based on "bag-of-words" models, supervised models are based on features extracted from "bag-of-words" and topic models are based on word co-occurrences. Hence, a mass of new words is a challenge for stance detection.

(2) The public often express their stances in a obscure way. For example, "韩寒的道德水平就和小四的身高一样。" (The degree of Han han's morality is as tall as Xiao si's height). Hereinto, the height of Xiao si (Guo jing-ming) is very short, so the aforementioned post tells us the level of Han han's morality is very low (Aganinst Han han).

(3) They also express their stances in a irony way, i.e., the public use the support (or against) words to express the contrary stance. For

example, "韩寒真是'天才'啊！" (Han han is a real 'genius'.-'Against Han Han'),"方舟子，真'打假'英雄" (Fang zhouzi is a real 'anti fraud' hero.-'Against Fang zhouzi').

(4) In some posts, they express contrary stances with similar words. It's difficult for our "bag-of-words" model to detect accurate stances in these posts. For example, "韩寒有无代笔虽尚待定论，但其表现令人心寒" (There exist no final conclusion about Han han's ghostwriting, but Han's behavior is particularly chilling.)-(Aganinst Han han), "韩寒有无代笔还尚待定论，网民表现令人心寒" (There exist no final conclusion about Han han's ghostwriting, but Netizen's behavior is particularly chilling)-(Support Han han).

The above-mentioned difficulties are a part of typical problems encountered during our experiments (Table 4).

**Table 4.** Example of topics extracted by SSTM

| FAVOR | (Han Han's Camp) | AGAINST | (Fang Zhouzi's Camp) |
|---|---|---|---|
| Topic 1 | Topic 2 | Topic 1 | Topic 2 |
| 方舟子(Fang Zhouzi) | 方舟子(Fang zhouzi) | 韩寒(Han han) | 韩寒(Han han) |
| 韩寒(Han Han) | 组织(Organize) | 带笔(Take pen) | 代笔(Ghostwritten) |
| 代笔(Ghostwritten) | 粉丝(Fan) | 马英九(Ma Ying-jeou) | 奖金(bonus) |
| 炒作(speculation) | 诋毁(slander) | 代笔(Ghostwritten) | 嘲笑(Mock) |
| 告(accuse) | 抹黑(smear) | 名誉(reputation) | 确有其事(real) |
| 污蔑(slander) | 五毛(propagandist) | 签名(Signature) | 缺乏(lack) |
| 劣迹(misdeed) | 辱骂(abuse) | 久仰(admired) | 诚意(sincerity) |
| 斑斑(much) | 咒骂(abuse) | 主席(president) | 粉丝(fan) |
| 恶意攻击(hostile attack) | 大肆(heavily) | 大声(loud) | 明明(obviously) |
| 半年(half year) | 脑残(brainless) | 喊(shout) | 声明(declaration) |

## 4.5   Visualization of Stance-Related Topics

To investigate the quality of subtopics towards the given issue discovered by SSTM, we selected four topics for visualization, i.e., two subtopics that support "Han Han" (FAVOR) and two subtopics that support "Fang zhouzi" (AGAINST). For each subtopics, we chose the top 10 words which can best represent the subtopic. In the camp of "Han Han", from the subtopic 1 and 2 in favor of "Han han", the fans or supporters of "Han Han" thought that "Fang zhouzi" deliberately slandered "Han han" and also maliciously attacked other people. From the subtopics in favor of "Fang zhouzi", the supporters of "Fang zhouzi" claimed "Han han" indeed ghostwrote and the supporters looked down upon this act. Obviously, in our results, there exist some bad subtopics which cannot represent the stances of the public. The reason we speculate is that, in our dataset, the numbers of posts belong to FAVOR subtopics, AGAINST subtopics and OTHER subtopics is unevenly distributed, but our model gave the topic numbers to the three kinds of subtopics. In fact, the public's stances towards most of controversial topics are one-sided, i.e., only one or few camps can get most of support.

## 5    Conclusion and Future Work

In this paper, we propose a semi-supervised topic model for modeling topics and stances from user-generated posts on the social media. Firstly, stance detection is a classification task, which is always conducted by a supervised classifier. However, only a few labeled data is available during training and a mass of unlabeled data is unexploited. In our work, we design a new stance-topic model (SSTM), where the supervised information is introduced with a Max-Ent classifier for stance classification and the generative process of SSTM is utilized for modeling the association between stances and topics. Moreover, for the short characteristic of posts, we also design SSTM to solve the context sparsity, i.e., we exploit the structural information on social media, time information of posts, the profile information of users who publish posts (location, age) to aggregate short posts as pseudo-documents to make up of context sparsity. Our quantitative evaluation showed that our model outperformed other models in stance classification. Moreover, we also chose four examples to visualize some stance-aware topics. In future work, we consider to use transfer learning methods to transfer available knowledge from similar tasks, i.e., sentiment analysis, to improve the performance of our task.

## References

1. Blei, D.M., Ng, A.Y., Jordan, M.I.: Latent dirichlet allocation. J. Mach. Learn. Res. **3**, 993–1022 (2003)
2. Bøhler, H., Asla, P., Marsi, E., Sætre, R.: IDI@NTNU at SemEval-2016 task 6: detecting stance in tweets using shallow features and glove vectors for word representation. In: Proceedings of SemEval@NAACL-HLT, pp. 445–450. ACL (2016)
3. Chieu, H.L., Ng, H.T.: Named entity recognition: a maximum entropy approach using global information. In: Proceedings of COLING (2002)
4. Diao, Q., Jiang, J., Zhu, F., Lim, E.-P.: Finding bursty topics from microblogs. In: Proceedings of ACL, pp. 536–544. ACL (2012)
5. Dias, M., Becker, K.: An heuristics-based, weakly-supervised approach for classification of stance in tweets. In Proceedings of WI, pp. 73–80. IEEE (2016)
6. Dong, Z., Dong, Q., Hao, C.: Hownet and its computation of meaning. In: Proceedings of COLING, pp. 53–56 (2010)
7. Ebrahimi, J., Dou, D., Lowd, D.: Weakly supervised tweet stance classification by relational bootstrapping. In: Proceedings of EMNLP, pp. 1012–1017. ACL (2016)
8. Geyer, C.J.: Practical Markov Chain Monte Carlo. Stat. Sci. 473–483 (1992)
9. Heinrich, G.: Parameter estimation for text analysis. Technical report (2005)
10. Igarashi, Y., Komatsu, H., Kobayashi, S., Okazaki, N., Inui, K.: Tohoku at SemEval-2016 task 6: feature-based model versus convolutional neural network for stance detection. In: Proceedings of SemEval@NAACL-HLT, pp. 401–407. ACL (2016)
11. Johnson, K., Goldwasser, D.: "All I know about politics is what I read in Twitter": weakly supervised models for extracting politicians' stances from Twitter. In: Proceedings of COLING, pp. 2966–2977 (2016)
12. Krejzl, P., Steinberger, J.: UWB at SemEval-2016 task 6: stance detection. In: Proceedings of SemEval@NAACL-HLT, pp. 408–412. ACL (2016)

13. Krizhevsky, A., Sutskever, I., Hinton, G.E.: Imagenet classification with deep convolutional neural networks. In: Proceedings of NIPS, pp. 1106–1114 (2012)
14. Lai, M., Farías, D.I.H., Patti, V., Rosso, P.: Friends and enemies of Clinton and Trump: using context for detecting stance in political tweets. CoRR, abs/1702.08021 (2017)
15. Li, C., Guo, X., Mei, Q.: Deep memory networks for attitude identification. In: Proceedings of WSDM, pp. 671–680. ACM (2017)
16. Lin, C., He, Y.: Joint sentiment/topic model for sentiment analysis. In: Proceedings of the 18th ACM Conference on Information and Knowledge Management. CIKM 2009, Hong Kong, 2–6 November 2009, pp. 375–384. ACM (2009)
17. Liu, C., Li, W., Demarest, B., Chen, Y., Couture, S., Dakota, D., Haduong, N., Kaufman, N., Lamont, A., Pancholi, M., Steimel, K., Kübler, S.: IUCL at SemEval-2016 task 6: an ensemble model for stance detection in Twitter. In: Proceedings of SemEval@NAACL-HLT, pp. 394–400. ACL (2016)
18. Ma, C., Wang, M., Chen, X.: Topic and sentiment unification maximum entropy model for online review analysis. In: Proceedings of WWW, pp. 649–654. ACM (2015)
19. Mikolov, T., Chen, K., Corrado, G., Dean, J.: Efficient estimation of word representations in vector space. CoRR, abs/1301.3781 (2013)
20. Miller, D.J., Uyar, H.S.: A mixture of experts classifier with learning based on both labelled and unlabelled data. In: Proceedings of NIPS, pp. 571–577 (1996)
21. Misra, A., Ecker, B., Handleman, T., Hahn, N., Walker, M.A.: NLDS-UCSC at SemEval-2016 task 6: a semi-supervised approach to detecting stance in tweets. In: Proceedings of SemEval@NAACL-HLT, pp. 420–427. ACL (2016)
22. Mohammad, S., Kiritchenko, S., Sobhani, P., Zhu, X.-D., Cherry, C.: SemEval-2016 task 6: detecting stance in tweets. In: Proceedings of SemEval@NAACL-HLT, pp. 31–41. ACL (2016)
23. Patra, B.G., Das, D., Bandyopadhyay, S.: JU_NLP at SemEval-2016 task 6: detecting stance in tweets using support vector machines. In: Proceedings of SemEval@NAACL-HLT, pp. 440–444. ACL (2016)
24. Ramage, D., Manning, C.D., Dumais, S.T.: Partially labeled topic models for interpretable text mining. In: Proceedings of KDD, pp. 457–465. ACM (2011)
25. Tutek, M., Sekulic, I., Gombar, P., Paljak, I., Culinovic, F., Boltuzic, F., Karan, M., Alagic, D., Snajder, J.: Takelab at SemEval-2016 task 6: stance classification in tweets using a genetic algorithm based ensemble. In: Proceedings of SemEval@NAACL-HLT, pp. 464–468. ACL (2016)
26. Vijayaraghavan, P., Sysoev, I., Vosoughi, S., Roy, D.: Deepstance at SemEval-2016 task 6: detecting stance in tweets using character and word-level CNNs. In: Proceedings of SemEval@NAACL-HLT, pp. 413–419. ACL (2016)
27. Wallach, H.M., Mimno, D.M., McCallum, A.: Rethinking LDA: why priors matter. In: NIPS, pp. 1973–1981 (2009)
28. Wei, W., Zhang, X., Liu, X., Chen, W., Wang, T.: pkudblab at SemEval-2016 task 6: a specific convolutional neural network system for effective stance detection. In: Proceedings of SemEval@NAACL-HLT, pp. 384–388. ACL (2016)
29. Williams, R.J., Zipser, D.: A learning algorithm for continually running fully recurrent neural networks. Neural Comput. 1(2), 270–280 (1989)
30. Xu, K., Qi, G., Huang, J., Wu, T.: A joint model for sentiment-aware topic detection on social media. In: Proceedings of ECAI, pp. 338–346. IOS Press (2016)
31. Yan, X., Guo, J., Lan, Y., Cheng, X.: A biterm topic model for short texts. In: Proceedings of WWW, pp. 1445–1456. ACM (2013)

32. Zarrella, G., Marsh, A.: MITRE at SemEval-2016 task 6: transfer learning for stance detection. In: Proceedings of SemEval@NAACL-HLT, pp. 458–463. ACL (2016)
33. Zhang, Z., Lan, M.: ECNU at SemEval 2016 task 6: relevant or not? Supportive or not? A two-step learning system for automatic detecting stance in tweets. In: Proceedings of SemEval@NAACL-HLT, pp. 451–457. ACL (2016)
34. Zhao, W.X., Jiang, J., Yan, H., Li, X.: Jointly modeling aspects and opinions with a MaxEnt-LDA hybrid. In: Proceedings of EMNLP, pp. 56–65. ACL (2010)
35. Zhao, W.X., Jiang, J., Weng, J., He, J., Lim, E.-P., Yan, H., Li, X.: Comparing Twitter and traditional media using topic models. In: Clough, P., Foley, C., Gurrin, C., Jones, G.J.F., Kraaij, W., Lee, H., Mudoch, V. (eds.) ECIR 2011. LNCS, vol. 6611, pp. 338–349. Springer, Heidelberg (2011). https://doi.org/10.1007/978-3-642-20161-5_34

# Mining Inverse and Symmetric Axioms in Linked Data

Rajeev Irny[✉] and P. Sreenivasa Kumar

Department of Computer Science and Engineering,
Indian Institute of Technology, Madras, India
{rajeeviv,psk}@cse.iitm.ac.in

**Abstract.** In the context of Linked Open Data, substantial progress has been made in mining of property subsumption and equivalence axioms. However, little progress has been made in determining if a predicate is symmetric or if its inverse exists within the data. Our study of popular linked datasets such as DBpedia, YAGO and their associated ontologies has shown that they contain very few inverse and symmetric property axioms. The state-of-the-art approach ignores the open-world nature of linked data and involves a time-consuming step of preparing the input for the rule-miner. To overcome these shortcomings, we propose a schema-agnostic unsupervised method to discover inverse and symmetric axioms from linked datasets. For mining inverse property axioms, we find that other than support and confidence scores, a new factor called predicate-preference factor (*ppf*) is useful and setting an appropriate threshold on *ppf* helps in mining quality axioms. We also introduce a novel mechanism, which also takes into account the semantic-similarity of predicates to rank-order candidate axioms. Using experimental evaluation, we show that our method discovers potential axioms with good accuracy.

**Keywords:** Ontology learning · Property axioms · Rule mining · Semantic similarity

## 1 Introduction

With the popularity of Linked Open Data (LOD), we are witnessing a rapid rise in the number and size of linked datasets. Advancements in Information Extraction have made creation of triples from web-resources very efficient [6]. The extracted triples can enhance existing knowledge bases like DBpedia or Freebase by using ontologies to ensure consistent mapping of entities, as in the case of DBpedia [2]. However, datasets in LOD often have incomplete or rudimentary ontology axioms and this hinders efforts towards semantic interoperability of data. We observe that unlike the *subsumption* property axiom, *inverse* and *symmetric* property axioms are very few in number in most ontologies in the LOD as shown in Table 1. Thus, the absence of such axioms motivated us to investigate if the discovery of latent symmetric and inverse axioms could be of

© Springer International Publishing AG 2017
Z. Wang et al. (Eds.): JIST 2017, LNCS 10675, pp. 215–231, 2017.
https://doi.org/10.1007/978-3-319-70682-5_14

**Table 1.** Property axiom count

| Property axiom | DBpedia | YAGO | GeoSpecies | SWRC |
|---|---|---|---|---|
| Inverse | 2 (1 pair) | 0 | 0 | 10 (5 pairs) |
| Symmetric | 0 | 1 | 0 | 0 |

any value to the community. We recognize some potential uses of such axioms in downstream applications like query re-writing and inconsistency detection.

For instance, consider the following axiom (in triple form) generated by our method: <dbo:subsidiary> <owl:inverseOf> <dbo:parentCompany>. Now, if we wish to obtain all subsidiaries of *dbr:Activision* (a company) we may have to use a SPARQL query. In the absence of above axiom, the SPARQL query one would write is shown in Listing 1.1.

```
SELECT distinct(?s) WHERE { dbr:Activision dbo:subsidiary ?s }
```

**Listing 1.1.** SPARQL Query 1

Owing to the inverse axioms above, we can now write a more complete version of the SPARQL Query 1 as shown in Listing 1.2. which returns 20 results as opposed to 14 results returned by SPARQL Query 1.

```
SELECT distinct(?s) WHERE {
{ ?s dbo:parentCompany dbr:Activision }
UNION
{ dbr:Activision dbo:subsidiary ?s }}
```

**Listing 1.2.** SPARQL Query 2

Additionally, on having identified *symmetric* or *inverse* axioms, we can impose restrictions on Domain and Range of a predicate and identify those triples which deviate from the semantics. For instance, our method discovers dbo:relative as a *symmetric* predicate. We can infer that the instances in the *subject* and *object* position of a triple with dbo:relative as the predicate must be instances of the same class-type. The triples that violate this restriction are the candidates for inconsistency check.

Thus, it is clear that the discovery of *inverse* and *symmetric* axioms helps us to advance towards a LOD that is more complete, consistent and richer in schema information.

## 1.1 Challenges and Contributions

Discovery of *inverse* and *symmetric* axioms is a challenging task because of the incomplete nature of linked datasets. We may opt to tide over this impediment by leveraging ontological knowledge like Domain/Range restrictions. However, work in [17] has shown that Domain and Range restrictions are often violated in LOD datasets and thus its use may lead to unreliable results. Also, as shown in Table 1, absence of *inverse* and *symmetric* axioms in most ontologies restrict us from applying Ontology Matching or Alignment techniques [15] to find such axioms. In fact, the output of our method can be used as a gold standard for training supervised relation alignment systems [11].

Moreover, triples in LOD do not always conform to the ontology axioms, thus resulting in data incoherency. For instance, DBpedia[1] mentions dbo:influenced and dbo:influencedBy as inverses of each other. But we observe that in only 15% of the cases both <$x$ dbo:influenced $y$> <$y$ dbo:influencedBy $x$> exists. This suggests existence of a bias is usage among predicates since dbo:influenced is the more preferred predicate in use. Such a bias in usage among predicates makes the discovery of *inverse* axioms a challenging task. To mitigate such a bias in use of predicates, we introduce predicate preference factor (*ppf*) in Sect. 4.1.

Thus, owing to a lack of useful ontology axioms, we opt to discover *inverse* and *symmetric* axioms independent of schema information. We believe that doing so will not affect the discovery of *inverse* and *symmetric* axioms as there are few such axioms to begin with. Our contributions in this paper are as follows:

1. We present a detailed method to identify *symmetric* and *inverse* axioms based on the evidence in datasets.
2. We propose a function to rank-order the candidate axioms based on the evidence in the dataset and context-based semantic similarity of predicates.
3. We introduce a measure called *predicate preference factor* to control the quality of axioms and estimate an upper-bound on the number of axioms that can be discovered by our method.
4. We also demonstrate that our method discovers more axioms compared to the state-of-the-art [7] at higher accuracy.

The rest of the paper is as follows - in Sect. 2 we give an overview of related works and the shortcomings of the state-of-the-art [7]. In Sect. 3 we discuss the relevant concepts and notations used throughout the paper. In Sect. 4 we provide the details of the proposed method to discover *symmetric* and *inverse* property axioms from the linked datasets. In Sect. 5 we evaluate the axioms generated by our method and discuss how the parameters influence the results. Finally we end the paper by discussing conclusions and scope for future work.

## 2   Related Work

While research for learning ontology axioms from text [13,14,16] has progressed, we focus on approaches that learn axioms from instances in linked datasets. Linked datasets contain triples from multiple sources that comply to different ontologies, many of which are poorly endowed with property axioms (see Table 1). One way to enrich these ontologies is to learn new axioms from the assertions in the datasets. Popular techniques include generating taxonomies using clustering techniques and Formal Concept Analysis (FCA) [4]. Others [12]

---

[1] Version 2015-10.

are Inductive Logic Programming (ILP) based approaches which generate only concept axioms. However, most of these approaches discuss the generation of concept axioms.

There have also been efforts in learning some property axioms like the work by Zhang et al. [19] discussed discovering locally equivalent relations (i.e. equivalent relations that hold only within a particular class and not across all classes) to reconcile synonymous predicates in heterogeneous linked datasets. [9] discovers relation subsumption and equivalence to align knowledge bases using an instance-based approach. Work in [7,18] introduces statistical induction of property and concept axioms and is closer to our approach. Section 2.1 discusses this in detail.

## 2.1    Limitations of Existing Work

Fleischhacker et al. [7] discuss a method to mine property axioms from linked datasets. They begin with a terminology acquisition phase where instances, predicates and concepts are extracted by querying a SPARQL endpoint and these are assigned unique identifiers and are stored in a MySQL database. This phase is followed by creation of *Transaction Tables*, where a row, representing a pair of instances $(i_1, i_2)$, contains a list of predicates $(r_i)$ such that $<i_1 \ r_i \ i_2>$ is a triple (if $<i_2 \ r_i \ i_1>$ is a triple, then $r_i^-$ is added to the list of predicates). Every such instance pair and the predicates that hold between them are obtained by issuing a series of SPARQL queries. An off-the-shelf association rule miner [3] takes these tables as input and produces association rules on predicates. A user defined confidence threshold is used to filter the rules. To obtain property axioms, these rules are parsed for specific patterns.

We note several drawbacks of this method and propose techniques to significantly improve upon it.

**Closed World Assumption.** Linked datasets are inherently incomplete in nature and follow the Open World Assumption (OWA). Under OWA, the absence of an assertion in the dataset does not mean it is false. Instead, it means that nothing is known about its truth value. Fleischhacker et al. [7] make use of traditional association rule miner [3] to mine axioms that does not account for the linked data setting and conforms to the Closed World Assumption (CWA). However, our method conforms to OWA (Sect. 3.2).

**Negative Examples.** Rule mining over transaction tables conforms to CWA and generates rules with a confidence score that doesn't account for Negative Examples (see Sect. 3.2). Under such a setting, every unknown assertion is assumed to be a negative assertion. Thus, axioms are discovered with a lower accuracy.

**Context of predicates.** The intuition that similar predicate pairs are candidates for finding equivalent predicates has been used as a starting point in discovery of equivalence axioms [19]. We extend this intuition to discover *inverse* axioms by introducing a context-based semantic-similarity measure for predicates in linked datasets. Existing methods do not account for this intuition. Details of this measure are discussed in Sect. 4.2.

**Transaction Tables.** Input to the rule miner in [7, 18] are *transaction tables* and they play a vital role in mining axioms. Multiple tables are created depending on the axiom being mined. Also, time to construct a transaction table corresponding to an axiom can vary widely between a couple of hours to well over a day. These tables are sparse and can consume anywhere between 4 MB to 56 GB of space (for DBpedia 3.7).

Thus the method involving transaction tables does not scale well. Instead, we use a specialized rule miner that consumes RDF triples. Additionally, [7, 18] have a cumbersome setup which involves a MySQL database and a local SPARQL endpoint for mining axioms. Our method is comparatively straightforward.

## 3  Preliminaries

In this section we discuss the concepts and terminologies which are necessary for understanding our work.

### 3.1  Property Axioms

Ontologies effectively capture and organize domain knowledge as Class and Property axioms. Ontologies offer definitions for a set of classes (or concepts) ($C$) and a set of predicates (or roles) ($P$). Axioms assert relationships between elements of $C$ and $P$. In the linked data setting, a *fact* can be categorized as belonging to either *T-Box*, which contains axioms or *A-Box*, which contains assertions. In addition to $C$ and $P$, linked datasets contain set of Instances (or Individuals) ($I$) which are instances of concepts in $C$. Linked datasets also contain Literals $L$ (string, integers etc.). We will be mining *A-Box* assertions to obtain *T-Box* axioms. The semantics of some property axioms are as shown below, where $r_i, r_j \in P$ and $r_i(x, y)$ represents a triple $<x\ r_i\ y>$.

**Note:** Every *symmetric* predicate is an *inverse* of itself and thus we can consider *symmetric* axioms as a special case of *inverse* axioms with $r_i \equiv r_j$.

| Semantics | T-Box Axiom |
|---|---|
| $\forall x, y : r_i(x, y) \iff r_i(y, x)$ | $r_i$ *rdf:type owl:SymmetricProperty* |
| $\forall x, y : r_i(x, y) \iff r_j(y, x)$ | $r_i$ *owl:inverseOf* $r_j$ |
| $\forall x, y : r_i(x, y) \Rightarrow r_j(x, y)$ | $r_i$ *rdfs:SubPropertyOf* $r_j$ |

### 3.2  Mining Logical Rules Under OWA

Association Rule Mining (ARM) is a technique used to make predictions over the occurrence of an item based on the occurrences of other items in a transaction database. Based on a *support* threshold, frequent items are identified and are subsequently used for discovering association rules of the form $A \Rightarrow B$. Using ARM, interesting rules are identified based on confidence scores, that conform to CWA. So to mine rules under the OWA, we use AMIE+ instead which introduces the notion of Partial Completeness Assumption (PCA) [8]. The logical rules generated by AMIE+ are of the form as shown below:

$$\overrightarrow{B} \Rightarrow r(x, y)$$
$$\overrightarrow{B} = B_1 \wedge B_2 \wedge .... \wedge B_n \tag{1}$$

Here, $B_i$ is a triple of the form: $<?x\, p_i\, ?y>$, $p_i$ is a predicate. $x$, $y$ are either variables or instances from the linked dataset. $\overrightarrow{B}$ is the *Body* of the rule. $r(x, y)$ is the *Head* of the rule and it represents *prediction* made by the rule.

Logical rules in Eq. (1) are closed, which means that every variable appears in the rule at least twice and even number of times. The rule in Eq. (1) is of size $n + 1$. In our work, we use AMIE+ to generate rules of size 2 ($n = 1$). So the rules would take one of the forms shown in Eq. (2)

$$B_1(x, y) \Rightarrow r(x, y) \quad or \quad B_1(y, x) \Rightarrow r(x, y) \tag{2}$$

**Support and Confidence.** In linked datasets *A-Box* assertions are in triple form and under such a setting, Support (*supp*) and Standard Confidence (*Conf$_{STD}$*) scores generated under ARM adapted for triples are shown in Eq. (3):

$$supp(\overrightarrow{B} \Rightarrow r(x, y)) = \{\#(x, y) : \overrightarrow{B} \wedge r(x, y)\}$$
$$Conf_{STD}(\overrightarrow{B} \Rightarrow r(x, y)) = \frac{supp(\overrightarrow{B} \Rightarrow r(x, y))}{\#(x, y) : \overrightarrow{B}} \tag{3}$$

Here, *supp* represents the number of valid instantiations of the rule in the dataset, while *Conf$_{STD}$* is the confidence score as computed under the CWA in ARM. The denominator represents count of triples that validate the body ($\overrightarrow{B}$) of the rule alone and is called the *Body Count*.

**Conforming to OWA.** Given any Semantic Knowledge Base ($\mathcal{K}_s$), the set of assertions ($F$) in it can belong to either the set of positive assertions ($F^+$) or set of negative assertions ($F^-$). Triples in linked datasets make positive assertions (like *Sam hasBrother Bob*) but do not make any negative assertions (like $\neg$ *Sam hasBrother Bob*). So in linked datasets we have $F^- = \emptyset$ and $F^+ = F$. For any rule of the form shown in Eq. (2) the *supp* value represents the count of correct predictions $\in F^+$, i.e. those $<subject, object>$ pairs that are the instantiations of both the *Body* and *Head* of the rule. These $<subject, object>$ pairs form the *PositiveExamples* of the rule. Under CWA, it is assumed that any $<subject, object>$ pair that is a valid instantiation of $\overrightarrow{B}$ but is an invalid instantiation of *Head* of a rule is considered as a negative assertion. This assumption is not in the spirit of OWA because such a $<subject, object>$ pair could either belong to $F^-$ or be an unknown assertion. To compensate for the absence of negative assertions in linked datasets, we use Partial Completeness Assumption.

**Partial Completeness Assumption** (PCA). States that for any rule $\overrightarrow{B} \Rightarrow r(x, y)$, given a predicate $r$ and its subject $x$, if we know some $r$-attribute of $x$ (i.e. $y$), then we know all $r$-attributes of $x$ [8]. Under this assumption, we say that the prediction $r(x, y')$ is a negative assertion of the rule, if $\exists y : r(x, y) \in F^+ \land r(x, y') \notin F^+ \land (y \neq y')$. Such negative assertions are called the *NegativeExamples* of the rule. PCA is explained in detail in [8].

$$Conf_{PCA}(\vec{B} \Rightarrow r(x, y)) = \frac{supp(\vec{B} \Rightarrow r(x, y))}{\#(x, y) : \exists y' : \vec{B} \land r(x, y')} \qquad (4)$$

With PCA, the confidence score is as shown in Eq. (4). The numerator of $Conf_{PCA}$ is *#PositiveExamples* for a rule and the denominator denotes the union of positive and negative examples of the rule. Alternatively, we have:

$$Conf_{PCA}(\vec{B} \Rightarrow r(x, y)) = \frac{\#PositiveExamples}{PCABodySize} \qquad (5)$$

$$PCABodySize = PositiveExamples + NegativeExamples$$

**Table 2.** Axiom patterns to identify rules of interest

| Pattern | Axiom (in triple form) |
|---------|------------------------|
| **RP₁**: ?a <prop1> ?b ⇒ ?b <prop1> ?a | <prop1> <rdf:type> <owl:SymmetricProperty> |
| **RP₂**: ?a <prop1> ?b ⇒ ?b <prop2> ?a | <prop1> <owl:inverseOf> <prop2> |
| **RP₃**: ?a <prop1> ?b ⇒ ?a <prop2> ?b | <prop1> <rdfs:subPropertyOf> <prop2> |

## 4 Proposed Method

In this section, we discuss the steps to obtain *symmetric* and *inverse* property axioms from the triples in a linked dataset. First, we mine the triples to obtain logical rules of size 2. Among these rules, we identify *Rules of Interest (roi)* by applying a threshold to predicate preference factor (*ppf*) and obtain an upper-bound on discoverable axioms. Each rule in *roi* is assigned a score called *R-Score* in Sect. 4.2 to rank-order rules in *roi*. In Sect. 4.3 we determine a cut-off threshold for *R-Score* in an unsupervised way to get top-*k* rules which will be converted to axioms (Sect. 4.4). Overview of the proposed method is shown in Fig. 1.

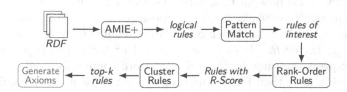

**Fig. 1.** Overview of method to generate axioms

## 4.1   Rules of Interest

A predicate in a triple can either be an Object Property or Data-type property. An Object property relates two instances of a class, like dbo:cousin which relates two persons, while a Data-type property relates an instance to a literal value like a string or a date, like dbo:length. We discard triples with Data-type properties as only Object properties can possess *inverse* and *symmetric* property axioms. We use AMIE+ to mine logical rules of size 2 as shown in Eq. (2). The sets of rules that satisfy rule-patterns $RP_1$ or $RP_2$ in Table 2 are classified as *Rules of Interest* (*roi*) and are named $roi_1$, $roi_2$ respectively such that $roi = roi_1 \cup roi_2$. Similar patterns ($RP_3$) were proposed in [9] to discover property equivalence and subsumption axioms to align knowledge bases. Note that $RP_1$, $RP_2$ and $RP_3$ collectively represent the entire spectrum of rules (of size 2) generated by AMIE+. Rules that conform to $RP_3$ also aid in obtaining an upper-bound on the number of discoverable axioms in a linked dataset, as discussed below:

**Predicate Preference Factor** (*ppf*). Given a linked dataset ($K$), let the set of all Object Properties in it be $P = \{p_1, p_2, ..., p_k\}$. For any $p_i \in P$, let $d_i =$ Domain($p_i$), $r_i =$ Range($p_i$). Let $shr(p_i, p_j) = \{(s, o) | (s, p_i, o) \in K \wedge (o, p_j, s) \in K\}$. Ideally, for any pair $p_i, p_j$ that are potentially inverses of each other, we have:

1. $(d_i = r_j) \wedge (d_j = r_i)$,
2. $shr(p_i, p_j) \neq \emptyset \wedge shr(p_j, p_i) \neq \emptyset$

Condition 1 is contained in Condition 2, so the number predicates pairs ($p_i, p_j$) that fulfill condition 2 can be considered as an approximate upper-bound on the number of *inverse* axioms. Enforcing condition 2 is equivalent to obtaining a rule conforming to $RP_2$ for the corresponding predicates ($p_i, p_j$) with a *support* $\geq 1$. Thus, for an instance-based method, we can say that the existence of a rule conforming to $RP_2$ is a necessary condition for $p_i, p_j$ to be inverses of each other. Incidentally, $roi_2$ contains such rules. So the size of $roi_2$ can be a naive upper-bound on the number of *inverse* axioms. Note that rules in $roi_2$ with a low $Conf_{STD}$ could mostly have inconsistent triples in their *support*, so we may filter these rules by imposing a cut-off on $Conf_{STD}$ and get a refined upper-bound. However, we observe that using $Conf_{STD}$ to filter rules can result in loss of prospective *inverse* axioms due to incomplete data. For instance, assume predicates $p_i, p_j$ are inverse of each other. So we could either have a rule *R:* $\{?x$ $p_i$ $?y \Rightarrow ?y$ $p_j$ $?x\}$ or *R:* $\{?x$ $p_j$ $?y \Rightarrow ?y$ $p_i$ $?x\}$ or both. The triples $<s\ p_i\ o>$, $<o\ p_j\ s>$ express the same fact and if $p_i$ is more commonly used compared to $p_j$ to express a fact, then fewer triples exist that have $p_j$ as a predicate (i.e. incomplete data). As a result, $Conf_{STD}(R)$ for the former rule will be lower than $Conf_{STD}(R)$ for the latter rule. Thus, we could overlook prospective inverse predicates involved in a rule of the latter form if we had used $Conf_{STD}$ to filter rules.

$$ppf(R) = \frac{supp(R)}{min\{|r|, |B|\}}$$

$$here, R \in roi, \; R : \overrightarrow{B} \Rightarrow r(x,y)$$

$$|r| = |\{(s,r,o)|(s,r,o) \in K)\}|, |B| = |\{(s,B,o)|(s,B,o) \in K)\}| \tag{6}$$

So, to avoid losing out on plausible axioms, we introduce a measure called predicate-preference factor (*ppf*). As shown in Eq. (6), given a rule containing two predicates, *ppf* measures the ratio of *support* of the rules to the count of the least popular of the two predicates in the rule. *ppf* ensures that we select those rules lead to potential axioms, which is a challenging task due to incomplete data. Unlike $Conf_{STD}$, *ppf* balances the bias of usage among predicates in an axiom by offsetting the influence of incomplete data in LOD. Thus, using *ppf* as a criterion for filtering rules in $roi_2$ resolves inequity in preference of predicates. To get an upper-bound, we apply a threshold ($\alpha$) on *ppf*, i.e. the rules in $roi_2$ with $ppf \geq \alpha$ are counted towards computing the upper-bound. In Sect. 5.3 we show how *ppf* influences the quality of discovered axioms.

Additionally, we also use the fact that a pair of distinct non-symmetric predicates can either be equivalent or inverse but not both. So if a predicate pair exists for which the support of the corresponding rule in $RP_3$ is greater than the support for the rule in $RP_2$, we do not consider this rule as a candidate axiom as it is more likely to be a candidate for subsumption axioms. We can now report the number of candidate axioms filtered from $roi_2$ as the upper-bound on the inverse axioms. For symmetric predicate pairs, $p_i = p_j$ and *ppf* is equivalent to $Conf_{STD}$, so any given rule in *roi* will have $ppf \geq Conf_{STD}$.

## 4.2 Ordering Rules

To order the rules in *roi*, we propose two measures namely Normalized $Conf_{PCA}$ (*NPCA*) and Context Similarity (*CS*) and combine them to obtain a score called *R-Score*. The details of the these measure are discussed below. Unlike [7], we look beyond just the confidence scores to select interesting rules.

**Normalized $Conf_{PCA}$.** Each rule in *roi* is associated with an OWA based confidence score i.e. $Conf_{PCA}$. $Conf_{PCA}$ scores are basically ratios and do not distinguish between rules with confidence scores like $\frac{40}{60}$ and $\frac{6000}{9000}$. Clearly, we are more interested in the rule with confidence score $\frac{6000}{9000}$ as it has a higher number of instances in its support and is less likely to have erroneous instances in its support. To distinguish between such cases we introduce *Normalized $Conf_{PCA}$* (*NPCA*), described in Eq. (7) which is a function of Positive examples and (assumed) Negative examples of a rule.

$$s = \#PositiveExamples \; b = PCABodySize$$

$$NPCA = \frac{X+Y}{2}, \; r' = \frac{s+100}{b+400} \; and \; \alpha = \sqrt{\frac{r'*(1-r')}{b+400}} \tag{7}$$

$$X = max(0, r' - 2*\alpha) \quad Y = min(1, r' + 2*\alpha)$$

Equation (7) represents a normalized version of the $Conf_{PCA}$ (see Eq. (5)). The normalization of $Conf_{PCA}$ as the average of $X, Y$ in Eq. (7) is adapted from the *improved Wald method* [1] which is a technique for approximating confidence intervals of a ratio. The numerical values 100 and 400 in $r'$ are empirically determined. The *max* and *min* expressions ensure that $NPCA \in [0, 1]$. Normalization of $Conf_{PCA}$ with this technique helps us in identifying statistically significant rules because it disproportionately affects the rules that have a high $Conf_{PCA}$ but are inadequately supported by *PositiveExamples*. As a result, $NPCA$ transforms the $Conf_{PCA}$ value $\frac{40}{60}$ to 0.3043 and $\frac{6000}{9000}$ to 0.6489 thereby assigning higher score to a likely rule.

In Sect. 5.2, we experimentally show that without the normalization of $Conf_{PCA}$, inferior rules are selected to generate axioms.

**Context Similarity** (CS). The objective of this measure is to obtain a *context* based semantic similarity between a pair of predicates in the property axiom. This similarity is based on the degree of overlap of the *contexts* of the predicates and is based on the intuition that semantically similar predicates have high degree of co-occurrence of *contexts*. We define *context* of a predicate as a set of class-types of the instances that occur in the subject and object position of a predicate. Formally, we define *context* for predicate $p$ as $C(p)$ in Eq. (8). Let $KB$ be the linked dataset in use and *(s, a, t)* means that subject $s$ is an instance of a class-type $t$.

$$C(p) = \{t \,|((s, p, o)) \in KB, ((s, a, t) or (o, a, t))\}$$
$$C(p, q) = \{t \,|(s, p, o) \in KB, (o, q, s) \in KB, ((s, a, t) or (o, a, t))\} \qquad (8)$$

We use $C(p), C(p, q)$ to model a context-based semantic distance measure called *Predicate Distance* (PD) as shown in Eq. (9).

$$f(p) = |C(p)| \; and \; f(p, q) = |C(p, q)|$$
$$PD(p, q) = \frac{max\{log\,f(p), log\,f(q)\} - log\,f(p, q)}{log\,T - min\{log\,f(p), log\,f(q)\}} \qquad (9)$$

Here $T$ is the total number of class-types that can be assigned to instances in a $KB$. Equation (9) measures dissimilarity between predicates $p, q$ over different semantic *contexts*. Intuitively, $PD$ is the conditional probability of the co-occurrence of predicates $p, q$ in the linked dataset. A high $PD$ value suggests low similarity and vice-versa. Note that $PD$ can have values $\geq 1$ and therefore, it is not a metric. For the sake of comparison, we scale $PD$ and use it as a similarity metric, as shown in Eq. (10). We name this similarity metric as Context similarity ($CS$). $CS$ generalizes the Normalized Semantic Web Distance ($NSWD$) [5].

$$CS(p, q) = \begin{cases} 1 - PD * (1 - \frac{1}{PD_{max}}), & if \; PD \in [0, 1] \\ (1 - \frac{PD}{PD_{max}}) * \frac{1}{PD_{max}}, & if \; PD \in (1, PD_{max}] \end{cases} \qquad (10)$$

Here, $PD_{max}$ is $log_2 \left(\frac{|T|}{2} + 1\right)$, an upper-bound on $PD$. The scaling applied is simple to follow from Eq. (8), $PD \in [0, 1]$ are scaled such that corresponding $CS$

$\in [\frac{1}{PD_{max}}, 1]$ and $PD \in (1, PD_{max})$ is scaled to $CS \in [0, \frac{1}{PD_{max}})$. As explained in [5], for the special case when $f(p, q) = 0$, we define $PD(p, q) = PD_{max}$. With this scaling applied to $PD$, we have $CS \in [0, 1]$. We get $CS(p, q) = 0$ when $PD(p, q)$ is $PD_{max}$, implying that $p, q$ are dissimilar as they do not occur in similar *contexts*. Similarly we have $CS = 1$ when $PD(p, q) = 0$ which means that $p, q$ have a perfect overlap of *contexts*. We assign a $CS$ score for the pair of predicates that appear in a rule in *roi*.

**Combined-Score.** The average of $NPCA$ and $CS$ measure is assigned as a score to each rule in *roi*. We name this score as *R-Score* short for rule-score.

$$R\text{-}Score = \frac{(NPCA + CS)}{2} \tag{11}$$

Rules are ordered in decreasing order of their *R-Score*. Rules with higher *R-Score* are the ones with a higher share of *PositiveExample* in the *PCABodySize* count and a greater overlap of context, such rules are thus desirable. Note that for symmetric predicates $CS(p, q) = 1$ as $p = q$. Section 4.3 explains the steps involved in determining a threshold on *R-Score* to obtain top-$k$ *symmetric* and *inverse* axioms.

## 4.3 Clustering

Association rule mining approaches normally have a predetermined, user selected confidence threshold to identify interesting rules. It is an empirical value and is not generalizable as it is characterized by the domain of the linked dataset. Our method determines the threshold in an unsupervised way. It is inspired by [19], where it was used to detect similarity threshold for equivalence relations. We use Jenks Natural Breaks [10], which is a clustering technique to determine an ordering of univariate values into different clusters. It maximizes the inter-cluster variance while minimizing the intra-cluster variance of the *R-Scores* of the rules. This technique efficiently serves our purpose as the number of rules in *roi* even for large linked datasets is observed to be less than 1000. Jenks Natural Breaks clusters the rules in *roi* into $R_A$ and $R_D$ based on their *R-Scores*. $R_A$ contains rules with higher *R-Score* and suggest that a substantial number of triples in the linked dataset validate the corresponding logical rule and the predicates in the rule have similar *contexts*. Rules in $R_D$ are discarded as we expect them to be a result of incoherent triples.

Threshold value $t_H$ is obtained from these clusters and is equal to the expression $max\{R\text{-}Score(r_i) \mid r_i \in R_D\}$, i.e. the maximum *R-Score* of the rules in $R_D$[2]. Threshold $t_H$ for *inverse* and *symmetric* axioms are found independent of each other. We use $t_H$ and $k = |\{R_A\}|$ to compare results against the state-of-the-art. The rules $r_i \in R_A$ form candidates for the top-$k$ property axioms. We use these top-$k$ axioms to compare against the axioms generated by other methods.

---

[2] It could very well be the minimum *R-Score* in $R_A$.

## 4.4  Axiom Generation

This is the final step in our method. At this point we have identified the rules in
*roi* which will be converted to axioms (i.e. top-$k$ axioms). For each rule $\in R_A$,
we form the corresponding axiom as shown in Table 2. The generated axioms are
then presented to the user in *descending* order of their likelihood as a potential
addition to the schema. To maximize recommendations, we also check for inferred
axioms i.e. if we assume prop1 is a symmetric predicate and is inverse of prop2
then we offer prop2 as a plausible *symmetric* predicate. Based on the preference
of the user, any of the generated axioms can be added to the schema, thereby
completing the schema enrichment process.

## 5  Experimental Results

In this section we evaluate the quantitative and qualitative performance of our
method against the existing state-of-the-art [7] and two naive baseline. We show
that our method performs better on both fronts. We also discuss the effects of
using semantic information (context) and *ppf* to discover axioms. All experiments were conducted on a system with 32 GB RAM with a 4-Core Intel Xeon
Processor. We set the threshold ($\alpha$) on *ppf* as 0.01 to find the upper-bound on
axioms and to discover top-$k$ property axioms.

### 5.1  Datasets

Our intention is to enrich schemas of popular datasets like DBpedia. We also choose the
GeoSpecies[3] and SW-DogFood[4] datasets because
together, these form a representative set of large,
medium and small sized datasets, as seen by the
triple count in Table 3. DBpedia is the most extensive dataset in LOD and is an extract of structured
information from Wikipedia. Specifically, we use

**Table 3.** Datasets sizes

| Dataset | Triples |
| --- | --- |
| DBpedia 3.7 | 26,988,054 |
| DBpedia 2015 | 67,054,254 |
| SW-DogFood | 132,963 |
| GeoSpecies | 2,201,532 |

the Mapping-based infobox properties dataset in DBpedia as it is cleaner and
more structured than the Raw infobox dataset [2]. We ran experiments with
DBpedia versions 3.7, 3.8, 3.9, 2014, 2015, 2016. This demonstrates that our approach suggests axioms with consistency for datasets of all sizes and thus it is
scalable.

### 5.2  Comparison with Naive Baselines

Consider a baseline (Naive₁) where we ignore *ppf*, *NPCA* and *CS* to discover
axioms. Instead, only $Conf_{PCA}$ is used to select interesting rules. A threshold of
0.5 on $Conf_{PCA}$ ensures a fair comparison. Table 4 compares the results of Naive₁

---

[3] https://datahub.io/dataset/geospecies.

[4] Semantic Web Dog Food: http://data.semanticweb.org/.

with our proposed method. The results are insightful and they demonstrate the need of $NPCA, CS$ and $ppf$ in discovering axioms. For *symmetric* axioms, the results of our method and $\text{Naive}_1$ are very similar. This is because, in our method $ppf$ and $CS$ do not contribute meaningfully in the discovery of *symmetric* axioms (since $p \equiv q$, thus $ppf = Conf_{STD}$ and $CS = 1$ for all predicates). In contrast, for *inverse* axioms where $ppf$ and $CS$ are relevant (since $p \not\equiv q$), our method clearly out-performs $\text{Naive}_1$. Also, use of $NPCA$ in our method, ensures better accuracy because $NPCA$ selects superior rules compared to $\text{Naive}_1$, thus showing that discovery of latent *inverse* axioms is more challenging and requires the use of $ppf$, $NPCA$ and $CS$ to discover axioms.

Consider another baseline for *inverse axioms* where we generate candidate axioms without learning any rules. Instead, the candidate axioms comprise of predicate pairs $(p, q)$ that satisfy the condition: $\frac{|\{(x,y):p(x,y)\in \text{KB} \wedge q(y,x)\in \text{KB}\}|}{max\{|p|,|q|\}} \geq t$. Here, KB refers to the linked dataset and $|p|$ is number of triples $p(x, y) \in$ KB and $t$ is a threshold. For *symmetric axioms*, we have $p \equiv q$. Note that the above expression is similar to $Conf_{STD} \geq t$ because the numerator is same as *support* of a rule and the denominator $\leq BodySize$ and thus will produce similar results like in [7]. In Sects. 5.3 and 5.4 we show that we out-perform a similar method [7].

**Table 4.** Count of inverse and symmetric axioms

| Datasets | #Correct/#Inverse | | | #Correct/#Symmetry | | | UB |
|---|---|---|---|---|---|---|---|
| | Proposed | Existing | $\text{Naive}_1$ | Proposed | Existing | $\text{Naive}_1$ | |
| DBpedia 3.7 | 23/39 | 12/34 | 16/36 | 13/13 | 3/25 | 11/11 | 128 |
| DBpedia 3.8 | 23/38 | 11/34 | 16/36 | 12/12 | 3/27 | 11/11 | 123 |
| DBpedia 3.9 | 25/44 | 12/30 | 16/41 | 13/13 | 4/22 | 12/13 | 132 |
| DBpedia 2014 | 24/44 | 11/29 | 15/38 | 12/12 | 3/20 | 13/14 | 137 |
| DBpedia 2015 | 24/45 | 2/11 | 16/38 | 14/15 | 3/16 | 13/16 | 122 |
| DBpedia 2016 | 25/45 | 12/30 | 16/38 | 12/13 | 4/22 | 13/16 | 122 |
| SW-DF | 7/8 | 8/18 | 6/10 | 0/0 | –/– | –/– | 9 |
| GeoSp | 12/16 | 10/21 | 10/12 | 1/2 | 1/27 | 1/2 | 17 |

## 5.3 Quantitative Performance

Table 4 compares the number of axioms discovered by our proposed method and the existing state-of-the-art [7] for various datasets. The column #Correct/#Inverse represents the proportion of correct *inverse* axioms contained in the top-$k$ *inverse* axioms discovered. To ensure optimum performance, a confidence threshold of 0.5 was set for rules in [7]. Fleischhacker et al. [7] mine axioms under the CWA by using $Conf_{STD}$ which over-compensates for negative assertions by considering unknown assertions as negative. This makes [7] susceptible to inconsistent triples and results in discovery of fewer correct axioms (Table 4). To discover axioms under OWA, we use $NPCA$ as it restricts the set of negative assertions for a rule by ignoring unknown assertions. $NPCA$ also offsets the presence of inconsistent triples by normalizing $Conf_{PCA}$. Unlike [7], we leverage semantic information like class-type statements in $CS$ to assign importance to predicate pairs that have a similar semantic-contexts. The use of *contexts* is modeled as a proxy to make-up for the absence of exact Domain and Range information. The results (Table 4) support our claim that *NPCA* as a confidence measure and semantic-context based similarity of predicates ($CS$) helps to discover greater number of correct axioms at better accuracy. Also, unlike [7] where transaction tables are generated in considerable time, our method generates axioms in under five minutes for each DBpedia datasets and under 10 s for GeoSpecies and SW-Dogfood.

Also, comparison of Naive$_1$ with [7] suggests that our decision to use AMIE+ and $Conf_{PCA}$ to mine rules under OWA produces better results.

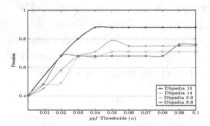

**Fig. 2.** Precision at different *ppf*

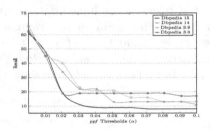

**Fig. 3.** Recall at different *ppf*

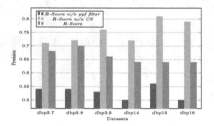

**Fig. 4.** Precision comparision (Color figure online)

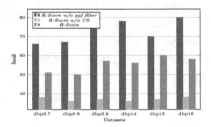

**Fig. 5.** Recall comparision (Color figure online)

***Effect of ppf.*** For each dataset in Table 4, we present the upper-bound (UB) on the number of *inverse* axioms that can be discovered by an instance-based method at $ppf = 0.01$ for different datasets. For the remaining paper, let us consider *recall* as the number of top-$k$ axioms discovered by our method and *precision* as the proportion of correct axioms contained in the top-$k$ axioms. Figures 2 and 3 show the *precision, recall* of *inverse* axioms discovered in some datasets at different threshold values for *ppf*. We observe that *ppf* controls the quality of the axioms, because at higher values of *ppf* we sacrifice *recall* to improve *precision* while at lower values of *ppf* we improve *recall* at the expense of lower *precision*. Further, we observe that for *ppf* ≥ 0.05 both precision and recall values level-out and we discover better quality axioms with higher precision but at reduced recall. Similarly, as *ppf* approaches 0, we discover the maximum probable *inverse* axioms that can be discovered by an instance-based method. Thus, we can use *ppf* as a measure to control the quality of axioms discovered and we can calibrate *ppf* based on nature of the application consuming these axioms.

## 5.4 Qualitative Performance

The correctness of the top-$k$ axioms is evaluated by a group of 5 evaluators because no gold standard exists. Evaluators queried the dataset to understand the context in which a predicate is used in the dataset. Given a predicate, the corresponding Wikidata pages[5] and the *rdf:type* of its *subjects*, *objects* were inspected. Finally, the correctness was decided based on the shared agreement of the evaluators.

We also evaluate our system under three different settings (at $\alpha = 0.01$), as shown in Figs. 4 and 5 to qualify the importance of *ppf* and *CS*. In the first setting, shown in red, we ignore the filtering of rules based on *ppf* to select *roi* but rest of the method remains unchanged. In the second setting, shown in orange, we compute *R-Score* without *Context similarity* (*CS*) measure and the last setting, shown in blue is the method proposed in this paper. We observe that without *ppf* to filter rules, axioms are discovered with least precision (see Figs. 4). This is because many predicate pairs that have a high *CS* score were selected, though they are similar, they are not ideal candidates for *inverse* axioms. In the second setting we ignore contribution of *CS* in *R-Score* and use only *NPCA* to order rules ∈ *roi*. Figures 4 and 5 show that we discover axioms with the highest precision albeit at least recall. Under this setting the recall falls 48% on an average compared to the first setting across the datasets. Thus such a setting is also not desirable because we fail to discover many latent axioms. Now, we contrast both these cases with the third case i.e. the proposed method, where we employ both *ppf* and *CS* to discover axioms. This setting provides a suitable balance of precision and recall where we have recall at values close to the maximum recall (in first setting) but at considerably improved precision. Thus, with this setting we maximize the possibility of discovering prospective latent axioms. This demonstrates that *CS* is helpful in discovering axioms that would otherwise have been overlooked and that *ppf* is useful in filtering inferior rules. Thus, these observations support our claim that the use of *ppf* and *CS* are indeed useful in mining axioms.

Tables 5 and 6 illustrate some of the top-$k$ *inverse* and *symmetric* axioms generated by our method for DBpedia 2016. Due to paucity of space in the paper, the details of axioms generated, their *R-Scores* and links to relevant resources are hosted on a web page[6]. Rules that we discard i.e. rules in R$_D$

---

[5] https://www.wikidata.org/wiki/.

[6] http://www.cse.iitm.ac.in/~rajeeviv/praxis/praxis.html.

**Table 5.** Some symmetric axioms

| Symmetric axioms | R-Score | Correct |
|---|---|---|
| dbo:spouse | 0.9162 | ✓ |
| dbo:neighboringMunicipality | 0.9010 | ✓ |
| dbo:sisterStation | 0.7973 | ✓ |
| dbo:formerPartner | 0.7191 | ✓ |
| dbo:associate | 0.5362 | ✓ |

**Table 6.** Some inverse axioms

| Inverse axioms | | R-score | Correct |
|---|---|---|---|
| dbo:subsequentWork | dbo:previousWork | 0.9160 | ✓ |
| dbo:previousEvent | dbo:followingEvent | 0.9009 | ✓ |
| dbo:doctoralStudent | dbo:doctoralAdvisor | 0.8804 | ✓ |
| dbo:lieutenant | dbo:governor | 0.8333 | ✗ |
| dbo:predecessor | dbo:successor | 0.8202 | ✓ |
| dbo:subsidiary | dbo:parentCompany | 0.7524 | ✓ |

indicate the presence of noisy triples. This suggests that the community can use these rules to refine the concepts/properties in the ontology. For instance, in DBpedia 2016, the upper-bound on *inverse* axioms is 122 (see Table 4) of which $77 \in R_D$ The predicates in these rules can now be considered as candidates for checking inconsistency. We aim to tackle this issue in our future work.

## 6   Conclusion

In this paper, we presented a method that suggests *symmetric* and *inverse* property axioms, based only on the linked datasets. We proposed the use of predicate preference factor (*ppf*) to obtain prospective rules and get an upper-bound on the axioms that can be discovered. We also introduced *NPCA* and *CS* measures to rank-order the *rules of interest* which ensure that our method conforms to OWA and also utilizes the semantic information available in linked datasets. We also proposed the use of an unsupervised approach to select top-$k$ *inverse* and *symmetric* axioms and thus avoid selecting an empirical confidence threshold. Through experimental evaluation we have shown that the proposed technique is effective in discovering axioms. Thus, in comparison to the state-of-the-art [7] and Naive₁, we are able to suggest more axioms at better precision.

## References

1. Agresti, A., Coull, B.A.: Approximate is better than "exact" for interval estimation of binomial proportions. Am. Statist. **52**(2), 119–126 (1998)
2. Bizer, C., Lehmann, J., Kobilarov, G., Auer, S., Becker, C., Cyganiak, R., Hellmann, S.: DBpedia - a crystallization point for the web of data. Web Semant. Sci. Serv. Agents World Wide Web **7**(3), 154–165 (2009)
3. Borgelt, C., Kruse, R.: Induction of association rules: apriori implementation. In: Härdle, W., Rönze, B. (eds.) Compstat, pp. 395–400. Physica, Heidelberg (2002)
4. Cimiano, P., Hotho, A., Staab, S.: Comparing conceptual, divisive and agglomerative clustering for learning taxonomies from text. In: ECAI, vol. 16, p. 435 (2004)
5. De Nies, T., et al.: Normalized semantic web distance. In: Sack, H., Blomqvist, E., d'Aquin, M., Ghidini, C., Ponzetto, S.P., Lange, C. (eds.) ESWC 2016. LNCS, vol. 9678, pp. 69–84. Springer, Cham (2016). https://doi.org/10.1007/978-3-319-34129-3_5
6. Etzioni, O., Fader, A., Christensen, J., Soderland, S.: Mausam: open information extraction: the second generation. In: IJCAI, vol. 11, pp. 3–10 (2011)

7. Fleischhacker, D., Völker, J., Stuckenschmidt, H.: Mining RDF data for property axioms. In: Meersman, R., et al. (eds.) OTM 2012. LNCS, vol. 7566, pp. 718–735. Springer, Heidelberg (2012). https://doi.org/10.1007/978-3-642-33615-7_18
8. Galárraga, L., Teflioudi, C., Hose, K., Suchanek, F.M.: Fast rule mining in ontological knowledge bases with AMIE+. VLDB J. **24**, 707–730 (2015). Springer
9. Galárraga, L.A., Preda, N., Suchanek, F.M.: Mining rules to align knowledge bases. In: Proceedings of the 2013 Workshop on Automated Knowledge Base Construction, pp. 43–48. ACM (2013)
10. Jenks, G.: The data model concept in statistical mapping. Int. Yearb. Cartogr. **7**, 186–190 (1967)
11. Koutraki, M., Preda, N., Vodislav, D.: Online relation alignment for linked datasets. In: Blomqvist, E., Maynard, D., Gangemi, A., Hoekstra, R., Hitzler, P., Hartig, O. (eds.) ESWC 2017. LNCS, vol. 10249, pp. 152–168. Springer, Cham (2017). https://doi.org/10.1007/978-3-319-58068-5_10
12. Lehmann, J.: DL-learner: learning concepts in description logics. J. Mach. Learn. Res. **10**, 2639–2642 (2009)
13. Petrucci, G., Ghidini, C., Rospocher, M.: Ontology learning in the deep. In: Blomqvist, E., Ciancarini, P., Poggi, F., Vitali, F. (eds.) EKAW 2016. LNCS (LNAI), vol. 10024, pp. 480–495. Springer, Cham (2016). https://doi.org/10.1007/978-3-319-49004-5_31
14. Ramakrishnan, C., Kochut, K.J., Sheth, A.P.: A framework for schema-driven relationship discovery from unstructured text. In: Cruz, I., et al. (eds.) ISWC 2006. LNCS, vol. 4273, pp. 583–596. Springer, Heidelberg (2006). https://doi.org/10.1007/11926078_42
15. Shvaiko, P., Euzenat, J.: Ontology matching: state of the art and future challenges. IEEE Trans. Knowl. Data Eng. **25**(1), 158–176 (2013)
16. Del Vasto Terrientes, L., Moreno, A., Sánchez, D.: Discovery of relation axioms from the web. In: Bi, Y., Williams, M.-A. (eds.) KSEM 2010. LNCS (LNAI), vol. 6291, pp. 222–233. Springer, Heidelberg (2010). https://doi.org/10.1007/978-3-642-15280-1_22
17. Tonon, A., Catasta, M., Demartini, G., Cudré-Mauroux, P.: Fixing the domain and range of properties in linked data by context disambiguation. In: LDOW@ WWW (2015)
18. Völker, J., Niepert, M.: Statistical schema induction. In: Antoniou, G., Grobelnik, M., Simperl, E., Parsia, B., Plexousakis, D., De Leenheer, P., Pan, J. (eds.) ESWC 2011. LNCS, vol. 6643, pp. 124–138. Springer, Heidelberg (2011). https://doi.org/10.1007/978-3-642-21034-1_9
19. Zhang, Z., Gentile, A.L., Blomqvist, E., Augenstein, I., Ciravegna, F.: An unsupervised data-driven method to discover equivalent relations in large linked datasets. Semant. Web **8**(2), 1–27 (2015)

# Enhancing Knowledge Graph Embedding from a Logical Perspective

Jianfeng Du[1(✉)], Kunxun Qi[1], Hai Wan[2], Bo Peng[2], Shengbin Lu[1], and Yuming Shen[1]

[1] School of Computer Science and Technology,
Guangdong University of Foreign Studies, Guangzhou 510006, China
jfdu@gdufs.edu.cn
[2] School of Data and Computer Science,
Sun Yat-sen University, Guangzhou 510006, China
wanhai@mail.sysu.edu.cn

**Abstract.** Knowledge graph embedding aims to represent entities and relations in a knowledge graph as low-dimensional real-value vectors. Most existing studies exploit only structural information to learn these vectors. This paper studies how logical information expressed as RBox axioms in OWL 2 is used for embedding. The involvement of RBox axioms could prevent existing methods from learning predictive vectors. For example, the symmetric, reflexive or transitive relations can be declared by RBox axioms, but popular translation-based methods are unable to learn distinguishable vectors for multiple these relations in the ideal case. To overcome these limitations introduced by the involvement of RBox axioms, this paper proposes to enhance existing translation-based methods by logical pre-completion and bi-directional projection of entities. Experimental results demonstrate that these enhancements improve the predictive performance in link prediction and triple classification.

## 1 Introduction

Knowledge graph, a kind of directed graph with vectors labeled by entities and edges labeled by relations, has become a popular formalism for knowledge representation. Meanwhile, knowledge graph embedding, a technology that typically represents entities and relations as low-dimensional real-value vectors, has shown promising results on a number of prediction tasks including link prediction [5,6], triple classification [19,22] and relation extraction [22,23]. Up to date, most existing studies for knowledge graph embedding e.g. [3–7,10–16,18,19,22,25] exploit only structural information of the knowledge graph to learn entity or relation vectors. Seldom work focuses on using other information in the learning process. It remains an open question if other information could be used in the learning process to improve the predictive performance.

Logical information is often attached to a knowledge graph. OWL 2 [8], the newest version of the Web Ontology Language, has become prevalent in

© Springer International Publishing AG 2017
Z. Wang et al. (Eds.): JIST 2017, LNCS 10675, pp. 232–247, 2017.
https://doi.org/10.1007/978-3-319-70682-5_15

knowledge representation. It introduces RBox, expressed as a set of axioms, to declare characteristics of relations and interactions among multiple relations. For example, a relation can be declared to be symmetric, reflexive or transitive, whereas one relation can be declared to be subsumed by another relation. Many existing knowledge graphs such as WordNet [17] and DBPedia [1] come with OWL 2 ontologies that have RBox axioms.

A knowledge graph is usually represented as a set of triples of the form $\langle h, r, t \rangle$, where $h$ is the head entity, $r$ the relation and $t$ the tail entity.[1] When a knowledge graph comes with RBox axioms, the embedding of triples in it should conform to the logical constraints enforced by these RBox axioms. For example, consider a knowledge graph that has a triple $\langle h, r, t \rangle$ where $r$ is declared to symmetric by an axiom attached to the graph. Knowledge graph embedding requires the embedding of $\langle h, r, t \rangle$ to satisfy a certain condition. According to logical inference, the knowledge graph entails the triple $\langle t, r, h \rangle$, thus the embedding of $\langle t, r, h \rangle$ should also satisfy the same condition.

Recently, translation-based methods become popular in knowledge graph embedding due to their high predictive performance and good applicability in sparse knowledge graphs. They represent entities and relations as low-dimensional real-value vectors and typically enforce that $\|f_r(\boldsymbol{h}) + \boldsymbol{r} - f_r(\boldsymbol{t})\|_{L_1/L_2}$ should be less than $\|f_r(\boldsymbol{h'}) + \boldsymbol{r} - f_r(\boldsymbol{t'})\|_{L_1/L_2}$ by a margin, where $\boldsymbol{x}$ denotes the vector representation for $x$, $\|\boldsymbol{x}\|_{L_1}$ (resp. $\|\boldsymbol{x}\|_{L_2}$) denotes the L1-norm (resp. L2-norm) of the vector $\boldsymbol{x}$, $\langle h, r, t \rangle$ (resp. $\langle h', r, t' \rangle$) is a triple in (resp. not in) the given knowledge graph, and $f_r(\cdot)$ is a projection function on entity vectors which is either an identity function (e.g. in TransE [5], TransM [7] and TransA [14]), or a $r$-specific function (e.g. in TransH [22] and TransR [16]) which may also depend on the given entity (e.g. in TransD [12]) or on the statistical triple distribution (e.g. in TranSparse [13]).

In the ideal case, $\|f_r(\boldsymbol{h}) + \boldsymbol{r} - f_r(\boldsymbol{t})\|_{L_1/L_2}$ declines to 0 for all triples $\langle h, r, t \rangle$ in the given knowledge graph. According to logical inference, $\|f_r(\boldsymbol{h}) + \boldsymbol{r} - f_r(\boldsymbol{t})\|_{L_1/L_2}$ should also decline to 0 for all triples $\langle h, r, t \rangle$ entailed by the knowledge graph. We find that, when the relation $r$ is symmetric, reflexive or transitive, the vector representation $\boldsymbol{r}$ declines to an all-0 vector in the ideal case, thus the vectors for multiple symmetric, reflexive or transitive relations are not distinguishable. Moreover, when the projection function $f_r(\cdot)$ is the identity function, for two relations $r_1$ and $r_2$ such that $r_1$ is subsumed by $r_2$, the vector representations $\boldsymbol{r_1}$ and $\boldsymbol{r_2}$ are also not distinguishable in the ideal case. To overcome these limitations, we propose to enhance existing methods that use $\|f_r(\boldsymbol{h}) + \boldsymbol{r} - f_r(\boldsymbol{t})\|_{L_1/L_2}$ as the loss function by logical pre-completion and bi-directional projection of entities. Logical pre-completion enlarges the given knowledge graph by all entailed triples according to the attached RBox, so that learnt vectors conform to the logical constraints enforced by the RBox. Bi-directional projection of entities revises the loss function to

---

[1] The head entity, relation and tail entity are respectively called the *subject*, *predicate* and *object* in OWL/RDF terminology. We use terminology in the field of knowledge graph embedding rather than OWL/RDF terminology throughout the paper.

$||f_r(\boldsymbol{h}) + \boldsymbol{r} - f'_r(\boldsymbol{t})||_{L_1/L_2}$ for $f_r(\cdot)$ and $f'_r(\cdot)$ two different functions, so that the above limitations are resolved.

We conducted experiments on four benchmark datasets that are widely used in the field of knowledge graph embedding. Experimental results show that the enhancements for TransE, TransH, TransR and TransD improve the predictive performance in two tasks namely link prediction and triple classification.

## 2    Preliminaries

### 2.1    OWL 2 RBox and Knowledge Graph

OWL 2 [8] is the newest version of the W3C proposed Web Ontology Language. It is a formal language with rigorous syntactics and semantics, underpinned by decidable fragments of first-order logic (FOL). Due to a good balance between expressivity and computational complexity, OWL 2 has become popular in both academy and industry. An OWL 2 ontology usually has an RBox and a TBox so as to express logical information in real-life applications. The relations in OWL 2 are mostly object properties or data properties. The RBox consists of axioms declaring characteristics of relations and interactions among multiple relations, whereas the TBox consists of axioms declaring interactions among multiple classes possibly with the help of relations.

A knowledge graph can often be expressed as a set of triples of the form $\langle h, r, t \rangle$, where $h$ is called the *head entity* (simply *head*), $r$ is called the *relation*, $t$ is called the *tail entity* (simply *tail*). Many publicly available knowledge graphs such as WordNet [17] and DBPedia [1] come with OWL 2 ontologies. In such a scenario, the knowledge graph and the attached OWL 2 ontology constitute a first-order logic program, where a triple $\langle h, r, t \rangle$ is translated to a unary fact $t(h)$ if $r$ is rdf:type, or to a binary fact $r(h, t)$ otherwise. It should be noted that only triples on rdf:type involve classes; i.e., for a triple $\langle h, r, t \rangle$, $t$ is a class if and only if $r$ is rdf:type. Since the number of triples on rdf:type is often much smaller than that of other triples in a knowledge graph, in this work we do not consider triples on rdf:type. Accordingly, since the triples that we consider do not involve classes, we do not consider TBox axioms either. We leave the consideration of classes in knowledge graph embedding with OWL 2 as our future work.

A relation in an OWL 2 RBox can be declared as a primary relation or an inverse relation. An inverse relation is of the form ObjectInverseOf($r$) for $r$ a primary relation. We assume that knowledge graphs use only primary relations. The syntax and semantics for RBox axioms are shown in Table 1. The semantics is given by translating axioms to rules in FOL under the *unique name assumption* (*UNA*). UNA ensures that two entities having different names are semantically different. It is often adopted in real-life applications. A rule $R$ in FOL is of the form $\mathbf{B} \rightarrow h$, where $\mathbf{B}$ is called the *body* of $R$, written body($R$), which is a conjunction of atoms or empty, and $h$ is called the *head* of $R$, written head($R$)), which is an atom or empty. The natural meaning of RBox axioms can easily be seen from their syntactic representation. To name only a few, SymmetricObjectProperty($r$) means that $r$ is symmetric, whereas

**Table 1.** The syntax and semantics for RBox axioms in OWL 2

| Functional-style syntax | Semantics via translation to FOL |
|---|---|
| SymmetricObjectProperty$(r)$ | $\mathsf{ar}(x, r, y) \rightarrow \mathsf{ar}(y, r, x)$ |
| AsymmetricObjectProperty$(r)$ | $\mathsf{ar}(x, r, y) \wedge \mathsf{ar}(y, r, x) \rightarrow$ |
| ReflexiveObjectProperty$(r)$ | $\rightarrow \mathsf{ar}(x, r, x)$ |
| IrreflexiveObjectProperty$(r)$ | $\mathsf{ar}(x, r, x) \rightarrow$ |
| TransitiveObjectProperty$(r)$ | $\mathsf{ar}(x, r, y) \wedge \mathsf{ar}(y, r, z) \rightarrow \mathsf{ar}(x, r, z)$ |
| FunctionalObjectProperty$(r)$ | $\mathsf{ar}(x, r, y_1) \wedge \mathsf{ar}(x, r, y_2) \wedge y_1 \neq y_2 \rightarrow$ |
| InverseFunctionalObjectProperty$(r)$ | $\mathsf{ar}(x_1, r, y) \wedge \mathsf{ar}(x_2, r, y) \wedge x_1 \neq x_2 \rightarrow$ |
| DisjointObjectProperties$(r, s)$ | $\mathsf{ar}(x, r, y) \wedge \mathsf{ar}(x, s, y) \rightarrow$ |
| SubObjectPropertyOf$(r, s)$ | $\mathsf{ar}(x, r, y) \rightarrow \mathsf{ar}(x, s, y)$ |
| SubObjectPropertyOf(ObjectPropertyChain$(p, q), r)$ | $\mathsf{ar}(x, p, y) \wedge \mathsf{ar}(y, q, z) \rightarrow \mathsf{ar}(x, r, z)$ |

Note: $x$, $y$ and $z$ are universally quantified variables; $\mathsf{ar}(x, r, y)$ denotes $s(y, x)$ if $r$ is ObjectInverseOf$(s)$, or $r(x, y)$ otherwise.

**Table 2.** The projection functions in different translation-based methods

| Method | Projection function |
|---|---|
| TransE | $f_r(\boldsymbol{x}) = \boldsymbol{x}$ for $\boldsymbol{x}, \boldsymbol{r} \in \mathbb{R}^n$ |
| TransH | $f_r(\boldsymbol{x}) = \boldsymbol{x} - \boldsymbol{w}_r^{\mathrm{T}} \boldsymbol{x} \boldsymbol{w}_r$ for $\boldsymbol{x}, \boldsymbol{r}, \boldsymbol{w}_r \in \mathbb{R}^n$ |
| TransR | $f_r(\boldsymbol{x}) = \boldsymbol{M}_r \boldsymbol{x}$ for $\boldsymbol{x} \in \mathbb{R}^n$, $\boldsymbol{r} \in \mathbb{R}^m$ and $\boldsymbol{M} \in \mathbb{R}^{m \times n}$ |
| TransD | $f_r(\boldsymbol{x}) = (\boldsymbol{p}_r \boldsymbol{p}_x^{\mathrm{T}} + \boldsymbol{D}) \boldsymbol{x}$ for $\boldsymbol{x}, \boldsymbol{p}_x \in \mathbb{R}^n$, $\boldsymbol{r}, \boldsymbol{p}_r \in \mathbb{R}^m$, and $\boldsymbol{D} \in \mathbb{R}^{m \times n}$ being a diagonal matrix whose principal diagonal is $\mathbf{1}^{\min(m,n)}$ |
| TranSparse (share or separate version) | $f_r(\boldsymbol{x}) = \boldsymbol{M}_{r,\theta_r} \boldsymbol{x}$ (share) or $f_r(\boldsymbol{h}) = \boldsymbol{M}_{r,\theta_{r1}} \boldsymbol{h}$ and $f_r(\boldsymbol{t}) = \boldsymbol{M}_{r,\theta_{r2}} \boldsymbol{t}$ (separate) for $\boldsymbol{x}, \boldsymbol{h}, \boldsymbol{t} \in \mathbb{R}^n$, $\boldsymbol{r} \in \mathbb{R}^m$, $\boldsymbol{M}_{r,\theta} \in \mathbb{R}^{m \times n}$ a sparse matrix having $\lfloor \theta mn \rfloor$ non-zero elements, where $\theta_r = 1 - (1 - \theta_{\min}) \frac{\#\{(h',t') \mid \langle h', r, t' \rangle \in \mathcal{G}\}}{\max_{r'} \#\{(h',t') \mid \langle h', r', t' \rangle \in \mathcal{G}\}}$, $\theta_{r1} = 1 - (1 - \theta_{\min}) \frac{\#\{h' \mid \langle h', r, t' \rangle \in \mathcal{G}\}}{\max_{r'} \#\{h' \mid \langle h', r', t' \rangle \in \mathcal{G}\}}$, $\theta_{r2} = 1 - (1 - \theta_{\min}) \frac{\#\{t' \mid \langle h', r, t' \rangle \in \mathcal{G}\}}{\max_{r'} \#\{t' \mid \langle h', r', t' \rangle \in \mathcal{G}\}}$, $\mathcal{G}$ is the given knowledge graph, $\theta_{\min}$ is a hyper-parameter, and $\#S$ denotes the cardinality of the set $S$ |

SubObjectPropertyOf$(r, s)$ means that $r$ is subsumed by $s$. The union of a knowledge graph $\mathcal{G}$ and an OWL 2 RBox $\mathcal{R}$ can be translated to a function-free Horn logic program, denoted by $\pi(\mathcal{G} \cup \mathcal{R})$, which inherits the standard FOL semantics. We say a triple $\langle h, r, t \rangle$ is *entailed by* by a knowledge graph $\mathcal{G}$ w.r.t. an OWL 2 RBox $\mathcal{R}$ if all models of $\pi(\mathcal{G} \cup \mathcal{R})$ are also models of $r(h, t)$.

## 2.2 Translation-Based Methods in Knowledge Graph Embedding

There have emerged a number of methods for embedding a knowledge graph into a continuous vector space, called *knowledge graph embedding*, to tackle prediction problems such as link prediction [5,6], triple classification [19,22] and relation extraction [22,23]. Among these methods, the translation-based methods, which represent entities as points and relations as translation from head entities to tail entities in the vector space, become very popular recently as they work well for sparse knowledge graphs with high predictive performance. TransE [5] is the pioneer translation-based method which defines translation directly on entity vectors. Most subsequent translation-based methods define translation on projection of entity vectors. TransH [22] defines projection on relation specific hyperplanes. TransR [16] defines projection by relation specific matrices. TransD [12] defines projection by relation-entity specific matrices. TranSparse [13] defines projection by sparse matrices that depend on relation specific triple distributions.

All the above methods finally learn entity vectors and relation vectors by minimizing a global margin-based loss function with the help of a loss function for triples. The loss function for a triple $\langle h, r, t \rangle$, written $\mathsf{loss}_r(h, t)$, can be uniformly written as $||f_r(\boldsymbol{h}) + \boldsymbol{r} - f_r(\boldsymbol{t})||_{L_1/L_2}$, where $\boldsymbol{x}$ denotes the vector representation $(x_1, \ldots, x_n)$ for $x$, $||\boldsymbol{x}||_{L_1}$ is the L1-norm of $\boldsymbol{x}$ defined as $\sum_{i=1}^{n} |x_i|$, $||\boldsymbol{x}||_{L_2}$ is the L2-norm of $\boldsymbol{x}$ defined as $\sqrt{\sum_{i=1}^{n} x_i^2}$, and $f_r(\cdot)$ is a projection function defined using the formulae given by Table 2. The global margin-based loss function to be minimized is defined on the set $\Delta$ of triples in the given knowledge graph and a randomly generated set $\overline{\Delta}$ of triples that are not in the given knowledge graph, where $\overline{\Delta}$ is constructed from $\Delta$ by randomly corrupting triples in either heads or tails; i.e., all elements of $\overline{\Delta}$ are randomly selected from

$$\{\langle h', r, t \rangle \notin \Delta \mid h' \neq h, \langle h, r, t \rangle \in \Delta\} \cup \{\langle h, r, t' \rangle \notin \Delta \mid t' \neq t, \langle h, r, t \rangle \in \Delta\}.$$

By introducing the margin $\gamma$ as a hyper-parameter, the global margin-based loss function to be minimized can be uniformly written as

$$\sum_{\langle h,r,t \rangle \in \Delta} \sum_{\langle h',r,t' \rangle \in \overline{\Delta}} \max(0, \gamma + \mathsf{loss}_r(h, t) - \mathsf{loss}_r(h', t')).$$

This function is commonly minimized, with all entity vectors, relation vectors and parameters on the projection function learnt, by stochastic gradient descent (SGD), where hyper-parameters are determined by a validation set.

Some other translation-based methods such as TransM [7] and TransA [14] are slightly extended from TransE. TransM defines the loss function for a triple $\langle h, r, t \rangle$ as $c_r ||\boldsymbol{h} + \boldsymbol{r} - \boldsymbol{t}||_{L_1/L_2}$, where $c_r$ is the $r$-specific weight for loss functions. TransA directly computes the margin $\gamma$ according to some statistics on the training set instead of determining $\gamma$ by a validation set.

## 3    Enhancements for Translation-Based Methods

When a knowledge graph comes with an OWL 2 RBox, the embedding of triples in it should conform to the logical constraints enforced by the RBox. In other words, what should be satisfied by all triples *in* the knowledge graph should also be satisfied by all triples *entailed by* the knowledge graph w.r.t. the attached

**Algorithm.** ComputeEntailedTripleSet($\mathcal{G}$, $\mathcal{R}$)
**Input:** A knowledge graph $\mathcal{G}$ and an attached OWL 2 RBox $\mathcal{R}$.
**Output:** The set of triples entailed by $\mathcal{G}$ w.r.t. $\mathcal{R}$.
1: Let $\Pi$ be obtained from $\pi(\mathcal{G} \cup \mathcal{R})$ by modifying rules to safe rules;
2: Let $\Delta' = \emptyset$ and $\Delta = \{h \mid \rightarrow h \in \Pi\}$;
3: **while** $\Delta \neq \Delta'$ **do**
4:     Let $\Delta' = \Delta$;
5:     **for** each $R \in \Pi$ and each ground substitution $\sigma$ for $\mathsf{body}(R)$ such that $\mathsf{head}(R)$ is not empty and $\mathsf{body}(R)\sigma \subseteq \Delta'$ **do**
6:         Let $\Delta = \Delta \cup \{\mathsf{head}(R)\sigma\}$;
7: **return** $\{\langle h, r, t \rangle \mid r(h, t) \in \Delta\}$;

**Fig. 1.** The algorithm for computing all entailed triples

RBox. As summarized in Subsect. 2.2, a translation-based method requires that $\gamma + \mathsf{loss}_r(h, t) - \mathsf{loss}_r(h', t')$ be as small as possible for a triple $\langle h, r, t \rangle$ in the given knowledge graph and a triple $\langle h', r, t' \rangle$ not in the given knowledge graph. Hence $\gamma + \mathsf{loss}_r(h, t) - \mathsf{loss}_r(h', t')$ should also be as small as possible for a triple $\langle h, r, t \rangle$ entailed by the given knowledge graph and a triple $\langle h', r, t' \rangle$ not entailed by the given knowledge graph w.r.t. the attached RBox.

Let $\mathcal{G}$ denote the given knowledge base, $\mathcal{R}$ the attached RBox and $\Delta^+$ the set of triples entailed by $\mathcal{G}$ w.r.t. $\mathcal{R}$. To enhance translation-based methods with RBox axioms, we only need to modify the global margin-based loss function to

$$\sum_{\langle h,r,t \rangle \in \Delta^+} \sum_{\langle h',r,t' \rangle \in \overline{\Delta^+}} \max(0, \gamma + \mathsf{loss}_r(h, t) - \mathsf{loss}_r(h', t')) \tag{1}$$

without modifying the loss function for triples, where $\overline{\Delta^+}$ is a set of triples not entailed by $\mathcal{G}$ w.r.t. $\mathcal{R}$, constructed from $\Delta^+$ by randomly corrupting triples in either heads or tails; i.e., all its elements are randomly selected from

$$\{\langle h', r, t \rangle \notin \Delta^+ \mid h' \neq h, \langle h, r, t \rangle \in \Delta^+\} \cup \{\langle h, r, t' \rangle \notin \Delta^+ \mid t' \neq t, \langle h, r, t \rangle \in \Delta^+\}.$$

To accomplish this enhancement, we give an algorithm in Fig. 1 for computing $\Delta^+$, the set of triples entailed by $\mathcal{G}$ w.r.t. $\mathcal{R}$. Simply speaking, the algorithm modifies all rules in $\pi(\mathcal{G} \cup \mathcal{R})$ to *safe rules*, which are rules such that all variables appearing in the head also appear in the body, then iteratively computes newly entailed facts according to these safe rules and previously entailed facts until a fix-point is reached, and finally returns the set of triples translated from entailed binary facts. A safe rule is computed from the original rule by adding the atom $\mathsf{dom}(x)$ to the body for every variable $x$ in the head but not in the body, and by adding the fact $\mathsf{dom}(e)$ to $\pi(\mathcal{G} \cup \mathcal{R})$ for every entity $e$ in $\mathcal{G}$, where $\mathsf{dom}$ is a new globally unique predicate symbol.

Since this enhancement computes the set of entailed triples once before applying a translation-based method, we call it *logical pre-completion*. The following theorem shows that ComputeEntailedTripleSet($\mathcal{G}$, $\mathcal{R}$) is correct.

**Theorem 1.** *Given a knowledge graph $\mathcal{G}$ and an OWL 2 RBox $\mathcal{R}$, Compute-EntailedTripleSet($\mathcal{G}$, $\mathcal{R}$) returns the set of triples entailed by $\mathcal{G}$ w.r.t. $\mathcal{R}$.*

*Proof (sketch).* Let $\Pi$ be obtained from $\pi(\mathcal{G} \cup \mathcal{R})$ by modifying rules to safe rules. Since $\pi(\mathcal{G} \cup \mathcal{R})$ is a function-free Horn logic program, $\Pi$ has a unique minimal model $M_{\min}$ while ComputeEntailedTripleSet($\mathcal{G}$, $\mathcal{R}$) returns $\{\langle h, r, t \rangle \mid r(h, t) \in M_{\min}\}$. Thus, a triple $\langle h, r, t \rangle$ is entailed by $\mathcal{G}$ w.r.t. $\mathcal{R}$ if and only if $r(h, t) \in M$ for all models $M$ of $\Pi$ if and only if $r(h, t) \in M_{\min}$. $\qquad\square$

Although logical pre-completion can guarantee knowledge graph embedding to meet the logical constraints enforced by the attached RBox, it also has some side effects. We find that this enhancement leads to undistinguishable learnt vectors in the ideal case where the loss value $\mathsf{loss}_r(h, t)$ declines to 0 for all entailed triples $\langle h, r, t \rangle$. The ideal case amounts to a sufficient condition under which the revised global margin-based loss function given by Formula (1) is minimized to 0.

The first side effect is that when multiple symmetric, reflexive, or transitive relations exist, the learnt vectors of these relations must decline to all-0 vectors and cannot be distinguished from each other in the ideal case, as shown in the following theorem.

**Theorem 2.** *For a knowledge graph $\mathcal{G}$, an OWL 2 RBox $\mathcal{R}$ and a relation $r$ appearing in $\mathcal{G}$ such that (1) SymmetricObjectProperty($r$) $\in \mathcal{R}$ and there exists $\langle h, r, t \rangle$ entailed by $\mathcal{G}$ w.r.t. $\mathcal{R}$, or (2) ReflexiveObjectProperty($r$) $\in \mathcal{R}$, or (3) TransitiveObjectProperty($r$) $\in \mathcal{R}$ and there exist $\langle e_1, r, e_2 \rangle$ and $\langle e_2, r, e_3 \rangle$ entailed by $\mathcal{G}$ w.r.t. $\mathcal{R}$, if $\mathsf{loss}_r(h, t) = \|\boldsymbol{h} + \boldsymbol{r} - \boldsymbol{t}\|_{L_1/L_2} = 0$ for all triples $\langle h, r, t \rangle$ entailed by $\mathcal{G}$ w.r.t. $\mathcal{R}$, then $\boldsymbol{r} = \boldsymbol{0}^m$ where $\boldsymbol{r}$ is the vector representation for $r$ and $m$ is the dimension of relation vectors.*

*Proof.* (1) Consider the case where SymmetricObjectProperty($r$) $\in \mathcal{R}$. For every triple $\langle h, r, t \rangle$ entailed by $\mathcal{G}$ w.r.t. $\mathcal{R}$, $\langle t, r, h \rangle$ is also entailed by $\mathcal{G}$ w.r.t. $\mathcal{R}$. Thus $f_r(\boldsymbol{h}) + \boldsymbol{r} - f_r(\boldsymbol{t}) = f_r(\boldsymbol{t}) + \boldsymbol{r} - f_r(\boldsymbol{h}) = \boldsymbol{0}^m$. It follows that $\boldsymbol{r} = \boldsymbol{0}^m$. (2) Consider the case where ReflexiveObjectProperty($r$) $\in \mathcal{R}$. The triple $\langle e, r, e \rangle$ must be entailed by $\mathcal{G}$ w.r.t. $\mathcal{R}$ for every entity $e$ in $\mathcal{G}$. Thus $f_r(\boldsymbol{e}) + \boldsymbol{r} - f_r(\boldsymbol{e}) = \boldsymbol{0}^m$, i.e., $\boldsymbol{r} = \boldsymbol{0}^m$. (3) Consider the case where TransitiveObjectProperty($r$) $\in \mathcal{R}$. For any entities $e_1$, $e_2$ and $e_3$ such that $\langle e_1, r, e_2 \rangle$ and $\langle e_2, r, e_3 \rangle$ are entailed by $\mathcal{G}$ w.r.t. $\mathcal{R}$, $\langle e_1, r, e_3 \rangle$ is also entailed by $\mathcal{G}$ w.r.t. $\mathcal{R}$. Hence $f_r(\boldsymbol{e_1}) + \boldsymbol{r} - f_r(\boldsymbol{e_2}) = f_r(\boldsymbol{e_2}) + \boldsymbol{r} - f_r(\boldsymbol{e_3}) = f_r(\boldsymbol{e_1}) + \boldsymbol{r} - f_r(\boldsymbol{e_3}) = \boldsymbol{0}^m$. It follows that $f_r(\boldsymbol{e_1}) + 2\boldsymbol{r} - f_r(\boldsymbol{e_3}) = f_r(\boldsymbol{e_1}) + \boldsymbol{r} - f_r(\boldsymbol{e_3}) = \boldsymbol{0}^m$ and thus $\boldsymbol{r} = \boldsymbol{0}^m$. $\qquad\square$

Since TransE [5], TransH [22], TransR [16], TransD [12], TranSparse(share) [13], TransM [7] and TransA [14] use projection function of the form $f_r(\cdot)$, these methods have the first side effect when enhanced by logical pre-completion.

The second side effect is that when a relation is subsumed by another relation, the vector representations for these two relations learnt by a translation-based method that uses the identity function as the projection function are not distinguishable in the ideal case, as shown in the following theorem.

**Theorem 3.** *For a knowledge graph $\mathcal{G}$, an OWL 2 RBox $\mathcal{R}$ and two relations $r_1$ and $r_2$ such that* SubObjectPropertyOf$(r_1, r_2) \in \mathcal{R}$ *and there exists $\langle h, r_1, t \rangle$ entailed by $\mathcal{G}$ w.r.t. $\mathcal{R}$, if* $\mathsf{loss}_r(h, t) = \|\boldsymbol{h} + \boldsymbol{r} - \boldsymbol{t}\|_{L_1/L_2} = 0$ *for all triples $\langle h, r, t \rangle$ entailed by $\mathcal{G}$ w.r.t. $\mathcal{R}$, then $\boldsymbol{r_1} = \boldsymbol{r_2}$ where $\boldsymbol{r_1}$ and $\boldsymbol{r_2}$ are respectively the vector representations for $r_1$ and $r_2$.*

*Proof.* For a triple $\langle h, r_1, t \rangle$ entailed by $\mathcal{G}$ w.r.t. $\mathcal{R}$, $\langle h, r_2, t \rangle$ is also entailed by $\mathcal{G}$ w.r.t. $\mathcal{R}$. Therefore $\boldsymbol{h} + \boldsymbol{r_1} - \boldsymbol{t} = \boldsymbol{h} + \boldsymbol{r_2} - \boldsymbol{t}$ and thus $\boldsymbol{r_1} = \boldsymbol{r_2}$.          □

Since TransE [5], TransM [7] and TransA [14] use the identity function as the projection function, these methods have the second side effect when enhanced by logical pre-completion.

To resolve the above two side effects, we further enhance existing translation-based methods by revising the projection function. We only consider enhancing translation-based methods whose projection function is of the form $f_r(\cdot)$ and is not the identity function. These methods include TransH [22], TransR [16], TransD [12] and TranSparse(share) [13]. It can be seen from the proof of Theorem 2 that, when the projection functions on head entities and on tail entities are different, the first side effect does not exist. Hence we modify the loss function for triples $\langle h, r, t \rangle$ to $\mathsf{loss}_r(h, t) = \|f_r(\boldsymbol{h}) + \boldsymbol{r} - f_r'(\boldsymbol{t})\|_{L_1/L_2}$, where $f_r(\cdot)$ and $f_r'(\cdot)$ are two projection functions composed of different sets of parameters to be learnt. For example, to enhance TransH (see Table 2) we define $f_r(\boldsymbol{x}) = \boldsymbol{x} - \boldsymbol{w}_r^{\mathrm{T}} \boldsymbol{x} \boldsymbol{w}_r$ and $f_r'(\boldsymbol{x}) = \boldsymbol{x} - \boldsymbol{w'}_r^{\mathrm{T}} \boldsymbol{x} \boldsymbol{w'}_r$, where $\boldsymbol{w}_r$ and $\boldsymbol{w'}_r$ are two different sets of parameters. We call this enhancement *bi-directional projection of entities*.

## 4     Experiments

We empirically evaluate the two enhancements for four most popular translation-based methods, i.e. TransE [5], TransH [22], TransR [16] and TransD [12], in link prediction and triple classification. By TransX+ we denote the method enhanced from TransX by logical pre-completion, and BiTransX+ the method enhanced from TransX+ by bi-directional projection of entities, where $X \in \{E, H, R, D\}$. Since TransE+ employs the identity function as the projection function on entity vectors, it cannot apply the bi-directional projection enhancement, thus we only consider BiTransX+ for $X \in \{H, R, D\}$. To guarantee a fair comparison, we implemented all TransX, TransX+ and BiTransX+ methods with multi-threads in Java, using standard SGD with fixed mini-batch size 1, and evaluated them in the RapidMiner platform[2] to ensure all methods to be compared in the same environment. The implementations of TransE/H/R were modified from the C++ code available at the KB2E repository[3] [16], whereas the implementation of TransD was based on [12] but employs standard SGD instead. The evaluation results obtained by our reimplementations will be slightly different from those reported in the literature, because a different implementation language was used here and the mini-batch size in SGD was tuned in original implementations.

---

[2] https://rapidminer.com/.
[3] https://github.com/thunlp/KB2E/.

**Table 3.** Datasets and their attached RBoxes

| Dataset | #relation | #entity | #train | #valid | #test | #sym | #ref | #trans | #subs |
|---------|-----------|---------|--------|--------|-------|------|------|--------|-------|
| WN18 | 18 | 40,943 | 141,442 | 5,000 | 5,000 | 3 | 0 | 0 | 0 |
| FB15k | 1,345 | 14,951 | 483,142 | 50,000 | 59,071 | 95 | 33 | 1,112 | 566 |
| WN11 | 11 | 38,696 | 112,581 | 2,609 | 10,544 | 0 | 0 | 0 | 0 |
| FB13 | 13 | 75,043 | 316,232 | 5,908 | 23,733 | 0 | 0 | 8 | 0 |

### 4.1    Data Sets

We collected four datasets commonly used in evaluating translation-based methods, which were built from two knowledge graphs WordNet [17] and Freebase [2]. Two datasets come from WordNet. They are WN18 provided by [5] for link prediction and WN11 provided by [19] for triple classification. Another two come from Freebase. They are FB13 provided by [19] for triple classification and FB15k provided by [5] for link prediction and triple classification. We composed RBoxes for these datasets by generating SymmetricObjectProperty (**sym**metricity), ReflexiveObjectProperty (**ref**lexivity), TransitiveObjectProperty (**trans**itivity) and SubObjectPropertyOf (**subs**umption) axioms. In more details, a candidate axiom $\alpha$ is generated from a dataset $\mathcal{G}$ if $\text{head}(R)\,\sigma$ is included in $\mathcal{G}$ for all ground substitutions $\sigma$ for $R$ that make $\text{body}(R)\,\sigma$ included in $\mathcal{G}$, where $R$ is the FOL rule translated from $\alpha$ according to Table 1. All the candidate axioms are then manually filtered by their natural meanings to form an RBox. The statistics for all the datasets and their attached RBoxes are listed in Table 3.[4]

### 4.2    Link Prediction

Link prediction aims to predict the missing head $h$ or the missing tail $t$ for a triple $\langle h, r, t \rangle$. This task focuses more on ranking a set of candidate entities from the knowledge graph for each missing position rather than computing only one best entity. As set up in previous work [5,7,12–14,16,22], we conduct experiments on two datasets WN18 and FB15k.

**Evaluation Protocol.** We follow a similar protocol originally given by [5]. For each test triple $\langle h, r, t \rangle$, we replace the tail $t$ with an arbitrary entity $e$ in the knowledge graph and calculate the loss value $\text{loss}_r(h, e)$ on the corrupted triple $\langle h, r, e \rangle$. By ranking the loss values in ascending order, we can get the rank of the original triple. Similarly, we can get another rank for $\langle h, r, t \rangle$ by corrupting the head $h$. Since a corrupted triple is also correct if it is entailed by the knowledge graph w.r.t. the attached RBox, ranking it before the original triple should not be considered wrong. To eliminate this factor, we remove those corrupted triples that are entailed by the union of the training set, the validation set and the test set w.r.t. the attached RBox before getting the rank of every test triple. In

---

[4] The test datasets and their attached RBoxes as well as the RapidMiner platform with test processes are available at http://www.dataminingcenter.net/JIST17/.

this setting two metrics are reported: the mean rank averaged by all test triples (denoted as *Mean*) and the proportion of ranks not larger than 10 (denoted as *Hits@10*). A lower Mean is better while a higher Hits@10 is better.

**Implementation.** We tune the hyper-parameters in a similar way presented in [5, 12, 16, 22]. We select the learning rate $\lambda$ in standard SGD among $\{0.001, 0.005, 0.01\}$, the margin $\gamma$ among $\{0.25, 0.5, 1, 2\}$, the dimensions of entity embedding $k$ and relation embedding $d$ among $\{20, 50, 100\}$, and the dissimilarity measure as either L1-norm or L2-norm. The optimal hyper-parameters are determined by seeking the lowest Mean in the validation set. Following [22], we adopt two strategies for sampling negative triples in every method. By unif we denote the traditional way of replacing head or tail with equal probability, and by bern we denote the way of replacing head or tail with different probabilities proportional to the average frequency for entities in head or entities in tail. The most prevalent optimal configurations are: $\lambda = 0.005$, $\gamma = 1$, $k = d = 100$ and using L2-norm on WN18 under both unif and bern strategies; $\lambda = 0.001$, $\gamma = 2$, $k = d = 100$ and using L1-norm on FB15k under the unif strategy; $\lambda = 0.001$, $\gamma = 1$, $k = d = 100$ and using L1-norm on FB15k under the bern strategy. The other optimal configurations include: $\lambda = 0.01$, $\gamma = 0.5$, $k = d = 100$ with L2-norm for training TransR+ (bern), BiTransR+ (bern), TransD (bern), TransD+ (bern) and BiTransD+ (bern) on WN18; $\lambda = 0.001$, $\gamma = 1$, $k = d = 100$ with L2-norm for training TransD+ (bern) on FB15k. For training TransE/H, TransE+/H+ and BiTransH+ on both datasets, we traverse all the training triples for 2000 rounds. For training TransR/D (resp. TransR+/D+ and BiTransR+/D+) on both datasets, in order to speed up the convergence and avoid overfitting, we initialize the entity vectors and the relation vectors with the results of TransE (resp. TransE+) and traverse all the training triples for another 500 rounds.

**Results.** The experimental results for all methods in terms of Mean and Hits@10 are reported in Table 4, where the best performance in each metric on each dataset is highlighted in bold. It can be seen that every TransX+ method consistently outperforms its baseline TransX method except for two cases, namely TransR+ (bern) on WN18 and TransD+ (bern) on FB15k. It shows that the unif strategy for sampling negative triples is more stable than the bern strategy in learning vectors. The performance improvements gained by TransX+ can be explained by that the logical pre-completion enhancement feeds up TransX+ with more training triples than TransX and that in TransX+ none of the entailed triples is treated as negative triples. Furthermore, in most cases the BiTransX+ method achieves at least the same good performance as its former TransX+ method. It shows that the bi-directional projection enhancement is effective in further improving the performance in link prediction.

To further demonstrate the effectiveness of both enhancements in link prediction, we separate the relations in FB15k into four categories, namely **symmetric**, **reflexive**, **transitive** and other relations that are not symmetric, reflexive, or transitive. The experimental results on different relation categories of FB15k in terms of macro average Hits@10, calculated as the mean of Hits@10 for every relation in a specific category, are reported in Table 5, where the highest Hits@10

**Table 4.** The performance for link prediction (H@10 is Hits@10 in percentage)

| Method | TransX | | | | TransX+ | | | | BiTransX+ | | | |
|---|---|---|---|---|---|---|---|---|---|---|---|---|
| Dataset | WN18 | | FB15k | | WN18 | | FB15k | | WN18 | | FB15k | |
| Metric | Mean | H@10 | Mean | H@10 | Mean | H@10 | Mean | H@10 | Mean | H@10 | Mean | H@10 |
| X = E(unif) | 320 | 90.7 | 53 | 75.4 | 314 | 91.4 | 52 | 75.7 | – | – | – | – |
| X = E(bern) | 324 | 91.1 | 97 | 75.7 | 316 | 91.8 | 91 | 75.8 | – | – | – | – |
| X = H(unif) | 324 | 91.1 | 52 | 75.5 | 323 | 91.7 | **51** | 75.6 | 318 | 91.7 | **51** | 75.7 |
| X = H(bern) | 314 | 91.2 | 99 | 75.7 | 300 | 91.8 | 97 | 76.2 | 273 | 90.4 | 97 | 76.2 |
| X = R(unif) | 330 | 91.4 | 68 | 75.5 | 319 | 91.6 | 67 | 76.1 | 325 | 94.0 | 68 | 76.2 |
| X = R(bern) | **27** | 90.8 | 98 | 76.2 | 143 | 86.4 | 64 | 77.1 | 118 | 87.6 | 60 | **77.8** |
| X = D(unif) | 322 | 90.7 | 55 | 75.9 | 320 | 91.2 | 54 | 76.1 | 317 | 91.2 | 56 | 76.1 |
| X = D(bern) | 395 | 92.9 | 96 | 76.1 | 46 | **98.6** | 149 | 62.9 | 45 | **98.6** | 65 | 76.8 |

**Table 5.** The performance for link prediction on FB15k separated by different relation categories (measured by macro average Hits@10 in percentage)

| Method | TransX | | | | TransX+ | | | | BiTransX+ | | | |
|---|---|---|---|---|---|---|---|---|---|---|---|---|
| Relation Category | sym | ref | trans | other | sym | ref | trans | other | sym | ref | trans | other |
| X = E(unif) | 100 | 100 | 74.1 | 79.3 | 100 | 100 | 74.2 | 80.1 | – | – | – | – |
| X = E(bern) | 100 | 100 | 76.5 | 80.2 | 100 | 100 | 76.7 | 80.2 | – | – | – | – |
| X = H(unif) | 100 | 100 | 74.8 | 78.8 | 100 | 100 | 75.0 | 79.0 | 100 | 100 | 75.0 | 80.1 |
| X = H(bern) | 100 | 100 | 76.5 | 79.6 | 100 | 100 | 76.9 | 80.8 | 100 | 100 | 76.9 | 80.8 |
| X = R(unif) | 100 | 100 | 75.2 | 80.0 | 100 | 100 | 76.4 | 80.4 | 100 | 100 | 75.2 | 82.1 |
| X = R(bern) | 100 | 100 | 77.7 | 80.8 | 100 | 100 | 86.3 | 86.7 | 100 | 100 | **87.3** | **87.1** |
| X = D(unif) | 100 | 100 | 75.5 | 80.1 | 100 | 100 | 75.2 | 80.4 | 100 | 100 | 75.5 | 80.5 |
| X = D(bern) | 100 | 100 | 77.8 | 80.0 | 100 | 100 | 80.0 | 77.0 | 100 | 100 | 85.7 | 85.7 |

on each category is highlighted in bold. This table reveals some wierd cases where the macro average Hits@10 in average category is larger than the mean Hits@10 displayed in Table 4. These cases are possible since the mean Hits@10 in Table 4 is triple-wise calculated and is not a weighted average of the four relation-wise calculated macro average Hits@10s. It can be seen that the TransX+ method generally outperforms its baseline TransX method while the BiTransX+ method also generally outperforms its former TransX+ method in both the transitive category and the other category. For the symmetric category and the reflexive category, all methods achieve the highest (100%) Hits@10.

### 4.3   Triple Classification

Triple classification aims to judge whether a triple $\langle h, r, t \rangle$ is correct or not. It is a binary classification task. We conduct experiments on all the four datasets.

**Evaluation Protocol.** We follow a similar protocol originally given by [19]. The evaluation of triple classification requires negative triples that are not entailed by the dataset w.r.t. the attached RBox. Although WN11 and FB13, released by [19], already contain negative triples that were built by corrupting the corresponding observed (positive) triples, these negative triples may be entailed by the dataset w.r.t. the attached RBox since the procedure for generating negative triples [19] does not consider RBox axioms. To guarantee generated negative triples to be logically valid, we construct negative triples for all the four datasets instead of reusing existing ones provided by [19]. The generation procedure is almost the same as the way presented in [19] except that a generated negative triple is required to be not entailed by the union of the training set, the validation set and the test set w.r.t. the attached RBox. The decision rule for triple classification is simple: given a triple $\langle h, r, t \rangle$, if its loss value $\mathsf{loss}_r(h, e)$ is not larger than a relation specific threshold $\theta_r$, the triple will be classified as positive; otherwise, it will be classified as negative. The relation specific threshold $\theta_r$ is determined by maximizing the classification accuracy on the validation set.

**Implementation.** We tune the hyper-parameters for all datasets in a similar way presented in [5,12,16,22]. The search space of hyper-parameters is identical to that in link prediction. The optimal hyper-parameters are determined by seeking the highest classification accuracy in the validation set. Following [22], we also adopt two strategies unif and bern for sampling negative triples in every method. The optimal configurations for FB15k are the same as those in link prediction. For other datasets, under both strategies the most prevalent optimal configurations are: $\lambda = 0.01$, $\gamma = 1$, $k = d = 100$ and using L2-norm on WN18; $\lambda = 0.01$, $\gamma = 2$, $k = d = 20$ and using L2-norm on WN11; $\lambda = 0.01$, $\gamma = 2$, $k = d = 100$ and using L2-norm on FB13. The other optimal configurations include: $\lambda = 0.005$, $\gamma = 1$, $k = d = 100$ with L2-norm for training TransE+ (unif), TransR (unif), TransR+ (unif), TransD (unif) and TransD+ (unif) on WN18; $\lambda = 0.005$, $\gamma = 2$, $k = d = 20$ with L2-norm for training TransE (unif/bern), TransE+ (unif/bern), TransR (unif), TransR+ (unif) and BiTransR+ (unif/bern) on WN11; $\lambda = 0.005$, $\gamma = 2$, $k = d = 20$ with L1-norm for training BiTransD+ (unif/bern) on WN11; $\lambda = 0.005$, $\gamma = 2$, $k = d = 100$ with L2-norm for training TransE (bern), TransE+ (unif/bern), TransH+ (unif/bern), BiTransH+ (bern) and BiTransR+ (unif) on FB13; $\lambda = 0.001$, $\gamma = 1$, $k = d = 100$ with L2-norm for training TransR (unif), TransR+ (unif), TransD (unif), TransD+ (unif) and BiTransD+ (unif) on FB13; $\lambda = 0.001$, $\gamma = 2$, $k = d = 100$ with L1-norm for training TransD (bern), TransD+ (bern) and BiTransD+ (bern) on FB13. On every dataset, we traverse all the training triples for the same number of rounds as in link prediction.

**Results.** The experimental results are reported in Table 6, where the highest accuracy on each dataset is highlighted in bold. It can be seen that every TransX+ method still outperforms its baseline TransX method except for three cases including TransE+ (bern) on WN18, TransE+ (bern) on FB15k, and TransD+ (bern) on FB13. The same accuracy on WN11 is due to the emptiness of the RBox attached to WN11, which implies that the set of entailed triples is the same as the set of observed triples on WN11. Moreover, for 13 out of 24 cases the BiTransX+ method achieves at least the same high accuracy as its former TransX+ method. Also, for three datasets the highest accuracy is achieved by a

**Table 6.** The performance for triple classification (measured by classification accuracy in percentage)

| Method | TransX | | | | TransX+ | | | | BiTransX+ | | | |
|--------|--------|--------|------|------|--------|--------|------|------|--------|--------|------|------|
| Dataset | WN18 | FB15k | WN11 | FB13 | WN18 | FB15k | WN11 | FB13 | WN18 | FB15k | WN11 | FB13 |
| X = E(unif) | 97.5 | 86.6 | 81.6 | 74.2 | 97.6 | 87.2 | 81.6 | 74.5 | – | – | – | – |
| X = E(bern) | 97.6 | 85.2 | 80.6 | 74.8 | 97.5 | 85.2 | 80.6 | 77.0 | – | – | – | – |
| X = H(unif) | 97.5 | 86.7 | 82.7 | 74.0 | 97.6 | 87.5 | 82.7 | 74.4 | 97.6 | 87.3 | **83.1** | 74.4 |
| X = H(bern) | 97.4 | 85.3 | 80.9 | 75.6 | 97.6 | 85.7 | 80.9 | **78.1** | 86.3 | 85.3 | 82.2 | 72.4 |
| X = R(unif) | 97.6 | 85.0 | 82.0 | 63.4 | **97.7** | 85.9 | 82.0 | 63.5 | 97.5 | **87.8** | 82.2 | 64.5 |
| X = R(bern) | 96.0 | 85.3 | 81.3 | 73.7 | 97.1 | 85.9 | 81.3 | 74.2 | 96.5 | 87.1 | 81.0 | 56.2 |
| X = D(unif) | 97.6 | 86.5 | 82.5 | 66.1 | **97.7** | 86.9 | 82.5 | 66.5 | **97.7** | 87.0 | 76.9 | 66.9 |
| X = D(bern) | 97.0 | 86.6 | 81.5 | 65.0 | 97.3 | 70.0 | 81.5 | 59.6 | 97.5 | 87.0 | 74.7 | 59.0 |

**Table 7.** The performance for triple classification on FB15k separated by different relation categories (measured by macro average accuracy in percentage)

| Method | TransX | | | | TransX+ | | | | BiTransX+ | | | |
|--------|--------|------|-------|-------|--------|------|-------|-------|--------|------|-------|-------|
| Relation Category | sym | ref | trans | other | sym | ref | trans | other | sym | ref | trans | other |
| X = E(unif) | 89.3 | 89.3 | 77.4 | 70.2 | 91.3 | 91.3 | 75.8 | 70.9 | – | – | – | – |
| X = E(bern) | 85.8 | 85.8 | 77.5 | 68.2 | 87.5 | 87.5 | 74.8 | 68.4 | – | – | – | – |
| X = H(unif) | 95.8 | 95.8 | 76.5 | 70.3 | 86.7 | 86.7 | 75.3 | 71.3 | 87.5 | 87.5 | 75.6 | 70.8 |
| X = H(bern) | **100** | **100** | 75.1 | 68.3 | **100** | **100** | 77.4 | 68.9 | 86.4 | 86.4 | 74.9 | 68.7 |
| X = R(unif) | 64.8 | 64.8 | 79.9 | 67.7 | 62.8 | 62.8 | 73.4 | 69.6 | 86.4 | 86.4 | **80.2** | **71.5** |
| X = R(bern) | 66.4 | 66.4 | 74.1 | 68.5 | 67.6 | 67.6 | 75.5 | 69.5 | 87.5 | 87.5 | 77.1 | 70.5 |
| X = D(unif) | 65.7 | 65.7 | 73.2 | 70.1 | 69.1 | 69.1 | 74.4 | 70.2 | 87.2 | 87.2 | 75.1 | 70.4 |
| X = D(bern) | 65.9 | 65.9 | 73.0 | 70.3 | 75.0 | 75.0 | 61.7 | 58.8 | 86.5 | 86.5 | 77.2 | 70.4 |

certain BiTransX+ method. These results show that the bi-directional projection enhancement is also effective in improving the classification accuracy.

To further demonstrate the effectiveness of both enhancements in triple classification, we also separate the relations in FB15k into four categories, namely **sym**metric, **ref**lexive, **trans**itive and other relations. The experimental results on different relation categories of FB15k in terms of macro average accuracy, calculated as the mean of classification accuracy for every relation in a specific category, are reported in Table 7, where the best accuracy on each category is highlighted in bold. It can be seen that, the TransX+ method generally outperforms its baseline TransX method while the BiTransX+ method also generally outperforms its former TransX+ method, although the improvement is not so significant as that for Hits@10 in link prediction.

## 5   Related Work

The translation-based method is a hot research direction for knowledge graph embedding. TransE [5] assumes $h + r \approx t$ for all triples $\langle h, r, t \rangle$ in the knowledge graph. To better embed 1-N, N-1 and N-N relations, TransH [22], TransR [16], TransD [12] and TranSparse [13] explore different projection functions on entity vectors. TransM [7] extends TransE by introducing relation specific weights on loss functions for triples. TransA [14] extends TransE by an automatic way to determine the optimal margin. We refer the interested reader for more details on the above method in Subsect. 2.2. There are also some other translation-based methods. Unstructured method [3,4] is a naive version of TransE where $r$ is fixed to an all-0 vector. PTransE [15] extends TransE by using reliable relation paths as translations between entities, and by developing a path-constraint resource allocation algorithm to measure the reliability of relation paths.

Another research direction for knowledge graph embedding is the energy-based method, which assigns low energies to plausible triples in the knowledge graph and employs neural network for learning. For example, Structured Embedding (SE) [6] defines two relation specific matrices for head entity and tail entity, and learns entity vectors in a neural network architecture. Other energy-based methods include Semantic Matching Energy (SME) [3,4], Latent Factor Model (LFM) [11], Single Layer Model (SLM) and Neural Tensor Network (NTN) [19].

There also exist some methods for knowledge graph embedding that are not in the above directions. RESCAL [18] is a typical matrix-factorization based method, which treats a knowledge graph as a 3-model tensor and learns the latent representation, namely an entity as a vector and a relation as a matrix, by reconstructing the original graph. KG2E [10] is density-based method which represents entities and relations by Gaussian distributions. TransG [25] is a generative method which leverages a mixture of relation specific component vectors to embed a triple so as to capture multiple relation semantics.

All the above methods only exploit the structure information on the knowledge graph to embed triples. External information such as text information and logical information can also be used to improve the embedding results. One kind of text information namely entity names and their corresponding Wikipedia anchors is used in the probabilistic TransE (pTransE) method [21] to embed both triples and words. Another kind of text information namely entity descriptions is used in the DKRL method [26] to embed both triples and descriptions.

Prior to this work, the logical information considered in knowledge graph embedding is restricted to rules. In [20] four kinds of rules are encoded into an integer linear programming problem to filter triples predicted by an embedding model; i.e., all rules are used in postprocess and will not impact embedding results. In [9] two kinds of rules of the form $b \rightarrow h$ and $b_1 \land b2 \rightarrow h$ are incorporated into an enhanced TransE method called KALE. This method instantiates given rules to true ground rules and introduces a loss function for ground rules, so that the global margin-based loss function to be minimized is summed up not only from the margin-based loss difference between positive triples and sampled negative triples but also from the margin-based loss difference between positive ground rules and sampled negative ground rules.

This work is the first work for exploiting OWL 2 RBox as logical information to embed triples. Although RBox axioms can be translated to rules, we enhance translation-based methods by computing all entailed triples rather than instantiating rules as in the KALE method [9]. In KALE a rule could be instantiated to quadratic number of ground rules in the number of entities since it only requires at least one instantiated body atom to appear in the knowledge graph. The instantiation of rules raises a scalability problem for large rule sets. In addition, the study in [9] has not considered side effects of the enhancement, while we consider side effects and propose solutions to resolve them.

## 6    Conclusions and Future Work

In order to use logical information represented by an OWL 2 RBox in knowledge graph embedding, we have proposed two enhancements to existing translation-based methods in this paper. The first enhancement called *logical pre-completion* enlarges the given knowledge graph by all entailed triples according to an attached RBox, so that the embedding results conform to the logical constraints enforced by the RBox. The second enhancement called *bi-directional projection of entities* allows the projection function on head entities to be different from the projection function on tail entities, so that the learnt relation vectors are guaranteed to vary with the different relations. Experimental results demonstrate the effectiveness of these enhancements in improving the predictive performance.

For future work, we plan to further extend translation-based methods to embed entity type triples, which are triples of the form $\langle h, \mathsf{rdf:type}, t \rangle$. We will study three sources of information for this extension. The first source is an OWL 2 ontology attached to the given knowledge graph. The second source is the heuristic information studied in the field of ontology learning [24] for predicting entity types. The last source is external links to other knowledge graphs, such as external links from entities to Wordnet [17] synsets.

**Acknowledgements.** This work was partly supported by National Natural Science Foundation of China under grants 61375056 and 61573386, Natural Science Foundation of Guangdong Province under grant 2016A030313292, Guangdong Province Science and Technology Plan projects under grant 2016B030305007, Sun Yat-sen University Cultivation Project (16lgpy40) and the Undergraduate Innovative Experiment Project in Guangdong University of Foreign Studies (201711846022).

## References

1. Bizer, C., Lehmann, J., Kobilarov, G., Auer, S., Becker, C., Cyganiak, R., Hellmann, S.: DBpedia - a crystallization point for the web of data. J. Web Semant. **7**(3), 154–165 (2009)
2. Bollacker, K.D., Evans, C., Paritosh, P., Sturge, T., Taylor, J.: Freebase: a collaboratively created graph database for structuring human knowledge. In: Proceedings of SIGMOD, pp. 1247–1250 (2008)
3. Bordes, A., Glorot, X., Weston, J., Bengio, Y.: Joint learning of words and meaning representations for open-text semantic parsing. In: Proceedings of AISTATS, pp. 127–135 (2012)
4. Bordes, A., Glorot, X., Weston, J., Bengio, Y.: A semantic matching energy function for learning with multi-relational data - application to word-sense disambiguation. Mach. Learn. **94**(2), 233–259 (2014)

5. Bordes, A., Usunier, N., García-Durán, A., Weston, J., Yakhnenko, O.: Translating embeddings for modeling multi-relational data. In: Proceedings of NIPS, pp. 2787–2795 (2013)
6. Bordes, A., Weston, J., Collobert, R., Bengio, Y.: Learning structured embeddings of knowledge bases. In: Proceedings of AAAI (2011)
7. Fan, M., Zhou, Q., Chang, E., Zheng, T.F.: Transition-based knowledge graph embedding with relational mapping properties. In: Proceedings of PACLIC, pp. 328–337 (2014)
8. Grau, B.C., Horrocks, I., Motik, B., Parsia, B., Patel-Schneider, P.F., Sattler, U.: OWL 2: the next step for OWL. J. Web Semant. **6**(4), 309–322 (2008)
9. Guo, S., Wang, Q., Wang, L., Wang, B., Guo, L.: Jointly embedding knowledge graphs and logical rules. In: Proceedings of EMNLP, pp. 192–202 (2016)
10. He, S., Liu, K., Ji, G., Zhao, J.: Learning to represent knowledge graphs with Gaussian embedding. In: Proceedings of CIKM, pp. 623–632 (2015)
11. Jenatton, R., Roux, N.L., Bordes, A., Obozinski, G.: A Latent factor model for highly multi-relational data. In: Proceedings of NIPS, pp. 3176–3184 (2012)
12. Ji, G., He, S., Xu, L., Liu, K., Zhao, J.: Knowledge graph embedding via dynamic mapping matrix. In: Proceedings of ACL, pp. 687–696 (2015)
13. Ji, G., Liu, K., He, S., Zhao, J.: Knowledge graph completion with adaptive sparse transfer matrix. In: Proceedings of AAAI, pp. 985–991 (2016)
14. Jia, Y., Wang, Y., Lin, H., Jin, X., Cheng, X.: Locally adaptive translation for knowledge graph embedding. In: Proceedings of AAAI, pp. 992–998 (2016)
15. Lin, Y., Liu, Z., Luan, H., Sun, M., Rao, S., Liu, S.: Modeling relation paths for representation learning of knowledge bases. In: Proceedings of EMNLP, pp. 705–714 (2015)
16. Lin, Y., Liu, Z., Sun, M., Liu, Y., Zhu, X.: Learning entity and relation embeddings for knowledge graph completion. In: Proceedings of AAAI, pp. 2181–2187 (2015)
17. Miller, G.A., Beckwith, R., Fellbaum, C.D., Gross, D., Miller, K.: WordNet: an online lexical database. Int. J. Lexicogr. **3**(4), 235–244 (1990)
18. Nickel, M., Tresp, V., Kriegel, H.: Factorizing YAGO: scalable machine learning for linked data. In: Proceedings of WWW, pp. 271–280 (2012)
19. Socher, R., Chen, D., Manning, C.D., Ng, A.Y.: Reasoning with neural tensor networks for knowledge base completion. In: Proceedings of NIPS, pp. 926–934 (2013)
20. Wang, Q., Wang, B., Guo, L.: Knowledge base completion using embeddings and rules. In: Proceedings of IJCAI, pp. 1859–1866 (2015)
21. Wang, Z., Zhang, J., Feng, J., Chen, Z.: Knowledge graph and text jointly embedding. In: Proceedings of EMNLP, pp. 1591–1601 (2014)
22. Wang, Z., Zhang, J., Feng, J., Chen, Z.: Knowledge graph embedding by translating on hyperplanes. In: Proceedings of AAAI, pp. 1112–1119 (2014)
23. Weston, J., Bordes, A., Yakhnenko, O., Usunier, N.: Connecting language and knowledge bases with embedding models for relation extraction. In: Proceedings of EMNLP, pp. 1366–1371 (2013)
24. Wong, W., Liu, W., Bennamoun, M.: Ontology learning from text: a look back and into the future. ACM Comput. Surv. **44**(4), 20:1–20:36 (2012)
25. Xiao, H., Huang, M., Zhu, X.: TransG: A generative model for knowledge graph embedding. In: Proceedings of ACL, pp. 992–998 (2016)
26. Xie, R., Liu, Z., Jia, J., Luan, H., Sun, M.: Representation learning of knowledge graphs with entity descriptions. In: Proceedings of AAAI, pp. 2659–2665 (2016)

# Knowledge Graphs

# Cross-Lingual Taxonomy Alignment
# with Bilingual Knowledge Graph Embeddings

Tianxing Wu[1]([✉]), Du Zhang[1], Lei Zhang[2], and Guilin Qi[1]

[1] School of Computer Science and Engineering, Southeast University, Nanjing, China
{wutianxing,duzhang,gqi}@seu.edu.cn
[2] FIZ Karlsruhe – Leibniz Institute for Information Infrastructure,
Karlsruhe, Germany
lei.zhang@fiz-karlsruhe.de

**Abstract.** Recently, different knowledge graphs have become the essential components of many intelligent applications, but no research has explored the use of knowledge graphs to cross-lingual taxonomy alignment (CLTA), which is the task of mapping each category in the source taxonomy of one language onto a ranked list of most relevant categories in the target taxonomy of another language. In this paper, we study how to perform CLTA with a multilingual knowledge graph. Firstly, we identify the candidate matched categories in the target taxonomy for each category in the source taxonomy. Secondly, we find the relevant knowledge denoted as triples for each category in the given taxonomies. Then, we propose two different bilingual knowledge graph embedding models called BTransE and BTransR to encode triples of different languages into the same vector space. Finally, we perform CLTA based on the vector representations of the relevant RDF triples for each category. Preliminary experimental results show that our approach is comparable and complementary to the state-of-the-art method.

**Keywords:** Taxonomy · Knowledge graph · Embedding

## 1 Introduction

One of the most crucial kinds of knowledge on the web is taxonomy, which refers to a hierarchy of categories that entities are classified to [1,7,8]. Taxonomies are prevalent on the web, such as product catalogues and Web site directories. To facilitate cross-lingual knowledge sharing, we aim to deal with the problem of cross-lingual taxonomy alignment (CLTA), which is the task of mapping each category in the source taxonomy of one language onto a ranked list of most relevant categories in the target taxonomy of another language.

The key step of aligning cross-lingual taxonomies is to measure the relevance between one category in the source taxonomy and another one in the target taxonomy. Recently, vector similarities depending on Bilingual Biterm Topic Model [8] has been introduced in CLTA and achieved the state-of-the-art performance. This method mainly leverages the textual context (extracted from

© Springer International Publishing AG 2017
Z. Wang et al. (Eds.): JIST 2017, LNCS 10675, pp. 251–258, 2017.
https://doi.org/10.1007/978-3-319-70682-5_16

the Web) of each category to learn its vector representation. However, as different knowledge graphs have become the essential components of many intelligent applications, we try to study CLTA in another direction, that is, performing CLTA with the structured knowledge in a multilingual knowledge graph.

In this paper, we propose a new approach to CLTA with bilingual knowledge graph embeddings. Firstly, we identify the candidate matched categories in the target taxonomy for each category in the source taxonomy, using the cross-lingual string similarity proposed in [8]. Secondly, for each category, we utilize XKnowSearch! (an entity-based information retrieval system) [9] to get its most relevant entities in DBpedia [4], which is the core multilingual knowledge graph in current Linking Open Data[1]. Then, we find each category's relevant triples (i.e. structured knowledge), which contain the above obtained entities. Afterwards, we propose two bilingual knowledge graph embedding models called BTransE and BTransR to encode triples of different languages into the same vector space. Finally, we compute the vector representation of each category and perform CLTA using the vector similarities between categories of different languages.

The rest of this paper is organized as follows. Section 2 introduces our proposed approach in detail. Section 3 presents the preliminary experimental results. Section 4 concludes this work and describes our future work.

## 2   The Proposed Approach

In this section, we introduce our approach by four main steps: (1) candidates identification, (2) acquiring relevant triples, (3) learning vector representations of triples with bilingual knowledge graph embedding models and (4) exact matching.

### 2.1   Candidates Identification

The purpose of this step is to get all possible matched categories in the target taxonomy for each category in the source taxonomy. We apply the cross-lingual string similarity proposed in [8] to finish this task. The key idea is *"two categories of different languages may be matched if they share the same or synonymous words"*. BabelNet [6] is used to find bilingual synonyms for each word.

### 2.2   Acquiring Relevant Triples

To acquire relevant knowledge for each category, we have two procedures: (1) searching the most relevant entities in DBpedia with XKnowSearch! and (2) finding the triples containing the above acquired entities.

XKnowSearch![2] is an entity-based information retrieval system, which can return the most relevant DBpedia entities according to the query. Here, we take

---

[1] http://lod-cloud.net/.
[2] http://km.aifb.kit.edu/sites/XKnowSearch/.

each category as the query and submit it to XKnowSearch!. As a result, each category corresponds to several relevant DBpedia entities. For example, a category in Yahoo! Directory named *Emergency Services*, we acquire its most relevant DBpedia entities [*Emergency service, Emergency control centre, Emerging nation, Social services, etc.*] using XKnowSearch!. After obtaining the most relevant DBpedia entities for each category, we find the knowledge denoted as triples containing the above obtained entities.

## 2.3   Learning Vector Representations

In this section, we propose two bilingual knowledge graph embedding models BTransE and BTransR, which can represent the triples of different languages in the same vector space. More specifically, we first learn vector representations of the triples in the same language with monolingual knowledge graph embeddings, in which BTransE adopts TransE [2] and BTransR adopts TransR [5], respectively. In order to make the triples of different languages in the same vector space, we then learn cross-lingual translations for the entities and relations of different languages by two strategies, i.e. non-relation-specific strategy (used in BTransE) and relation-specific strategy (used in BTransR). Finally, we compute the vector representation for each triple leveraging the trained vector representations of entities and relations.

**Symbol Specification:** In a multilingual knowledge graph $MKG$, we use $L$ to denote the set of languages, and $L^2$ to denote the 2-combination of $L$ (i.e. the set of *unordered* language pairs). For a language $\ell \in L$, $G_\ell$ denotes the language-specific part of $\ell$ in $MKG$, and $E_\ell$ and $R_\ell$ respectively denote the corresponding vocabularies of entity expression and relation expression. $T = (h, r, t)$ denotes a triple in $G_\ell$ such that $h, t \in E_\ell$ and $r \in R_\ell$. Boldfaced $\boldsymbol{h}, \boldsymbol{r}, \boldsymbol{t}$ respectively represent the embedding vectors of head entity $h$, relation $r$ and tail entity $t$. For a language pair $(\ell_1, \ell_2) \in L^2$, $\delta(\ell_1, \ell_2)$ denotes the alignment set which contains pairs of aligned triples of the same meaning between $\ell_1$ and $\ell_2$. For example, for the languages Chinese and English, we have the aligned triples [["法国农业信贷银行", "工业", "金融服务"], ["Cariparma", "industry", "Financial services"]]$\in \delta(Chinese, English)$

### BTransE

**Monolingual Knowledge Graph Embedding.** In this part, we apply TransE, which is the most efficient translation-based monolingual knowledge graph embedding model. For each language $\ell \in L$, a dedicated $k$-dimensional embedding space $\mathbb{R}_L^k$ is assigned for vectors of $E_\ell$ and $R_\ell$, where $\mathbb{R}$ is the field of real numbers. The score function is

$$f_m(h, t) = \|\boldsymbol{h} + \boldsymbol{r} - \boldsymbol{t}\|_2^2 \tag{1}$$

**An Alignment between Vector Spaces.** Here, the alignment is to construct the translation between the vector space of $\ell_i$ and that of $\ell_j$. The objective of the alignment is to learn three $k \times k$ matrices $\boldsymbol{M}_{i,j}^{h}$, $\boldsymbol{M}_{i,j}^{r}$ and $\boldsymbol{M}_{i,j}^{t}$ as linear translations from source language $\ell_i$ to target language $\ell_j$ on the vectors of head entities, relations and tail entities, respectively. The score function is defined as,

$$f_a(T_i, T_j) = \left\| \boldsymbol{M}_{i,j}^{h} \boldsymbol{h}_i - \boldsymbol{h}_j \right\|_2^2 + \left\| \boldsymbol{M}_{i,j}^{r} \boldsymbol{r}_i - \boldsymbol{r}_j \right\|_2^2 + \left\| \boldsymbol{M}_{i,j}^{t} \boldsymbol{t}_i - \boldsymbol{t}_j \right\|_2^2 \qquad (2)$$

where $T_i = (h_i, r_i, t_i)$ and $T_j = (h_j, r_j, t_j)$ are the triples in language $i$ and $j$, respectively. In practice, we enforce constraints on the norms of the mapping matrices, i.e. $\forall h, t$, we have $\left\| \boldsymbol{M}_{i,j}^{h} \boldsymbol{h}_i \right\|_2 \leq 1$, $\left\| \boldsymbol{M}_{i,j}^{r} \boldsymbol{r}_i \right\|_2 \leq 1$ and $\left\| \boldsymbol{M}_{i,j}^{t} \boldsymbol{t}_i \right\|_2 \leq 1$.

We define the margin-based score function as the objective for training as follows:

$$L = \sum_{(T_i, T_j) \in \delta(\ell_i, \ell_j)} \sum_{(T_i', T_j') \in \delta'(\ell_i, \ell_j)} \max(0, f_a(T_i, T_j) + \gamma - f_a(T_i', T_j')) \qquad (3)$$

where $\delta(\ell_i, \ell_j)$ is the aligned triples in language $i$ and $j$, $\delta'(\ell_i, \ell_j)$ is the corrupted triples in language $i$ and $j$, and $\gamma$ is the margin.

To sample enough aligned triples of language $i$ and $j$ for training, we use *interlanguage_links* in DBpedia to get the synonymous set of each head entity, relation and tail entity. Thus, a triple in language $i$ can also be represented as $T_i = (\mathbf{h_i}, \mathbf{r_i}, \mathbf{t_i})$ where $\mathbf{h_i}, \mathbf{r_i}$ and $\mathbf{t_i}$ are the synonymous sets of head entity, relation and tail entity, respectively. Similarly, a triple in language $j$ can be represented as $T_j = (\mathbf{h_j}, \mathbf{r_j}, \mathbf{t_j})$. The criterion to determine whether two triples $T_i$ and $T_j$ are aligned is defined as follows:

*if* $\mathbf{h_i} \cap \mathbf{h_j}$ *and* $\mathbf{r_i} \cap \mathbf{r_j}$ *and* $\mathbf{t_i} \cap \mathbf{t_j} \neq \emptyset$,

*then* $T_i$ *and* $T_j$ *are aligned triples.*

After deriving the aligned triples of different languages in a multilingual knowledge graph, we design a method to corrupt the aligned triples following the convention of translation-based knowledge graph embedding models. In detail, given a pair of correctly aligned triples $(T_i, T_j)$, it is corrupted by (i) randomly replacing one of the six elements (i.e. $h_i$, $r_i$, $t_i$, $h_j$, $r_j$ and $t_j$) in the two triples with another corresponding element in some triple of the same language, or (ii) randomly substituting either $T_i$ or $T_j$ with another triple of the same language. We corrupt aligned triples by (i) or (ii) with the same probability.

**Computing Vector Representations for Triples.** The vector of a triple $T$ can be acquired by integrating the vectors of its head entity $h$, relation $r$ and tail entity $t$. If the dimensions of the vectors for its head entity, relation and tail entity are respectively $k$, the dimension of the vectors for $T$ is $3k$, $0 \sim k - 1$ dimension is the vector of the head entity, $k \sim 2k - 1$ dimension is the vector of the relation, $2k \sim 3k - 1$ dimension is the vector of the tail entity. We use symbol $<>$ to represent this integration procedure. The vector of the triples in language $i$ can be represented as

$$T_i = < \boldsymbol{M}_{i,j}^{h} \boldsymbol{h}_i, \boldsymbol{M}_{i,j}^{r} \boldsymbol{r}_i, \boldsymbol{M}_{i,j}^{t} \boldsymbol{t}_i > \qquad (4)$$

where boldfaced $T_i$ represents the vector of the triple. The vector of the triples in language $j$ can be represented as $T_j = < h_j, r_j, t_j >$.

## BTransR

BTransE assumes the embeddings of entities and relations are in the same vector space. Enlightened by TransR, we propose a relation-specific bilingual knowledge graph embedding model BTransR, which models entities and relations in different spaces and performs translations in relation spaces.

**Monolingual Knowledge Graph Embedding.** We use TransR in this part. For each language $\ell \in L$, we model entities and relations in different spaces, i.e. an entity space and multiple relation spaces, and perform translations in relation spaces. The score function is given as follows:

$$f_m(h, t) = \| M_r h + r - M_r t \|_2^2 \tag{5}$$

where $M_r$ is the projection matrix for each relation $r$.

**An Alignment between Vector Spaces.** In the second step of BTransR, we need to learn a $k \times k$ matrix $M_{i,j}^r$ for each relation $r$ in language $i$. Since the number of relations in language $i$ and that of language $j$ are not the same, the result of learning translations from language $i$ to language $j$ is different from learning translations from language $j$ to language $i$. Here, we use the language with fewer relations as the source language, as we may sample more aligned triples per relation for better training.

With the relation-specific translation matrix $M_{i,j}^r$ from source language $i$ to target language $j$, we define the projected vectors $h_{i,j}^r$ and $t_{i,j}^r$ for head entities and tail entities as

$$h_{i,j}^r = M_{i,j}^r M_i^r h_i, \; t_{i,j}^r = M_{i,j}^r M_i^r t_i \tag{6}$$

where $M_i^r$ (trained in TransR) is relation-specific projection matrix in language $i$. The score function is correspondingly defined as

$$f_a(T_i, T_j) = \left\| h_{i,j}^r - M_j^r h_j \right\|_2^2 + \left\| M_{i,j}^r r_i - r_j \right\|_2^2 + \left\| t_{i,j}^r - M_j^r t_j \right\|_2^2 \tag{7}$$

where $M_j^r$ is relation-specific projection matrix in language $j$. Margin-based score function is the same as Eq. 3.

**Computing Vector Representations for Triples.** Similar to BTransE, we denote the vector of a triple in language $i$ and the vector of a triple in language $j$ as $T_i = < M_{i,j}^r M_i^r h_i, M_{i,j}^r r_i, M_{i,j}^r M_i^r t_i >$ and $T_j = < M_j^r h_j, r_j, M_j^r t_j >$, respectively.

### 2.4   Exact Matching

The purpose of this step is to use the vector representations of triples to compute the relevance score between each category in the source taxonomy and its candidate matched categories in the target taxonomy.

Given a category $c_i$ in the source taxonomy of language $i$ and a category $c_j$ in target taxonomy of language $j$, $c_i$ and $c_j$ correspond to $N$ triples $\{T_{i,u}\}_{u=1}^{N}$ and $M$ triples $\{T_{j,v}\}_{v=1}^{M}$, respectively. We compute the vector representation of each category as

$$c_i = \frac{1}{N} \sum_{u=1}^{N} T_{i,u}, c_j = \frac{1}{M} \sum_{v=1}^{M} T_{j,v} \tag{8}$$

where boldfaced $c_i$ is the vector of category $c_i$, boldfaced $c_j$ represents the vector of category $c_j$ and boldfaced $T$ denotes the vector of triple $T$.

The final relevance score between each category in the source taxonomy and its candidate matched categories in the target taxonomy is computed as the cosine similarity between the vector of each category.

## 3    Preliminary Experiments

In this section, we conduct preliminary experiments on two real-world datasets to evaluate our approach for CLTA. The source code is publicly available[3].

### 3.1    Experiment Settings

**Datasets:** Each dataset[4] (also used in [8]) consists of a Chinese taxonomy, an English taxonomy and a set of labeled cross-lingual alignments from the Chinese taxonomy to the English one. The taxonomies in one dataset are two product catalogues respectively extracted from JD.com (one of the largest Chinese B2C online retailers with 7,741 Chinese categories) and eBay.com (7,782 English categories), and those in another dataset are two Web site directories: Chinese Dmoz.org (the largest Chinese Web site directory with 2,084 Chinese categories) and Yahoo! Directory (2,353 English categories).

**Evaluation Metrics:** Similar to these works [1,7,8], we used MRR (Mean Reciprocal Rank) and P@1 (precision for the top 1 ranking result) as the evaluation metrics because CLTA is seen as a ranking problem.

**Baseline:** We took the state-of-the-art method [8] on CLTA as the baseline. The difference between our approach and the baseline is the resource used for learning vector representations for categories. The baseline relies on the textual context of each category and utilizes Bilingual Biterm Topic Model (BiBTM) to learn topic vectors for categories, but our approach depends on the structured knowledge in a multilingual knowledge graph and uses a knowledge graph embedding model to learn embeddings of categories. Hence, the baseline is denoted as BiBTM, and our approach is denoted as BTransE or BTransR.

---

[3] https://github.com/Jason101616/BTransX.
[4] https://github.com/jxls080511/080424.

## 3.2   Preliminary Results

In BTransE (or BTransR), we ran 1,000 epochs of TransE (or TransR) and 1,000 epochs of learning an alignment between vector spaces. Table 1 (except the last row) gives the results of our approach and the baseline. It can be observed that:

- BTransE and BTransR achieve better p@1 on Web site directories, which means that we can get more correct results with BTransE or BTransR than BiBTM on this dataset. It shows that the structured knowledge in a knowledge graph can indeed benefit CLTA.
- On product catalogues, BTransE and BTransR are not better than (but are comparable to) BiBTM. This phenomenon can be explained as the lack of relevant triples for the categories in product catalogues seriously affects the performance of CLTA. This problem may be alleviated after performing knowledge graph completion.
- Another interesting phenomenon is that BTransE always performs better than BTransR. Since TransR is an important component in BTransR, the weakness of TransR also exists in BTransR. As mentioned in [3], complex relations (i.e. the relations link many entity pairs) may be overfitting and simple relations (i.e. the relations link few entity pairs) may be underfitting in TransR because each relation (no matter complex or simple) has the same number of parameters to learn.

## 3.3   Combining BTransE with BiBTM

Since BTransE is better than BTransR on both datasets and we also found that the overlap of correct alignment results obtained by our approach using BTransE and those obtained by the baseline using BiBTM is low, we felt it is necessary to validate whether BTransE and BiBTM are highly complementary for CLTA. We combined BTransE with BiBTM using the equation as follows:

$$RS_{final}(c_i, c_j) = \alpha \cdot RS_{BTransE}(c_i, c_j) + (1 - \alpha) \cdot RS_{BiBTM}(c_i, c_j) \qquad (9)$$

where $\alpha \in [0, 1]$, $c_i$ and $c_j$ are two categories from two taxonomies of different languages, $RS_{BTransE}(c_i, c_j)$ and $RS_{BiBTM}(c_i, c_j)$ are respectively the relevance scores generated by our approach using BTransE and the baseline using BiBTM.

The last row in Table 1 shows the performance of such a combination, which gets the best MRR and p@1 on each dataset. This reflects that BTranE is complementary to BiBTM for CLTA. Due to the space limit, we only gave the value of the optimal parameter (omitted the details of training), i.e. $\alpha = 0.1$ for product catalogues and $\alpha = 0.4$ for Web site directories.

**Table 1.** Preliminary results

| Approach | Product catalogues | | Web site directories | |
|---|---|---|---|---|
| | MRR | p@1 | MRR | p@1 |
| BiBTM | 0.597 | 0.440 | 0.719 | 0.520 |
| BTransE | 0.541 | 0.440 | 0.695 | 0.550 |
| BTransR | 0.532 | 0.420 | 0.647 | 0.530 |
| BTransE+BiBTM | **0.628** | **0.490** | **0.783** | **0.620** |

## 4    Conclusion and Future Work

In this paper, we proposed a new approach to CLTA, utilizing the structured knowledge in a multilingual knowledge graph to learn the vector representation for each category. We show that our knowledge-based approach using bilingual knowledge graph embedding models is comparable and complementary to the state-of-the-art method (i.e. the text-based method). In the future, to complement the relevant knowledge for each category, we will study how to generate vector representations of categories using multiple knowledge graphs.

**Acknowledgements.** This work is supported in part by the National Natural Science Foundation of China (Grant No. 61672153), the 863 Program (Grant No. 2015AA015406), the Fundamental Research Funds for the Central Universities and the Research Innovation Program for College Graduates of Jiangsu Province (Grant No. KYLX16_0295).

## References

1. Boldyrev, N., Spaniol, M., Weikum, G.: Across: a framework for multi-cultural inter-linking of web taxonomies. In: WebSci, pp. 127–136 (2016)
2. Bordes, A., Usunier, N., Garcia-Duran, A., Weston, J., Yakhnenko, O.: Translating embeddings for modeling multi-relational data. In: NIPS, pp. 2787–2795 (2013)
3. Ji, G., Liu, K., He, S., Zhao, J.: Knowledge graph completion with adaptive sparse transfer matrix. In: AAAI, pp. 985–991 (2016)
4. Lehmann, J., Isele, R., Jakob, M., Jentzsch, A., Kontokostas, D., Mendes, P.N., Hellmann, S., Morsey, M., Van Kleef, P., Auer, S., et al.: Dbpedia-a large-scale, multilingual knowledge base extracted from wikipedia. Semantic Web **6**(2), 167–195 (2015)
5. Lin, Y., Liu, Z., Sun, M., Liu, Y., Zhu, X.: Learning entity and relation embeddings for knowledge graph completion. In: AAAI, pp. 2181–2187 (2015)
6. Navigli, R., Ponzetto, S.P.: Babelnet: Building a very large multilingual semantic network. In: ACL, pp. 216–225 (2010)
7. Prytkova, N., Weikum, G., Spaniol, M.: Aligning multi-cultural knowledge taxonomies by combinatorial optimization. In: WWW, pp. 93–94 (2015)
8. Wu, T., Qi, G., Wang, H., Xu, K., Cui, X.: Cross-lingual taxonomy alignment with bilingual biterm topic model. In: AAAI, pp. 287–293 (2016)
9. Zhang, L., Färber, M., Rettinger, A.: Xknowsearch!: exploiting knowledge bases for entity-based cross-lingual information retrieval. In: CIKM, pp. 2425–2428 (2016)

# KG-Buddhism: The Chinese Knowledge Graph on Buddhism

Tianxing Wu[1]([✉]), Cong Gao[1], Guilin Qi[1], Lei Zhang[2], Chuanqi Dong[1], He Liu[1], and Du Zhang[1]

[1] School of Computer Science and Engineering, Southeast University, Nanjing, China
{wutianxing,cgao,gqi,cdong,heliu,duzhang}@seu.edu.cn
[2] FIZ Karlsruhe - Leibniz Institute for Information Infrastructure, Karlsruhe, Germany
lei.zhang@fiz-karlsruhe.de

**Abstract.** One of the most important elements in human society is religion, which provides moralities to help regulate human behaviours. However, the Web lacks a specialized knowledge graph on religion. To facilitate religious knowledge sharing, we aim to build KG-Buddhism, i.e. the Chinese knowledge graph on Buddhism, which is the most widely spread religion in China. In this paper, we contribute to the development of the first version of KG-Buddhism, containing the knowledge of Buddhist figures and temples, extracted from existing encyclopedic knowledge graphs and unstructured Web text. KG-Buddhism is linked to DBpedia and can be accessed by our online API.

**Keywords:** Knowledge graph · Buddhism · Knowledge on religion

## 1 Introduction

With the development of the Semantic Web, structured knowledge in different domains has been created and published as the linked data on the Web. However, current Linking Open Data (LOD) lacks a specialized knowledge graph on religion, which is one of the most important elements in human society and provides moralities to help regulate human behaviours. In China, Buddhism is the most widely spread religion, so we aim to build the first Chinese knowledge graph on Buddhism, called KG-Buddhism[1]. Knowledge on Buddhism does exist in some large-scale encyclopedic knowledge graphs, such as DBpedia [1], Yago [2] and Zhishi.me [3], but the problem of how to collect and integrate such knowledge as well as further complement them is worthy to study.

In this paper, we present the first version of KG-Buddhism, containing the knowledge of Buddhist figures and temples. Besides the dataset KG-Buddhism 1.0 itself, there are three main contributions of this work summarized as follows:

---

[1] http://www.kg-buddhism.com.

© Springer International Publishing AG 2017
Z. Wang et al. (Eds.): JIST 2017, LNCS 10675, pp. 259–267, 2017.
https://doi.org/10.1007/978-3-319-70682-5_17

- We propose a development architecture which can (1) collect and integrate the knowledge of Buddhist figures (or temples) from different encyclopedic knowledge graphs; (2) extract more knowledge of Buddhist figures and temples from unstructured Web text.
- We preliminarily evaluate KG-Buddhism 1.0 and the results show the high quality of our built knowledge graph.
- We provide an online API to help access the whole dataset.

The rest of the paper is organized as follows. Section 2 introduces the development method of KG-Buddhism 1.0. Section 3 reports the statistical data and preliminary evaluation results of KG-Buddhism 1.0. Section 4 describes the usage of our online API and we conclude in the last section.

## 2   Development Method

Figure 1 shows the architecture of developing KG-Buddhism 1.0. The first development module is *Knowledge Collection*, which is utilized to collect the entities representing Buddhist figures (or temples) and their corresponding knowledge (denoted as RDF triples) from *three encyclopedic knowledge graphs* in Zhishi.me. The knowledge graphs in Zhishi.me are actually the structured versions of the three largest Chinese online encyclopedias: Chinese Wikipedia[2], Baidu Baike[3] and Hudong Baike[4], respectively. Here, we denote these three knowledge graphs as *zhishi.me:zhwiki*, *zhishi.me:baidubaike* and *zhishi.me:hudongbaike*. Then, we integrate the collected RDF triples with the module *Knowledge Fusion*. Finally, to further complement the knowledge of Buddhist figures and temples, we extract more property values for the collected entities from unstructured Web text using the module *Knowledge Completion*.

### 2.1   Knowledge Collection

In this module, we first collect the entities representing Buddhist figures (or temples) by two strategies as follows:

- **Collecting entities from specific categories**. For the category "佛教 (Buddhism)" in a given knowledge graph (e.g. *zhishi.me:zhwiki*), we find some of its direct sub-categories used for classifying Buddhist figures (e.g. 佛教圣者 (Buddhas) and 佛教相关人物 (Buddhists)) or temples (e.g. 佛寺 (Buddhist temples)) Thus, we collect the entities classified to these sub-categories themselves or their descendant categories.
- **Collecting entities with specific naming rules**. We observe that there are some naming rules for the labels of the entities representing Buddhist figures (or temples). For example, the labels of the entities representing Buddhist

---

[2] https://zh.wikipedia.org/.

[3] https://baike.baidu.com/.

[4] http://www.baike.com/.

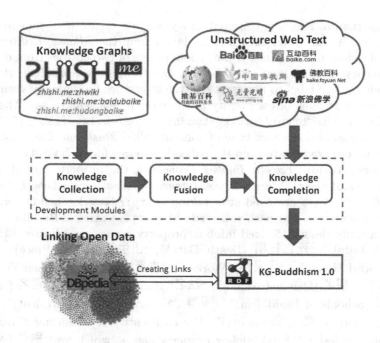

**Fig. 1.** Development architecture

deities often containing the suffix "菩萨 (Bodhisattva), 罗汉 (Arhat), etc. ". We manually summarize 60 naming rules for Buddhist figures and 10 naming rules for Buddhist temples. If the label of some entity satisfies one of these rules, this entity will be collected.

After applying the above strategies to each knowledge graph in Zhishi.me, we collect all the RDF triples taking the collected entities as subjects. These triples are divided into 10 types, i.e. *Abstracts, Categories, Aliases, Infobox Properties, External Links, Images, Internal Links, Labels, Redirects* and *Related Pages* (the details are given in [3]).

## 2.2  Knowledge Fusion

In this module, we first transform the URIs of the subjects and predicates in the collected RDF triples into new IDs in KG-Buddhism 1.0. The namespace "http://zhishi.me/" in their corresponding URIs is directly replaced with "http://www.kg-buddhism.com/". For objects, if an object is the entity collected by the module *Knowledge Collection*, we apply the same way to replace its namespace, otherwise we keep the original literal string or URI of the entity in *zhishi.me:zhwiki* or *zhishi.me:baidubaike* or *zhishi.me:hudongbaike*. Since all the collected RDF triples are from three knowledge graphs, we need to fuse the entities (may be subjects or objects), properties (i.e. predicates) and triples.

- **Entity fusion**. For each collected entity, we first create its synonym set by extracting its label and the synonym labels from the *Aliases*, *Redirects* and *sameAs Links* in Zhishi.me. If the synonym sets of two entities from different online encyclopedias have overlap, these two entities are equivalent. Here, we only choose one label to represent equivalent entities by majority voting (i.e. the most used label of each entity in three online encyclopedias). The ID of fused entities is "http://www.kg-buddhism.com/entity/[label]". In this part, we also create a new type of content called *Zhishi.me Links* using the predicate owl:sameAs to keep the collection source for each entity.

- **Property fusion**. Except *Infobox Properties*, the properties in *Abstracts*, *Categories*, *Aliases*, *External Links*, *Images*, *Internal Links*, *Labels*, *Redirects* and *Related Pages* are fixed after taking "http://www.kg-buddhism.com/" as the namespace. For each collected entity representing a Buddhist figure, we manually design 15 fixed infobox property names, which are "出生日期 (Birth Date)", "死亡日期 (Death Date)", "出生地 (Birth Place)", "籍贯 (Ancestral Home)", "代表作 (Works)", "主要成就 (Achievement)", "身份 (Title)", "俗名 (Original Name)", "法名 (Buddhist Name)", "别名 (Alias)", "宗派 (Schools of Buddhism)", "师傅 (Advisor)", "弟子 (Student)", "时代 (Era)", and "国籍 (Nationality)". For each entity representing a Buddhist temple, we design 9 fixed infobox property names, which are "别名 (Alias)", "创建人 (Founder)", "地址 (Address)", "供奉神明 (Deity)", "建筑 (Building)", "建筑风格 (Architectural Style)", "始建时间 (Founding Time)", "住持 (Buddhist Abbot)", and "宗派 (Schools of Buddhism)". Original property names collected from Zhishi.me are replaced with these designed fixed property names by synonym matching. For example, "出生时间 (Birth Time)" is replaced with "出生日期 (Birth Date)". Synonymous property names are gathered manually. Finally, each predicate in *infobox properties* is denoted as "http://www.kg-buddhism.com/property/[infoboxPropertyName]".

- **Triple fusion**. After the above entity fusion and property fusion, we need to fuse the RDF triples containing the same subject and predicate. For each single-valued property, we choose one object by majority voting (i.e. the most used object for the RDF triples containing the same subject and predicate). As shown in Fig. 2, "1927-08-19"@zh is the most used object for the given three triples, so we take it as the final property value. For each multivalued properties, we simply remove duplicated triples.

## 2.3   Knowledge Completion

*Infobox Properties* is a kind of important knowledge, which is used to summarize the characteristics of entities. However, the amount of this knowledge is limited in *zhishi.me:zhwiki*, *zhishi.me:baidubaike* and *zhishi.me:hudongbaike*, so we decide to extract the property value from unstructured Web text when given an entity and a property name. To acquire the corpus regarding Buddhist figures

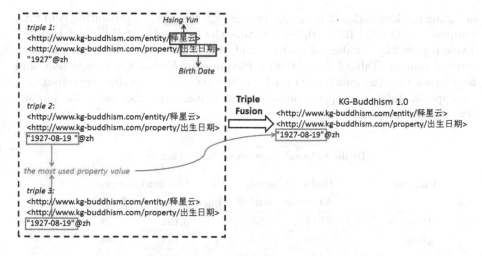

**Fig. 2.** An example of triple fusion

and temples, we crawl large-scale documents from online encyclopedias (i.e. Chinese Wikipedia, Baidu Baike, Hudong Baike) and famous Buddhist Web sites in China, such as Sina Buddhism[5], Chinese Buddhism[6], Buddhist Baike[7], etc. When given an entity and a property name, if the corresponding property value is missing, we will try to extract the property value by manually written patterns, which are actually regular expressions. For all of the 15 infobox property names of Buddhist figures, we manually write 66 patterns in total. For the 9 infobox property names of Buddhist temples, we write 19 patterns. The details of these patterns are shown in "http://www.kg-buddhism.com/pattern".

## 2.4 Linking to DBpedia

DBpedia, the structured Wikipedia, is one of the central knowledge graphs in LOD. Since a part of entities in KG-Buddhism 1.0 are actually from Chinese Wikipedia, linking such entities in KG-Buddhism 1.0 to those in DBpedia is a straightforward way to add our knowledge graph to LOD. We represent the links between KG-Buddhism 1.0 and DBpedia using the predicate owl:sameAs.

## 3 Dataset Statistics

In this section, we first give a general statistics about KG-Buddhism 1.0, and then report a preliminary evaluation on this new knowledge graph. KG-Buddhism 1.0 totally contains 3,152 entities representing Buddhist figures, 136,162 RDF triples

---

[5] http://fo.sina.com.cn/master/.
[6] http://ddks.zgfj.cn/.
[7] http://baike.foyuan.net/.

denoting the knowledge of Buddhist figures, 5,509 entities representing Buddhist temples and 188,829 RDF triples denoting the knowledge of Buddhist temples. Table 1 gives the number of entities and the number of RDF triples in each type of content. Table 2 shows the effectiveness of *Knowledge Completion*, and we can see that the number of entities that have property values corresponding to a specific infobox property name obviously increases after using the module *Knowledge Completion*.

**Table 1.** Data statistics of `KG-Buddhism` 1.0

| Content | Buddhist figures | | Buddhist temples | |
|---|---|---|---|---|
| | #Entities | #RDF triples | #Entities | #RDF triples |
| *Abstracts* | 2,017 | 3,235 | 3,191 | 4,779 |
| *Aliases* | 180 | 290 | 124 | 144 |
| *Categories* | 3,149 | 15,216 | 5,508 | 22,201 |
| *External links* | 1,009 | 3,390 | 2,652 | 5,953 |
| *Images* | 1,204 | 4,894 | 2,604 | 13,962 |
| *Internal links* | 2,446 | 34,887 | 3,511 | 94,633 |
| *Infobox properties* | 2,632 | 17,540 | 5,509 | 16,646 |
| *Labels* | 3,152 | 3,152 | 5,509 | 5,509 |
| *Redirects* | 422 | 2,169 | 656 | 1,278 |
| *Related pages* | 1,220 | 45,067 | 1,630 | 15,514 |
| *Zhishi.me links* | 3,152 | 4,904 | 4,707 | 6,345 |
| *DBpedia links* | 1,418 | 1,418 | 1,684 | 1,865 |

We analysed the quality of `KG-Buddhism` 1.0 by evaluating the domain relevance of entities and accuracy of RDF triples as follows:

– **Domain Relevance of Entities.** We asked five graduate students to annotate each entity representing a Buddhist figure using the labels *"100% Buddhist figure"*, *"related to Buddhism"* and *"unrelated to Buddhism"*. Each of the annotators can only choose one label for each entity. Since each entity got five labels from five annotators, we totally got 15,760 labels, in which 88% are *"100% Buddhist figure"*, 12% are *"related to Buddhism"* and 0% is *"unrelated to Buddhism"*. Similarly, for each entity representing a Buddhist temple, each annotator can only choose one label from *"100% Buddhist temple"*, *"related to Buddhism"* and *"unrelated to Buddhism"*. As a result, we got 27,545 labels in total from the five annotators, in which 94% are *"100% Buddhist temple"*, 6% are *"related to Buddhism"* and 0% is *"unrelated to Buddhism"*. It reflects that the module *Knowledge Collection* can really identify the entities representing Buddhist figures (or temples) from existing encyclopedic knowledge graphs.

**Table 2.** Number of the entities that have property values corresponding to a specific infobox property name **before/after** using the module *Knowledge Completion*

| Buddhist figures | | |
|---|---|---|
| Property Name | #Entities (before) | #Entities (after) |
| 出生日期 (Birth Date) | 394 | 936 |
| 死亡日期 (Death Date) | 319 | 842 |
| 出生地 (Birth Place) | 275 | 634 |
| 籍贯 (Ancestral Home) | 139 | 227 |
| 代表作 (Works) | 107 | 211 |
| 主要成就 (Achievement) | 113 | 832 |
| 身份 (Title) | 0 | 151 |
| 俗名 (Original Name) | 542 | 1,261 |
| 法名 (Buddhist Name) | 170 | 536 |
| 别名 (Alias) | 215 | 1,608 |
| 宗派 (Schools of Buddhism) | 126 | 898 |
| 师傅 (Advisor) | 16 | 356 |
| 弟子 (Student) | 10 | 615 |
| 时代 (Era) | 132 | 132 |
| 国籍 (Nationality) | 233 | 1,891 |
| Buddhist temples | | |
| Property Name | #Entities (before) | #Entities (after) |
| 别名 (Alias) | 280 | 5,509 |
| 创建人 (Founder) | 244 | 356 |
| 地址 (Address) | 275 | 634 |
| 供奉神明 (Deity) | 283 | 2,697 |
| 建筑 (Building) | 235 | 399 |
| 建筑风格 (Architectural Style) | 27 | 104 |
| 始建时间 (Founding Time) | 329 | 1,253 |
| 住持 (Buddhist Abbot) | 2 | 26 |
| 宗派 (Schools of Buddhism) | 451 | 630 |

– **Accuracy of RDF Triples**. We randomly selected 100 RDF triples from each type of contents (i.e. *Abstracts, Infobox Properties, DBpedia Links*, etc.), and also asked five graduate students to label whether these selected RDF triples are correct or not. Then, we computed the average accuracy of the RDF triples in each type of contents. As shown in Table 3, we can conclude that the RDF triples in KG-Buddhism 1.0 have quite high accuracy. This fully demonstrates the effectiveness of our proposed development modules for acquiring the knowledge of Buddhist figures and temples. The accuracy of RDF triples in *Infobox Properties* is the lowest (i.e. 0.9) compared with others. The main reason is that the quality of extracting the property values corresponding to some specific infobox property name from unstructured Web Text cannot always be ensured. This is the part that we plan to improve.

**Table 3.** Accuracy of RDF triples in KG-Buddhism 1.0

| Content | Accuracy |
|---|---|
| *Abstracts* | 1 |
| *Aliases* | 0.98 |
| *Categories* | 1 |
| *External links* | 1 |
| *Images* | 0.95 |
| *Internal links* | 1 |
| *Infobox properties* | 0.9 |
| *Labels* | 1 |
| *Redirects* | 0.95 |
| *Related pages* | 1 |
| *Zhishi.me links* | 1 |
| *DBpedia links* | 1 |

## 4   Online API

To make `KG-Buddhism` 1.0 publicly available, we provide an online API. Users can query the knowledge graph through an HTTP interface that returns JSON. Http requests should be executed using the `GET` method. Now we have two kinds of queries: (1) querying all the RDF triples w.r.t. a specific entity, and the querying format is "`http://www.kg-buddhism.com/entity/[label]`"; (2) querying all the RDF triples w.r.t. a specific entity and a specific property with the format "`http://www.kg-buddhism.com/entity/[label]&[property]`". Here, "`[label]`" means the entity label and "`[property]`" refers to the name of a specific content (the details are given in "`http://www.kg-buddhism.com/api`"). KG-Buddhism 1.0 is licensed under the `Creative Commons Attribution-Non Commercial-Share Alike 3.0 License`[8].

## 5   Conclusions and Future Work

In this paper, we presented the first version of the first Chinese knowledge graph on Buddhism, called `KG-Buddhism`. We first collected and fused RDF triples representing the knowledge of Buddhist figures (or temples) from Zhishi.me consisting of three Chinese encyclopedic knowledge graphs. We then performed knowledge completion by extracting more property values from unstructured Web text. With the above methods, we obtained 8,661 entities and 324,991 RDF triples. The preliminary evaluation results show the high quality of our built knowledge graph. KG-Buddhism 1.0 provides an online API to help users freely access the whole dataset. In the future, we will identify more entities representing Buddhist

---

[8] https://creativecommons.org/licenses/by-nc-sa/3.0/.

figures (or temples) from unstructured Web text and improve our development method in property value extraction with machine learning techniques. We also plan to design an ontology to better control the quality of acquired knowledge.

**Acknowledgements.** This work is supported in part by the National Natural Science Foundation of China (Grant No. 61672153), the 863 Program (Grant No. 2015AA015406), the Fundamental Research Funds for the Central Universities and the Research Innovation Program for College Graduates of Jiangsu Province (Grant No. KYLX16_0295).

# References

1. Lehmann, J., Isele, R., Jakob, M., Jentzsch, A., Kontokostas, D., Mendes, P.N., Hellmann, S., Morsey, M., Van Kleef, P., Auer, S., et al.: Dbpedia-a large-scale, multilingual knowledge base extracted from wikipedia. Semantic Web **6**(2), 167–195 (2015)
2. Mahdisoltani, F., Biega, J., Suchanek, F.: Yago3: a knowledge base from multilingual wikipedias. In: CIDR (2014)
3. Niu, X., Sun, X., Wang, H., Rong, S., Qi, G., Yu, Y.: Zhishi.me - weaving chinese linking open data. In: Aroyo, L., Welty, C., Alani, H., Taylor, J., Bernstein, A., Kagal, L., Noy, N., Blomqvist, E. (eds.) ISWC 2011. LNCS, vol. 7032, pp. 205–220. Springer, Heidelberg (2011). https://doi.org/10.1007/978-3-642-25093-4_14

# Semantic Graph Analysis for Federated LOD Surfing in Life Sciences

Atsuko Yamaguchi[1]([⊠]), Kouji Kozaki[2], Yasunori Yamamoto[1],
Hiroshi Masuya[3,4], and Norio Kobayashi[3,4]

[1] Database Center for Life Science (DBCLS), Research Organization of Information
and Systems, 178-4-4 Wakashiba, Kashiwa, Chiba 277-0871, Japan
{atsuko,yy}@dbcls.rois.ac.jp
[2] The Institute of Scientific and Industrial Research (ISIR), Osaka University,
8-1 Mihogaoka, Ibaraki, Osaka 567-0047, Japan
kozaki@ei.sanken.osaka-u.ac.jp
[3] RIKEN BioResource Center (BRC), 3-1-1, Koyadai, Tsukuba, Ibaraki
305-0074, Japan
hiroshi.masuya@riken.jp
[4] Advanced Center for Computing and Communication (ACCC), RIKEN,
2-1 Hirosawa, Wako, Saitama 351-0198, Japan
norio.kobayashi@riken.jp

**Abstract.** Currently, Linked Open Data (LOD) is increasingly used when publishing life science databases. To facilitate flexible use of such databases, we employ a method that uses federated query search along a path of class–class relationships. However, an effective method for federated query search requires analysis of the structure the relationships form for LOD datasets. Therefore, we constructed a graph of class–class relationships among 43 SPARQL endpoints and analyzed the connectivity of the graph. As a result, we found that (1) the sizes of connected components follow a power law; thus we should deal with the classes separately according to the size of connected components, (2) only the largest and second largest connected components have paths among classes from two or more SPARQL endpoints, and the datasets of each of the two connected components share ontologies, and (3) key classes that connect SPARQL endpoints are primarily upper-level concepts in the biological domain.

**Keywords:** Linked Open Data · Class–class relationships · Data integration · Federated query search

## 1 Introduction

Life sciences data are huge and heterogeneous, and holistic interpretation of such data is crucial to discover new biological knowledge. However, data are developed and maintained by various institutions; therefore, accessing such data in an integrated manner is difficult. Recently, increasing numbers of institutions have

© Springer International Publishing AG 2017
Z. Wang et al. (Eds.): JIST 2017, LNCS 10675, pp. 268–276, 2017.
https://doi.org/10.1007/978-3-319-70682-5_18

published their data as Linked Open Data (LOD) and have provided SPARQL endpoints to facilitate data access by both researchers and machines. To use such LOD effectively, a method to extract data from multiple SPARQL endpoints according to user requirements is needed.

Semantics can be translated from a sequence of links across different classes of data. In other words, semantics can be obtained by traversing the paths of the class–class relationships in the LOD. While some studies have investigated methods to find paths between resources in LOD, such as RelFinder [1], such methods have difficulty obtaining paths among large numbers of instances in classes that often come from experimental results in the life sciences research. Thus, our approach focuses on paths of class–class relationships. Such relationships can provide a foundations for the semantics in RDF databases.

Based on this approach, in a previous study [2], we developed a system called SPARQL Builder (http://www.sparqlbuilder.org/) that allows users to build a SPARQL query for a SPARQL endpoint without thorough understanding of the RDF. In the development of SPARQL Builder, we recognized that the essential data discovery depends on extracting data by traversing LOD over various life sciences SPARQL endpoints rather than constructing a single SPARQL query. With this new perspective, we have recently begun to develop a search system, which we refer to as LOD Surfer, based on the class–class relationships among SPARQL endpoints. We have been collecting metadata for SPARQL Builder called SPARQL Builder Metadata (SBM) that describes a data schema for each SPARQL endpoint, and we employ the SBM to handle class–class relationships in LOD Surfer.

In this study, to design a practical federated query search over life sciences datasets, we constructed a graph of class–class relationships from using SBMs from 43 SPARQL endpoints and analyzed the graph to determine the feasibility of employing federated queries among classes from two or more SPARQL endpoints.

## 2   LOD Surfer

LOD Surfer is a search system that discovers data along paths of class–class relationships over life sciences LOD provided by different SPARQL endpoints. A user can obtain desired data interactively by selecting two classes and paths between the classes.

SPARQL Builder Metadata (SBM) is used to deal with paths of class–class relationships efficiently. In this section, we briefly describe the SBM and introduce a labeled multigraph that describes class–class relationships constructed by a collection of SBM. Then, we describe how we process federated query searches from a path of the graph.

### 2.1   SPARQL Builder Metadata

The SBM (http://www.sparqlbuilder.org/doc/sbm_2015sep/) comprise a SPARQL endpoint catalogue that describes LOD schema. The SBM are defined as an extension of VoID (https://www.w3.org/TR/void/) and the SPARQL 1.1 service description (https://www.w3.org/TR/sparql11-service-description/) with our original vocabulary whose name space is `sbm:`. The SBM describe the classes, properties, class–class relationships, and statistics summaries of individual RDF graphs provided by each SPARQL endpoint. Note that some classes and properties, such as `owl:Class` and `rdfs:subClassOf`, are removed from the SBM to simplify the description.

By applying our SPARQL endpoint crawler program to the SPARQL endpoints obtained from YummyData (http://yummydata.org/), we collected SBM for 76 graphs from 43 SPARQL endpoints that provides life sciences datasets (as of July 2017). Note that some SPARQL endpoints from YummyData are not included due to a long response time. Because SBM can be used to deal with class-class relationships efficiently over SPARQL endpoints, we employ SBM to develop LOD Surfer.

### 2.2   Merged Class Graph for LOD

As described in our previous work [2], we used a specialized graph called a class graph for SPARQL Builder, where the nodes of a class graph correspond to classes. A directed edge from class $C_1$ to $C_2$ with label $p$ represents the existence of a triple $(s, p, o)$ such that $s$ and $o$ are instances of $C_1$ and $C_2$, respectively.

A class graph can be constructed from SBM efficiently because the SBM include a list of all classes and a list of all class–class relationships for a SPARQL endpoint. To deal with classes and class–class relationships from different SPARQL endpoints, we merged class graphs for individual SPARQL endpoints into a single graph. By overlapping nodes that correspond to the classes with the same class URIs and adding SPARQL endpoint information and a statistics summary to each edge, class graphs can be naturally merged into a single graph, which we refer to as a *merged class graph*.

Because each edge of a merged class graph has information about the SPARQL endpoints, each path of a merged class graph can easily translate to a federated SPARQL query with SERVICE patterns. In addition, because SBM includes statistics such as the number of triples for each class–class relationship, an efficient evaluation plan for a federated SPARQL query can be determined by starting from the edge with the minimum number of triples. Therefore, if a user can find paths between classes of interest, using the translated federated SPARQL queries with SERVICE patterns from the paths can obtain desired integrated data from LOD.

## 3   Graph Analysis for Merged Class Graph

To realize a federated query search system using paths of class–class relationships, an analysis of the structures of a merged class graph that focuses on paths between classes is required. To achieve this, we first constructed a merged class graph with 14,147 nodes and 18738 edges based on SBM obtained from 43 SPARQL endpoints with 76 life sciences datasets (Sect. 2.1). The list of SPARQL endpoints includes 31 eagle-i projects [3], 2 DBCLS projects [4], Bio2RDF [5], CiteSeer [6], DisGeNet [7], DrugBank@FU Berlin [1], the EBI RDF platform [8], LinkedCT [9], MeSH [2], NBDC LSDB Archive[3], Organic.Edunet[4], and WikiPathways[5]. Note that Bio2RDF and the EBI RDF platform include multiple datasets. Here, we used SBM of 30 and five datasets for Bio2RDF and the EBI RDF platform, respectively.

First, we computed the connected components of the constructed graph. Note that a path of class–class relationships between the classes exists if and only if two classes are in the same connected components.

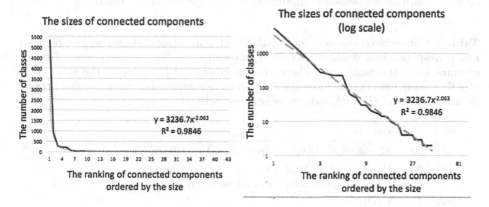

**Fig. 1.** Size distribution of 43 connected components. 6814 singletons are omitted from this figure. The solid black line shows the sizes of connected components. The dashed gray line shows $f(x) = cx^k$ for $c = 3236.7$ and $k = -2.063$ fit by least squares with $R^2 = 0.9846$

Figure 1 shows the size distribution of 43 connected components with more than two nodes in the merged graph. In addition to the 43 connected components, 6814 singletons were found in the merged graph. The left and right graphs show the distribution at normal and log scales, respectively. Here, the number of classes is 14147 and the number of singletons in the merged graph is 6814; thus, 7333 classes are distributed into 43 connected components. The distribution appears to follow a power law, as shown by the dashed gray line, which indicates $f(x) = 3236.7x^{-2.063}$ in the Fig. 1 because $R^2 = 0.9846$ is close sufficiently to 1.

---

[1] http://www4.wiwiss.fu-berlin.de/drugbank/sparql.
[2] https://id.nlm.nih.gov/mesh/sparql.
[3] https://dba-rdf.biosciencedbc.jp/sparql.
[4] http://data.organic-edunet.eu/sparql.
[5] http://sparql.wikipathways.org.

Then, we computed a list of SPARQL endpoints for each connected component and found that only the two largest connected components come from two or more SPARQL endpoints. Table 1 shows the SPARQL endpoint information for the five largest connected components. For these five connected components, the largest and second largest connected components come from 32 and six SPARQL endpoints, respectively. Therefore, for classes in these two connected components, paths can traverse two or more SPARQL endpoints. On the other hand, each of the remaining 41 connected components comes from a single SPARQL endpoint. A federated query search is not possible for classes in the connected components; therefore, generally, federated query searches can be applied to 44% of the classes, 48% of the classes are singletons, and the remaining classes comprise connected components in a single SPARQL endpoint.

**Table 1.** Connected component information for top five sizes of the merged graph. Each row corresponds to a single connected component. The "Size", "#ep" and "Details" columns show the numbers of classes in the connected component, the numbers of SPARQL endpoints that includes the classes of the connected component, and detailed information about the SPARQL endpoints, respectively.

| Size | #ep | Details |
|------|-----|---------|
| 5327 | 32 | 31 eagle-i SPARQL endpoints and EBI RDF Platform |
| 925 | 6 | Bio2RDF, EBI RDF Platform, WikiPathways, LinkedCT, DisGeNet and Organic Edunet |
| 269 | 1 | Bio2RDF |
| 220 | 1 | Bio2RDF |
| 219 | 1 | NBDC LSDB Archive |

## 3.1 Cut Classes for Federated Search

As discussed in the previous section, only the two largest connected components have paths that traverse two or more SPARQL endpoints. The largest connected component comes from 32 SPARQL endpoints comprising 31 eagle-i SPARQL endpoints and the EBI RDF platform. The second largest connected components comes from 6 SPARQL endpoints including Bio2RDF, EBI RDF platform, WikiPathways, LinkedCT, DisGeNet, and Organic Edunet.

A point of these two connected components that differs from the other connected components should be the existence of classes that connect two or more classes from different SPARQL endpoints. Therefore, to clear connecting points in these two connected components, we compute a set of cut vertices connecting two or more SPARQL endpoints for each connected component. Note that a vertex $v$ of a connected graph $G = (V, E)$ is a cut vertex if the vertex-induced subgraph with $V - \{v\}$ of $G$ is disconnected. In other words, if vertex $v$ is a cut vertex with two or more SPARQL endpoints, the vertex corresponds to a class connecting two datasets from different SPARQL endpoints. Note that we refer to a class of a cut vertex as a *cut class*.

Table 2 lists cut classes shared by two or more SPARQL endpoints for the largest connected components. As can be seen, the VIVO-ISF Ontology (VIVO) and the eagle-i Research Resource Ontology (ERO) are commonly used for datasets from 32 SPARQL endpoints, and these two ontologies connect classes from different SPARQL endpoints.

**Table 2.** Cut classes shared by two or more SPARQL endpoints for the largest connected components. Each row corresponds to a cut class. The "label", # ep, and # cp columns show the class label, the number of SPARQL endpoints that shares the class, and the number of connected components of the vertex-induced subgraph obtained by removing the class, respectively.

| Class URI | label | # ep | # cp |
|---|---|---|---|
| http://vivoweb.org/ontology/core#CoreLaboratory | Core laboratory | 27 | 16 |
| http://purl.obolibrary.org/obo/ERO_0000071 | software | 23 | 11 |
| http://vivoweb.org/ontology/core#Laboratory | Laboratory | 19 | 20 |
| http://purl.obolibrary.org/obo/OBI_0000272 | protocol | 13 | 3 |
| http://purl.obolibrary.org/obo/NCBITaxon_10090 | Mus musculus | 11 | 4 |
| http://purl.obolibrary.org/obo/ERO_0000229 | monoclonal antibody reagent | 8 | 4 |
| http://purl.obolibrary.org/obo/ERO_0001839 | algorithmic software suite | 5 | 4 |
| http://purl.obolibrary.org/obo/ERO_0001964 | human subject | 3 | 4 |
| http://purl.obolibrary.org/obo/ERO_0000565 | Technology transfer office | 2 | 3 |

The second largest connected component comes from six SPARQL endpoints, including Bio2RDF, the EBI RDF platform, WikiPathways, LinkedCT, Dis-GeNet and Organic Edunet. As shown in Table 3, classes from these SPARQL endpoints are connected through the Biological Pathway Exchange (BioPax).

**Table 3.** Cut classes shared by two or more SPARQL endpoints for the secondly largest connected components. Each row corresponds to a cut class. The "label", # ep, and # cp columns show the class label, the number of SPARQL endpoints that shares the class, and the number of connected components of the vertex-induced subgraph obtained by removing the class, respectively.

| Class URI | label | # ep | # cp |
|---|---|---|---|
| http://www.biopax.org/release/biopax-level3.owl#BiochemicalReaction | - | 2 | 3 |
| http://www.biopax.org/release/biopax-level3.owl#SmallMoleculeReference | - | 2 | 3 |
| http://www.biopax.org/release/biopax-level3.owl#ModificationFeature | - | 2 | 3 |

From these two connected components, we see that sharing ontologies plays a key role in integrating datasets using the class–class relationships from two or more SPARQL endpoints.

## 4   Discussion

The graph analysis of the merged class graph revealed that there are roughly two groups of connected components. The first group contains large components that involves "hub" classes, such as cut classes, which are essential for federated queries across multiple SPARQL endpoints. The second group is composed of small-size components with fewer links that do not contribute to finding federated search paths.

We found only upper-level concepts in the biological domain (e.g., software, laboratory, protocol, clinical trial, etc.) in the lists of cut classes of large connected components (Tables 2 and 3). In contrast, classes that represent lower-level individual concepts (e.g., the names of a disease, phenotype, or gene) never appeared in the lists. Paths composed of upper concepts of the domain give an overview of the meaning of the paths (e.g., "Laboratories have their own protocol for clinical trials") rather than instantiated paths (e.g., "John's lab has a protocol $X$ for specific clinical trial"). This fact is useful in the development of a high-performance path exploration system for federated search where the user interface shows an overview of the paths.

On the other hand, further analysis of small connected components would be required. In our preliminary observation, two groups of classes are involved in the small components. The first group is composed of middle or lower classes of ontologies, such as disease names. The second group is composed of classes that are not well established as a common vocabulary and are used rather locally. Further grouping with upper classes using `rdfs:subClassOf` for the first group and applying a method to match the local ontoloies used in the second group to major ontologies may be beneficial to realize greater usability of LOD surfer.

## 5   Conclusion

In this study, we constructed a merged graph from 43 life sciences SPARQL endpoints and analyzed the characteristics of the graph. to further the development of a practical LOD federated query search system based on class–class relationships, To investigate the existences of paths across multiple SPARQL endpoints, we computed connected the components of the merged class graph. In addition, for the largest and second largest connected components, which come from 32 and six SPARQL endpoints, respectively, we computed the classes of the cut vertices, i.e., the connecting points between classes in the merged class graph.

We conclude the following from these analysis. (1) The sizes of the connected components follow a power law; thus, we should deal with classes in large and small connected components separately. (2) Only the largest and second largest connected components have links among classes from two or more SPARQL endpoints, and each of the two connected components shares ontologies. (3) The classes of the cut vertices that connect classes from different SPARQL endpoints are primarily upper-level concepts in the biological domain.

These results are significant to realize LOD Surfer with high usability and performance. For example, for classes in large connected components, we can realize greater usability when finding paths across two or more SPARQL endpoints by focusing on the cut classes because cut classes are junctions of datasets from different SPARQL endpoints. For classes in small connected components, by grouping or matching classes according to their features, as discussed in Sect. 4, we can extend paths across different connected components. Further analysis for extension of paths using inference, such as `rdfs:subClassOf` and `owl:sameAs`, will be the focus of future work.

**Acknowledgments.** This work was supported by JSPS KAKENHI grant numbers 17K00434, 17K00424 and 17H01789, and by the National Bioscience Database Center (NBDC) of the Japan Science and Technology Agency (JST).

# References

1. Heim, P., Hellmann, S., Lehmann, J., Lohmann, S., Stegemann, T.: RelFinder: revealing relationships in RDF knowledge bases. In: Chua, T.-S., Kompatsiaris, Y., Mérialdo, B., Haas, W., Thallinger, G., Bailer, W. (eds.) SAMT 2009. LNCS, vol. 5887, pp. 182–187. Springer, Heidelberg (2009). https://doi.org/10.1007/978-3-642-10543-2_21
2. Yamaguchi, A., Kozaki, K., Lenz, K., Yamamoto, Y., Masuya, H., Kobayashi, N.: Semantic data acquisition by traversing class-class relationships over linkedopen data. In: 6th Joint International Conference (JIST 2016), LNCS 10055, pp. 136-151(2016)
3. Vasilevsky, N., Johnson, T., Corday, K., Torniai, C., Brush, M., Segerdell, E., Wilson, M., Shaffer, C., Robinson, D., Haendel, M.: Research resources: curating the new eagle-i discovery system. Database **2012**, bar067 (2012). https://doi.org/10.1093/database/bar067
4. Yamamoto, Y., Yamaguchi, A., Bono, H., Takagi, T.: Allie: a database and a search service of abbreviations and long forms. Database **2011**, bar013 (2011). https://doi.org/10.1093/database/bar013
5. Belleau, F., Nolin, M.A., Tourigny, N., Rigault, P., Morissette, J.: Bio2RDF: towards a mashup to build bioinformatics knowledge systems. J. Biomed. Inform. **41**(5), 706–716 (2008)
6. Gile, C.L., Bollacker, K.D., Lawrence, S.: CiteSeer: an automatic citation indexing system. In: Proceedings of the Third ACM Conference on Digital Libraries (DL 98), pp. 89–98 (1998)
7. Piñero, J., Queralt-Rosinach, N., Bravo, À., Deu-Pons, J., Bauer-Mehren, A., Baron, M., Sanz, F., Furlong, L.I.: DisGeNET: a discovery platform for the dynamical exploration of human diseases and their genes. Database (2015). https://doi.org/10.1093/database/bav028
8. Jupp, S., Malone, J., Bolleman, J., Brandizi, M., Davies, M., Garcia, L., Gaulton, A., Gehant, S., Laibe, C., Redaschi, N., Wimalaratne, S.M., Martin, M., Le Novére, N., Parkinson, H., Birney, E., Jenkinson, A.M.: The EBI RDF platform: linked open data for the life sciences. Bioinformatics **30**(9), 1338–1339 (2014)

9. Hassanzadeh, O., Miller, R.J.: Automatic Curation of Clinical Trials Data in LinkedCT. In: Arenas, M., Corcho, O., Simperl, E., Strohmaier, M., d'Aquin, M., Srinivas, K., Groth, P., Dumontier, M., Heflin, J., Thirunarayan, K., Staab, S. (eds.) ISWC 2015. LNCS, vol. 9367, pp. 270–278. Springer, Cham (2015). https://doi.org/10.1007/978-3-319-25010-6_16

# Development of Semantic Web-Based Imaging Database for Biological Morphome

Satoshi Kume[1,2,3]($\boxtimes$), Hiroshi Masuya[4,5], Mitsuyo Maeda[2], Mitsuo Suga[2], Yosky Kataoka[1,2,3], and Norio Kobayashi[2,4,5]

[1] RIKEN Center for Life Science Technologies (CLST), RIKEN,
6-7-3 Minatojima-minamimachi, Chuo-ku, Kobe, Hyogo 650-0047, Japan
{satoshi.kume,kataokay}@riken.jp
[2] RIKEN CLST-JEOL Collaboration Center, RIKEN,
6-7-3 Minatojima-minamimachi, Chuo-ku, Kobe, Hyogo 650-0047, Japan
mitsuyo.maeda@riken.jp, msuga@jeol.co.jp
[3] RIKEN Compass to Healthy Life Research Complex Program,
RIKEN Cluster for Science and Technology Hub, RIKEN,
6-7-1 Minatojima-Minamimachi, Chuo-ku, Kobe, Hyogo 650-0047, Japan
[4] RIKEN BioResource Center (BRC),
3-1-1, Koyadai, Tsukuba, Ibaraki 305-0074, Japan
hmasuya@brc.riken.jp
[5] Advanced Center for Computing and Communication (ACCC), RIKEN,
2-1 Hirosawa, Wako, Saitama 351-0198, Japan
norio.kobayashi@riken.jp

**Abstract.** We introduce the RIKEN Microstructural Imaging Meta-database, a semantic web-based imaging database in which image metadata are described using the Resource Description Framework (RDF) and detailed biological properties observed in the images can be represented as Linked Open Data. The metadata are used to develop a large-scale imaging viewer that provides a straightforward graphical user interface to visualise a large microstructural tiling image at the gigabyte level. We applied the database to accumulate comprehensive microstructural imaging data produced by automated scanning electron microscopy. As a result, we have successfully managed vast numbers of images and their metadata, including the interpretation of morphological phenotypes occurring in sub-cellular components and biosamples captured in the images. We also discuss advanced utilisation of morphological imaging data that can be promoted by this database.

**Keywords:** Microscopy ontology · Open microscopy environment · Imaging metadatabase · Morphome · Morphological phenotype · Scanning electron microscopy

## 1 Introduction

Imaging data are an important fundamental aspect of contemporary life sciences. They facilitate the understanding of detailed morphological changes related to

Z. Wang et al. (Eds.): JIST 2017, LNCS 10675, pp. 277–285, 2017.
https://doi.org/10.1007/978-3-319-70682-5_19

cellular functions and various diseases. We focus on microstructural images obtained using electron microscopes (EM), which provide detailed morphological information about tissues and cells at a nano-scale level [1]. Typically, EM images have two significant problems. (1) Because the target area of an EM is quite small (sub-cellular level: less than a micrometre square), a wide-range overview (tissue level: millimetre square) is required to screen phenotypes. (2) Interpreting an EM image requires an advanced understanding of histology and histopathology.

To solve the first problem, we have developed a technology to obtain comprehensive microstructural imaging data of biotissues using scanning electron microscopy (SEM) with an autofocus function (Maeda, et al., under preparation). This imaging technology provides wide-range and high-resolution images of a sub-millimetre square area in biosamples. This type of comprehensive imaging is referred to as the morphome analysis, especially micro-morphomics in order to identify the totality of the microstructural morphological features and is considered to be an omic research. While such imaging data are accumulating rapidly as big data, a common procedure for sharing, integrating and analysing such data has not yet been developed. In addition, specialised experience is required to interpret morphological phenotypes in the images. Recently, Williams et al. launched the Image Data Resource (IDR), a prototype public database and repository for imaging data [2]. The IDR is the first general repository for image data in life sciences. Thus, the potential impact of accessing, sharing and referencing imaging data in experimental results and published research has increased. However, issues associated with metadata, image formats and vocabulary associated with imaging experiments remain unresolved. To ensure reproducibility of experimental data and results, including expert image analysis, a common data publication framework and a multi-faceted ontology that describes various imaging experiments and experimental conditions are desired. Moreover, to complement specialised histology and histopathology experience and realise further automatic imaging data analysis, such as machine learning, metadata that describe related experimental conditions and phenotypes should be machine readable.

Our goal is to generate standardised machine-readable metadata that include related information, such as experimental conditions and phenotype data, for life sciences research. Currently, we are constructing an imaging metadata system for optical and electron microscopy that employs the Resource Description Framework (RDF). We are also developing an ontology to describe RDF metadata. Relative to the development of an ontology to describe imaging metadata, we have previously proposed general microscopy ontology concepts that involved translating XML-based Open Microscopy Environment (OME) metadata and the extension of incomplete vocabularies [3]. The RDF ontology provides vocabularies and semantics to describe metadata, including information about electron microscopy imaging conditions, biosamples and sample preparation. In this paper, we describe the proposed RDF-based RIKEN Microstructural Imaging Metadatabase, which provides imaging metadata integrated with biosample and phenotype metadata and an imaging viewer that takes advantage of such metadata.

# 2    Methods

## 2.1    Extension of Microscopy Ontology for Morphomics Data

The OME is an open-source interoperable toolset for biological imaging data to manage multi-dimensional and heterogeneous optical microscopy imaging data [4]. Previously, using the latest version of the OME data model, we translated the XML based OME data model into a base concept for an RDF schema [3]. At this point, we identified missing concepts in the OME data model and extended the vocabularies to describe imaging metadata integrated with biosample and phenotype metadata.

To improve the ability to describe experimental situations, we further expanded vocabularies in the microscopy ontology. Note that phenotypic annotations often differ between observation of microscopic images and direct observation of biosamples because their magnication levels are different (i.e. sub-cellular and individual levels). Therefore, in the microscopy ontology, we enabled separate descriptions of phenotype data obtained from imaging analysis (i.e. subcellular phenotypes) and direct observation of biosamples (i.e. phenotype data of biosamples obtained from an individual).

In comprehensive microstructural imaging at the sub-tissue level, it is necessary to manage image data continuously photographed in units of several thousands of images or huge image data at several gigabyte levels using a tiling process. Recently, such huge image data have been converted and visualised using a pyramidal image format called the Deep Zoom Image (DZI) format. Therefore, we added new vocabularies (image directory, image overlap, tiling method, tiling map, etc.) to describe the metadata for tiling image data and DZI visualisation. In addition, the class of the DZI dataset is related to `rdfs:subClassOf` of the dataset class, and it has a structure that can link individual image data.

## 2.2    Microstructural Bioimaging and Image Processing

To prepare a biosample, we initially prepared wide-range ultra-thin (70 nm) tissue sections obtained from a rat liver (600 μm × 500 μm). Following the electron staining of the liver sections, we performed large-scale microstructural imaging at high spatial resolution (7.87 nm/pixel) covering the entire area of the section using SEM. As a result, 288 16-bit images (5120 × 3840 pixels) were obtained, and the amount of data was approximately 10 GB.

To construct a tiling image, overlaps of resized images were computed using the Grid/Collection stitching plug-in in ImageJ/Fiji [5] and stitched to a single tiling image. Then, the tiling image was converted to the DZI format using the Python-based code deepzoom.py (initially developed by Kapil Thangavelu) for visualisation by the electron microscopy viewer.

## 2.3    Development of Electron Microscopy Viewer and the Annotation of Phenotype Data

To display many microstructural images as a single large image, we developed the JavaScript-based RIKEN CLST Electron Microscopy Viewer, in which an interactive multi-resolution visualisation function was implemented using OpenSeadragon (https://openseadragon.github.io). In addition, overview, metadata linkage, snapshot, annotation, measurement and importing and exporting annotation file functions were implemented. The metadata of the viewer are a part of metadata based on the microscopy ontology and are contained in the taxon of National Center for Biotechnology Information (NCBI), animal strain, derives from, imaging method and staining method used to describe tiled images. We then annotated the phenotype data of the images, such as the binuclear cell phenotype (CMPO:0000213) [6] in the liver, using the electron microscopy viewer.

## 2.4    Construction of Metadatabase for Microstructural Imaging Data of Biotissue

RIKEN has been making efforts to publish metadata which is suitable for the cutting-edge research communities in life sciences by discussions with experts in various fields such as ontology, informatics and biology [7]. Furthermore, RIKEN has developed a database platform called the RIKEN MetaDatabase (http://metadb.riken.jp) [7] with the help of two virtual machines on the RIKENs private cloud platform using OpenLink Virtuoso [https://virtuoso.openlinksw.com/], and have already started conducting pilot operation as a metadata publishing service for biologists. In this study, we introduced an RDF-based metadatabase for large-scale tissue microstructural images. Then we generated the metadata and published the images and their metadata in the Imaging Metadatabase (http://metadb.riken.jp/metadb/db/clstMultimodalMicrostruct).

## 3    Results and Discussion

### 3.1    Vocabulary Extension for Experimental Description and Morphome Data Description

We developed an ontology to define the upper-level concepts in microscope imaging [3] as descriptions of Project, Experimenter, Biosample, Image, Instrument and Screening experiments. This ontology is described in OWL (OWL 2 DL) and has 324 classes and 225 properties. The representative classes are shown in Table 1. For example, in the image descriptions, the classes, such as dataset and image, were translated directly from the OME data model. In addition, we added new classes, including a DZI dataset, phenotype data of biosample and image, bioformat and tiling image (Table 1).

We defined a high-level concept so as to realise the common description of the description vocabulary of the microscopy imaging and the difference among

**Table 1.** Overview of classes in the microscopy ontology

| Mainly used for | URI | Label | Newly added |
|---|---|---|---|
| Description of project | http://metadb.riken.jp/db/clstMultimodalMicrostruct/ome/Project | Project | No |
| | http://metadb.riken.jp/db/clstMultimodalMicrostruct/ome/Experiment | Experiment | No |
| Description of experimenter | http://metadb.riken.jp/db/clstMultimodalMicrostruct/ome/ExperimenterGroup | Experimenter group | No |
| Description of bio-samples | http://metadb.riken.jp/db/clstMultimodalMicrostruct/riken/SamplePreparation | Sample preparation | Yes |
| | http://metadb.riken.jp/db/clstMultimodalMicrostruct/riken/BioSample | BioSample | Yes |
| | http://metadb.riken.jp/db/clstMultimodalMicrostruct/riken/PhenotypeDataOfBioSample | Phenotype data of bioSample | Yes |
| Description of image | http://metadb.riken.jp/db/clstMultimodalMicrostruct/ome/Dataset | Dataset | No |
| | http://metadb.riken.jp/db/clstMultimodalMicrostruct/ome/Image | Image | No |
| | http://metadb.riken.jp/db/clstMultimodalMicrostruct/ http://metadb.riken.jp/db/clstMultimodalMicrostruct/riken/DeepZoomImageDataset | Deep zoom image dataset | Yes |
| | http://metadb.riken.jp/db/clstMultimodalMicrostruct/ http://metadb.riken.jp/db/clstMultimodalMicrostruct/riken/PhenotypeDataOfImages | Phenotype data of image | Yes |
| | http://metadb.riken.jp/db/clstMultimodalMicrostruct/riken/TilingImage | Tiling Image | Yes |
| Description of instrument | http://metadb.riken.jp/db/clstMultimodalMicrostruct/riken/ImagingCondition | Imaging condition | Yes |
| | http://metadb.riken.jp/db/clstMultimodalMicrostruct/ome/Instrument | Instrument | No |
| | http://metadb.riken.jp/db/clstMultimodalMicrostruct/ome/Microscope | Microscope | No |
| | http://metadb.riken.jp/db/clstMultimodalMicrostruct/riken/ElectronMicroscope/ | Electron microscope | Yes |
| | http://metadb.riken.jp/db/clstMultimodalMicrostruct/riken/ElectronDetector | Electron detector | Yes |
| | http://metadb.riken.jp/db/clstMultimodalMicrostruct/riken/ElectronGun | Electron gun | Yes |
| Description of screening experiments | http://metadb.riken.jp/db/clstMultimodalMicrostruct/riken/SampleContainer | Sample container | Yes |

various types of microscopies at the same time, by examining the concept in the microscopy ontology. As a specific example, in the microscopy imaging conditions, one of the difference between the optical microscope and the electron microscope is the apparatus and the initial parameters. Although they are individualised as subclasses, those that can be integrated using the higher concept ImagingCondition.

Additionally, we focused on the description of the phenotype data because phenotypic observations relating to the whole cell, cellular components and cellular processes are a noteworthy research topic in life sciences [6]. Then, we performed discrimination of the phenotype data relative to what was imaged (subject) and what was captured (image) as Phenotype data of sample and Phenotype data of image classes under the upper Phenotype data concept.

Further, we integrated such concepts which are having 15 super classes described using `rdfs:subClassOf` and 27 classes which defines choices using `rdf:type`. In this way, reusing data has progressed by handling such super classes and choice classes in a large scale, which was not organised by the XML-based OME data schema. Resultantly, imaging metadata for the rat liver tissue, such as imaging method, SEM parameters, experimental conditions, biological sample (a rat) and phenotype data, obtained from observations of the biosample and images were successfully described, as shown in Fig. 1.

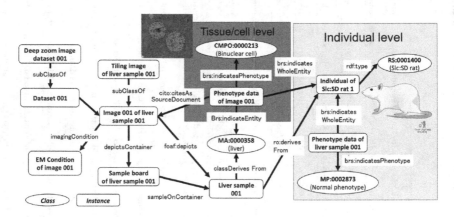

**Fig. 1.** Example graph structure showing that an abnormal binuclear phenotype of a liver cell is detected from the imaging analysis of a Slc:SD rat with a normal phenotype. The rounded rectangles are instances of classes, while ovals are classes. An EM image (Image 001 of liver sample 001) shows a liver sample (Liver sample 001) derived from a Slc:SD rat (Individual of Slc:SD rat 1). Referencing Image 001 of liver sample 001, phenotype data (Phenotype data of image 001) were produced. The graph structure can link bioresource information in RIKEN to an external database from RS:0001400.

On the other hand, we designed the semantics of the DZI dataset such that they can be handled properly by the image viewer application. We defined a DZI dataset to link to a DZI. By referring to the DZI URL, this viewer provides

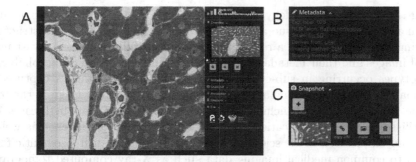

**Fig. 2.** Interactive visualisation of large tiling image in the RIKEN CLST Electron Microscopy Viewer. (A) DZI of the tiling image visualised using the viewer and images around the interlobular vein of the liver. (B) The viewer contains metadata such as NCBI taxon, animal strain and derives from. (C) The viewer includes a snapshot function to capture a certain part of the image. These viewer data are published on the RIKEN web site (http://clst.multimodal.riken.jp/CLST_ViewerData/EMV_025_161114SEM_RatLiverNormal/)

a graphical user interface to visualise the tiling image combined with hundreds of microstructural biotissue images (Fig. 2A). The viewer also functions successfully for the metadata card view for the tiling image (Fig. 2B) and generates snapshots and a direct URI for a part of an image (Fig. 2C). Furthermore, using the viewer's annotation function, users can define a region of interest (ROI) and insert phenotype annotation of the cells or cellular components (i.e. binuclear cell phenotype (CMPO:0000213) of liver cells) (Fig. 3). Users can search the ROIs for a specific phenotype in the search mode by designating a phenotype term via a text search. Thus, the electron microscopy viewer is a powerful and a useful tool to accumulate phenotype information in morphome data. Also, the CMPO provided the description of the general phenotypic observations in various kinds of biological samples and utilised for phenotype annotation in several image repositories including the IDR, allowing for future cross-species integration of phenotypic data [6].

### 3.2 Imaging Metadatabase and Comprehensive Morphome Analysis on the Semantic Web

Practically, a microscopy imaging database is required to provide various imaging metadata corresponding to various experimental results and conditions such as the disease phenotype of biotissues. Such metadata are a great benefit when users search images as linked data for research and education purposes. In this study, we conducted a series of histological experiments from the sample preparation of biotissue (rat liver) to imaging experiments by SEM. In addition, we described RDF-based metadata of these experimental processes using the microscopy ontology. We also verified that images, experimental conditions and phenotype data are linked and can be implemented as a practical metadatabase.

Using the microscopy ontology, we have described metadata for approximately 300 images of rat liver tissue and have published the RIKEN Microstructural Imaging Metadatabase. As a result, we have successfully managed a vast number of images and their metadata, including the interpretation of morphological phenotypes occurring in sub-cellular components and biosamples captured in the images. Also, the accumulation of machine-readable knowledge in the morphome, e.g. metadata that include morphological phenotype data in diseases, can be utilised as a higher educational system for specialised experience as well as an advanced research in life sciences. Also by introducing such a semantic foundation to common medical imaging data such as X-ray computed tomography (CT), magnetic resonance imaging (MRI), and positron emission tomography (PET), which are done in the clinical field, the management and sharing of these imaging data will be considered to be beneficial.

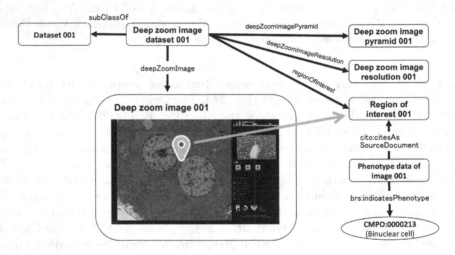

**Fig. 3.** Metadata of a DZI based on the microscopy ontology to control the image viewer. The deep zooming function of the image viewer accesses the DZI dataset, which links the class of deep zoom image and other metadata components. The DZI dataset can have an ROI that can be linked to image phenotype annotations.

## 4  Conclusions

We have developed a semantic web-based imaging database, i.e. the RIKEN Microstructural Imaging Metadatabase (http://metadb.riken.jp/metadb/db/clstMultimodalMicrostruct). As far as we know, a combination application of the ontology-based imaging metadatabase and its viewer is the first attempt in the world, and it has also made data integration with bioresources and other data in the life science possible. In this paper, we have shown an imaging dataset in the metadatabase that consists of approximately 300 images of rat liver tissue. To describe the imaging-related information, such as the experimental conditions of the EM and biosamples, we extended the original OME vocabularies for optical microscopy and used them to describe our EM imaging metadata. Moreover, we have developed an interactive EM image viewer that can visualise a tiling image in the DZI format by referring to the image metadata. We believe that these results will lead to the evolution of morphomics and emerging bioimaging research fields. Additionally, these results will be particularly useful for practical biomedical research and education.

**Acknowledgements.** This paper is dedicated to RIKEN's centennial. This work was partially supported by the management expenses grant for RIKEN CLST-JEOL Collaboration Center, the RIKEN ageing project, the RIKEN engineering network project, and the JSPS KAKENHI grant number 15K16536 and JP17K00424.

# References

1. Kasthuri, N., et al.: Saturated reconstruction of a volume of neocortex. Cell **162**(3), 648–661 (2015)
2. Williams, E., et al.: Image data resource: a bioimage data integration and publication platform. Nat. Methods **14**, 775–781 (2017)
3. Kume, S., et al.: Development of an ontology for an integrated image analysis platform to enable global sharing of microscopy imaging data. In: Proceedings of the ISWC 2016 Posters & Demonstrations Track (2016)
4. Goldberg, I.G., et al.: The open microscopy environment (OME) data model and XML file: open tools for informatics and quantitative analysis in biological imaging. Genome Biol. **6**, R47 (2005)
5. Preibisch, S., et al.: Globally optimal stitching of tiled 3D microscopic image acquisitions. Bioinformatics **25**(11), 1463–1465 (2009)
6. Jupp, S., et al.: The cellular microscopy phenotype ontology. J. Biomed. Semant. **1**, 7–28 (2016)
7. Kobayashi, N., Lenz, K., Masuya, H.: RIKEN metadatabase: a database platform as a microcosm of linked open data cloud in the life sciences. In: Li, Y.-F., Hu, W., Dong, J.S., Antoniou, G., Wang, Z., Sun, J., Liu, Y. (eds.) JIST 2016. LNCS, vol. 10055, pp. 99–115. Springer, Cham (2016). https://doi.org/10.1007/978-3-319-50112-3_8

# Applications of Semantic Technologies

# User Participatory Construction of Open Hazard Data for Preventing Bicycle Accidents

Ryohei Kozu$^{(\boxtimes)}$, Takahiro Kawamura, Shusaku Egami, Yuichi Sei,
Yasuyuki Tahara, and Akihiko Ohsuga

Graduate School of Informatics and Engineering,
The University of Electro-Communications, Tokyo, Japan
kozu.ryohei@ohsuga.lab.uec.ac.jp

**Abstract.** Recently, bicycle-related accidents, e.g., collision accidents at intersection increase and account for approximately 20% of all traffic accidents in Japan; thus, it is regarded as one of the serious social problems. However, the Traffic Accident Occurrence Map released by the Japanese Metropolitan Police Department is currently based on accident information records, and thus there are a number of near-miss events, which are overlooked in the map but will be useful for preventing the possible accidents. Therefore, we detect locations with high possibility of bicycle accidents using user participatory sensing and offer them drivers and government officials as Open Hazard Data (OHD) to prevent future bicycle accident. This paper uses smartphone sensors to obtain data for acceleration, location, and handle rotation information. Then, by classifying those data with convolutional neural networks, it was confirmed that the locations, where sudden braking occurred can be detected with an accuracy of 80%. In addition, we defined an RDF model for OHD that is currently publicly available. In future, we plan to develop applications using OHD, e.g., notifying alerts when users are approaching locations where near-miss events have occurred.

**Keywords:** Bicycle accident · Participatory sensing · Deep learning · Open data

## 1  Introduction

Recently, bicycle-related accidents have become one of the serious social problems. According to the bicycle accident analysis data[1] released in 2008 by the Japanese Metropolitan Police Department, more than 90,000 accidents related to bicycles occurred, which account for approximately 20% of all traffic accidents in Japan. In the bicycle-related accidents, it has been reported that collision accidents at the intersection correspond a very high proportion. Thus, we found that

---

[1] http://www.keishicho.metro.tokyo.jp/about_mpd/jokyo_tokei/tokei_jokyo/bicycle. files/002_28.pdf (in Japanese).

© Springer International Publishing AG 2017
Z. Wang et al. (Eds.): JIST 2017, LNCS 10675, pp. 289–303, 2017.
https://doi.org/10.1007/978-3-319-70682-5_20

the surrounding environment, e.g., low visibility at the intersection is a cause of such bicycle accidents.

The environment for bicycles and pedestrians is drawing attention from the viewpoint of accident prevention and/or universal design. Hazard map has been used as a mean to inform locations with high risk of bicycle-related accidents to car drivers and urban planner. For example, the Traffic Accident Occurrence Map[2] is released by the Japanese Metropolitan Police Department. However, it is currently based on accident information records, and thus there are several near-miss events, which are overlooked in the map but will be useful for preventing the possible accidents. As Heinrich, who is an American industrial safety pioneer claimed as Heinrich's Law in 1931, it is said for each accident that causes a major injury there are 300 accidents that cause no injuries and 29 accidents that cause minor injuries, since many accidents share common root causes [1]. Therefore, addressing more commonplace accidents that cause no injuries is so much important in order to prevent future accidents that cause major injuries. Also, the existing hazard maps relating to bicycle accidents are based on the questionnaire-based social investigation, and thus there is a problem that creation of such maps costs many human resources. Then, the Ministry of Land, Infrastructure, and Transport just started a public offering for "local project on pedestrian mobility support service utilizing open data"[3] for the establishment of a universal society.

On the other hand, with the spread of smartphones equipped with sensors, such as acceleration sensors and GPS, user participatory sensing has attracted attention as a method of social information gathering. The user participatory sensing is an approach for collecting sensor data mainly from general public carrying smartphones. If a number of users participate, it can be regarded as a social sensor that covers a large area with low cost. In order to collect information on social infrastructure such as roads on which bicycles run about, a method that has high coverage will be suitable.

Therefore, we detect locations with high possibility of bicycle accidents using the user participatory sensing and offer them bicycle drivers and government officials as Open Hazard Data (OHD) to prevent future bicycle accident. In this paper, we used smartphone sensors to obtain data for acceleration, location, and handle rotation information. Based on the above theory of behavior-based safety, we here focused on the sudden braking as the commonplace accidents of the whole bicycle-related traffic accidents. Then, by classifying sensor data with convolutional neural networks (CNN), we confirmed that the locations, where sudden braking occurred could be detected with 80% accuracy. In addition, we defined a Resource Description Framework (RDF) model for OHD that is publicly available.

The remainder of this paper is organized as follows. In Sect. 2, related works concerning the environment and behavior sensing, and user participation methods are described. In Sect. 3, our system for detecting sudden braking using user

---

[2] http://www3.wagamachi-guide.com/jikomap/ (in Japanese).
[3] https://www.mlit.go.jp/common/001191847.pdf (in Japanese).

participatory sensing is presented. In Sect. 4, we evaluate our results through preliminary experiments and actual deployment of the system. In Sect. 5, Linked Open Data (LOD) for OHD is described. Finally, Sect. 6 concludes this paper with future works.

## 2   Related Works

### 2.1   Environmental Sensing Methods

Due to increased health awareness and energy conservation awareness by the project[4] of Ministry of Health, Labor and Welfare, the number of Japanese bicycles owners increased by 2.6 times from 1954 to 1999 [2]. As described above, however, the bicycle-related accidents have accounted for a large proportion of all traffic accidents in Japan. Thus, sensing of bicycle driving has been studied. Shane et al. [3] collected bicycle speed, position, heart rate, road inclination, etc. by multiple sensors attached to bicycles and bicycle drivers, against the problem that the running and the running environment of the bicycle are not quantitatively evaluated. Then, the senser data can be shared within a group on their website. However, it was necessary to attach multiple dedicated sensors to bicycles and drivers, and thus it was hard to collect data from a large number of users on a daily basis.

To lower user's participation barriers, Saito et al. [4] extracted several driving conditions, e.g., straight-ahead driving, turning movement, and stops using sensors built into users' smartphones. However, the obtained data are not disclosed as open data and not applicable for third parties.

### 2.2   Behavior Recognition Methods

According to miniaturization and performance improvement of various sensors, those are utilized in the field of user behavior recognition. There are methods using Hidden Markov Model (HMM) based on finite state machine for time-series pattern recognition. Several studies tried to recognize human behaviors [5], such as standing, sitting, walking, and also behaviors of bicycles [4] using acceleration sensor data as the input of the HMM. HMM is often used as time-series recognition methods, since the algorithm is relatively fast and parameter setting is easy. However, in the previous research [4], Saito et al. only tried to recognize the running condition under the experimental environment, and it was not clear whether the condition can be detected accurately on the actual road.

Kiyohara et al. [6] applied a Time Series Data retrieval method called Dynamic Time Warping (DTW) to behavior recognition of dogs by attaching acceleration sensors to their necks. DTW uses the degree of similarity focused on the waveform as an index, and it is possible to recognize behaviors by considering the phase shift between waveforms. However, their study showed that the accuracy of behavior recognition using data acquired from multiple dogs is not satisfiable.

---

[4] http://www.kenkounippon21.gr.jp/ (in Japanese).

### 2.3    User Participation Methods

Umezu et al. [7] proposed a system for notifying information on obstacles that hinder the aged people and/or handy-capped persons, such as steps and crowds. FixMyStreet[5] is a famous platform for reporting regional problems, such as road conditions and illegal dumping. Moreover, to solve the illegally-parked bicycle problem, Egami et al. [8] collected the number and location information of bicycles parked in the street using social sensors. They also estimated missing data using Bayesian networks that consist of information of time, weather, and location information. These studies used crowdsourcing as a mean to collect local problems and obstacles to be solved. Therefore, data collection and recognition heavily rely on intentional reporting from the general public.

To efficiently and continuously collect data through user participation, it is known that incentive design for users is important. Arakawa et al. [9] studied that it was effective to give incentives to users through gamification. As an example, they built a game, in which points are added by reporting the position and intensity of the streetlight to collect the street lamp information [10]. We also plan to build smartphone applications based on the collected location data to be an efficient and sustainable framework to collect data without active engagement such as reporting, which is described again in the future work.

## 3    Detection System of Sudden Braking

To lower users' participation barriers, we decided to use the sensors built into users' smartphones to detect bicycle-related (possibly) accident locations. Although such information was not provided to bicycle drivers and local governments, we aim to share the knowledge of hazardous locations extracted from users' sensor data as open data. In this paper, we detect the locations of sudden braking occurrence.

However, smartphones installed several different sensors, and data qualities vary. Thus, quantitative use of them to recognize users' behaviors is more difficult than dedicated sensor data. Also, data characteristics highly depend on hazardous occasions, locations, users, and bikes. Therefore, we adapted a recent advanced neural network technique to recognize the users' behavior.

Many studies on deep learning have been done in recent years. In particular, in the field of image classification using CNN, since Hinton [11] dramatically improved the accuracy of image classification with the 2012 ImageNet Large-scale Visual Recognition Challenge (ILSVRC)[6], the accuracy of various image classification tasks has largely improved. Therefore, we used the image classification technique using CNN to enable more robust detection of sudden braking from several different sensor data. As in Fig. 5, the vertical line shows Y axis acceleration (acceleration with respect to the bicycle moving direction), and the horizontal line shows seconds. Then, this graph is used as input for CNN.

---

[5] http://fixmystreet.org/.
[6] http://www.image-net.org/challenges/LSVRC/.

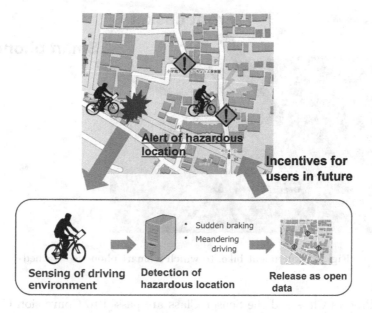

**Fig. 1.** Design of user participatory system

We considered that the increases and the decreases of Y axis acceleration, which are time-series elements, can be taken into a graph form and will work effectively for sudden braking detection. Also, since the CNN provides robustness against positional displacement and deformation of objects to be classified by performing pooling processing, we considered that the CNN is suitable for detecting hazardous locations under various road environments. Figure 1 shows the workflow of sudden braking detection.

### 3.1  Sensing User Data

In the following preliminarily experiments, we attached a smartphone based on Android OS on the left-side center of the handlebar of the bicycle with a smartphone holder. Figure 2 shows an experimental bike.

We built a system for user participatory sensing to collect users' moving information, which consists of an application on Android OS and a server-side process that correspond Sensing Agent collecting sensor data. When the application starts up, the Server Agent collects and extracts hazardous locations from the raw sensor data. Figure 3 shows the system configuration.

In detail, Sensing Agent collects three axes acceleration, three axes geomagnetic intensity, and latitude and longitude data with Sensor Class and passes these data to Process Class. In the Process Class, the time information for given sensor data and flags on whether the sudden braking has been performed are added, and then rotation of the smartphone is calculated from the three axes acceleration and the three axes geomagnetic intensity. The information obtained

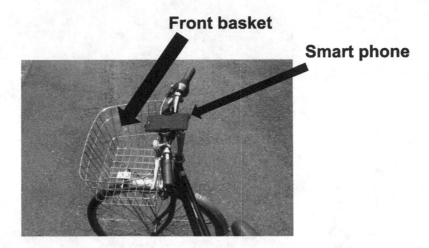

**Front basket**

**Smart phone**

**Fig. 2.** Experiment bike, to which a smart phone is attached

by the Process Class and the Sensor Class are passed to Conversion Class and stored in the smartphone as a text file in CSV format, which will be uploaded to the Server Agent on a daily basis.

Table 1 shows details of sensor data obtained from a smartphone attached on the left side center of a handlebar of a bicycle using a smartphone holder, and thus the moving direction was positive direction of the acceleration sensor Y axis. In the three axes acceleration sensors, the acceleration in the X axis represents the right and left with respect to the moving direction, the acceleration in the Y axis is the forward direction and the acceleration in the Z axis is the moving direction, respectively. In the rotation data calculated from the three axes geomagnetic sensors and the three axes acceleration sensors, pitch, roll, and azimuth represent the X, Y, and Z axis, respectively.

### 3.2 Sudden Braking Recognition

CNN in the Server Agent has two major processes. In the first convolution processing, a high dimensional feature map is obtained by the number of convolution filters by performing a convolution operation on the input image. This convolution filter is also called a kernel and has a weight as a fixed size matrix. In this paper, the trials were repeated, and the kernel size was empirically set to 3.

In the second pooling process, the universality of the position is obtained through collecting a plurality of features. In this paper, the max pooling which outputs the maximum value in the rectangle by using a fixed size rectangular filter was used. Repeating these processes leads to a drastic reduction in learning parameters making learning easier.

At each CNN layer, a value is an input to the activation function. In this case, the activation function determines how the sum of the CNN outputs is

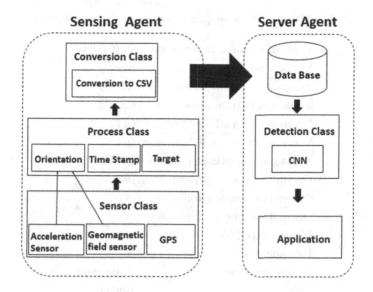

**Fig. 3.** System configuration

activated and plays a role of adjusting a value to be passed to the next layer. We used the ReLU (ramp) function as the activation function. As shown below, the ReLU function outputs 0 when the input is 0 or less, but when the input is larger than 0 the function outputs a value as it is.

$$y = max(0, x) \tag{1}$$

Besides, we used an algorithm called Adaptive Moment Estimation (Adam) that optimizes the gradient descent method used for parameter determination of neural network. For the purpose of preventing divergence of the loss function, the learning rate $lr$ for determining the scale for updating is set to 0.0001.

The image of graph, in which the vertical line shows Y axis acceleration (acceleration with respect to the bicycle moving direction) and the horizontal line shows seconds is used as input for our CNN-based image classifier. The collected data showing the braking events about three times is represented in a graph, which was a time duration to easily visualize and understand the fluctuation of the acceleration with respect to the bicycle moving during actual driving, and then each braking event was extracted. The graph is also converted to the fixed size and then classified by the CNN. The original graph includes 3500 sensor records (in the following experiments, one record was obtained in 0.01 s). We then normalized the acceleration values in the moving direction from 15 to $-15\,\mathrm{m/s}^2$. For each braking event, approximately 4 s before and after the peak of the Y axis acceleration and 10 to $-10\,\mathrm{m/s}^2$ on the Y axis was extracted and converted to a graph. Then, finally, it was resized to $100 \times 100$ pixels and used as input to the CNN classifier. Table 2 shows the model setting.

**Table 1.** Informations acquired by the sensing application

| Types of acquired information | Unit symbol |
|---|---|
| X-axis acceleration sensor | $(m/s^2)$ |
| Y-axis acceleration sensor | $(m/s^2)$ |
| Z-axis acceleration sensor | $(m/s^2)$ |
| X-axis geomagnetic field sensor | $(\mu T)$ |
| Y-axis geomagnetic field sensor | $(\mu T)$ |
| Z-axis geomagnetic field sensor | $(\mu T)$ |
| Orientation sensor (Pitch) | $(\theta)$ |
| Orientation sensor (Roll) | $(\theta)$ |
| Orientation sensor (Azimuth) | $(\theta)$ |
| Longitude (GPS) | $(\theta)$ |
| Latitude (GPS) | $(\theta)$ |
| Acquisition time | Timestamp |
| Target | Binary |

**Table 2.** Variables of analysis system of driving environment

| Variables | Values |
|---|---|
| Input size | $100 \times 100$ |
| Kernerl size | $3 \times 3$ |
| Activation | ReLU |
| Optimizer | $Adam(lr = 0.0001)$ |
| Epoch | 20 |
| Batch size | 25 |

# 4    Experiment Results on Sudden Braking Detection

Section 4.1 describes the results of preliminary experiments that aims to obtain sudden braking and normal braking data, which are in turn training data of the CNN classification model. Experimental results for actual sensor data on street using the trained model are described in Sect. 4.2.

## 4.1    Preliminary Experiment

In the preliminary experiment, we first collect data for training a CNN classification model. Three collaborators (students in our university) drove a bike in an experimental environment and intentionally made sudden braking and normal braking 25 times, respectively. The experimental environment is a concrete road with a flat length of 26 m as shown in Fig. 4. In the case of sudden braking operation, the collaborators accelerate the bike and perform constant speed exercise

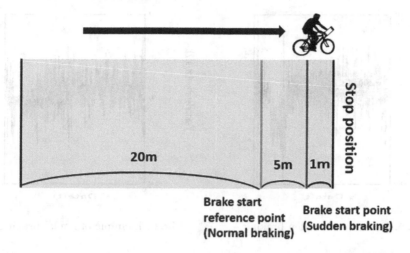

**Fig. 4.** Experimental environment in preliminary experiment (Color figure online)

at the first 20 m, and brake from only 1 m (the red line in Fig. 4) before the stop position. In this study, the sudden braking is defined as "a braking suddenly to avoid or mitigate the imminent danger," since we wanted to detect driving that decelerates without notice. In the case of the normal braking operation, the collaborators accelerate the bike and perform constant speed exercise at the first 20 m, and then they perform braking as usual from 6 m (the blue line in Fig. 4) before the stop position. In this study, the normal brake is defined as "brakes, drivers understand the necessity of stopping in advance, semi-unconscious and daily braking." Aiming to detect brakes which are normally done unconsciously, we gave the collaborators their discretion.

First, we constructed a CNN classifier that can classify sudden braking and normal braking obtained by the preliminary experiments. We then build a system that can detect sudden braking even on public streets, where users actually drive.

Figures 5 and 6 show example data for the sudden braking and the normal braking. As a result, 75 images representing sudden braking and 75 images representing normal braking were input to the classifier, and the accuracy was verified using 10-fold cross validation (10 times tests with 90% images for learning and 10% for testing). The trained classifiers classified sudden braking and normal braking data with 98.7% precision, 84.3% recall, and an F-measure of 90.7% (Table 3).

## 4.2 Actual Experiment on Public Streets

In the actual run experiment, we asked collaborators to use our smartphone application at bicycle driving. Then, we collected bicycle driving data from six collaborators (again, students in our university) for one month and tried to detect sudden braking (and its locations as a result) that occurred during actual bicycle driving on the street using the trained classifier. The six collaborators

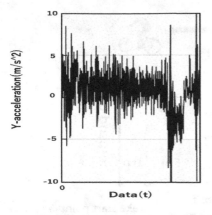

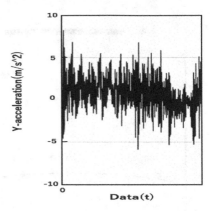

**Fig. 5.** Example of sudden braking    **Fig. 6.** Example of nomal braking

**Table 3.** Classification results when sudden braking is positive, normal braking is negative

| Accuracy | % |
|---|---|
| Precision | 98.7 |
| Recall | 84.3 |
| F-measure | 90.7 |

drove their own bikes with their own smartphones on a daily basis around Chofu station, Tokyo.

In detail, the six collaborators were divided into two groups. Group A is a group that uses a smartphone holder attached on the left side center of a handlebar of a bicycle as well as the preliminary experiments. Therefore, the moving direction was fixed to positive direction of the acceleration sensor Y axis. On the contrary, Group B put their smartphones into their own bags without using the holders, and then placed the bags on the bicycle front basket. The reason is that for user participatory sensing the number of users is very much important; thus, we intended to lower the barriers to participate and use the system as much as possible.

During the experiment, the collaborators were asked to notify sudden braking by pushing a button on our application (after confirming their safety). This button is used only for obtaining correct data to evaluate sudden braking detection. In actual user participatory sensing, hazardous locations will be automatically extracted without human intervention, such as pushing a sudden brake notification button.

Figures 7 and 8 show example data for the sudden braking observed in Group A and B. As a result, six sensor data representing sudden braking in Group A and five sensor data representing sudden braking in Group B were verified. In Group

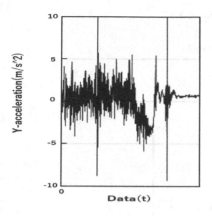

**Fig. 7.** Example of sudden braking in Group A

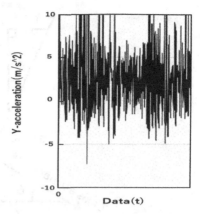

**Fig. 8.** Example of sudden braking in Group B

A, five out of six actual sudden braking events were detected as sudden braking (83.3% accuracy).

On the contrary, in Group B, two out of five events were correctly detected (20% accuracy). This poor result was because in Group A and the preliminary experiment the Y always axis represents the direction of the bicycle moving, but in Group B, the moving direcation was the three accelerations due to the directions of smartphones in the bag. Therefore, we corrected the moving direction using the rotation data that can be calculated from the geomagnetic sensors and the acceleration sensors before inputting the graphs to the CNN classifier.

The smartphones in the bag have nave arbitrary directions. Figure 9 shows an example of installation and a method to correct the acceleration in the bicycle moving direction. Since the smartphone is a rectangle, the rotation in the Y axis direction can be excluded from the correction. The acceleration with respect to the bicycle moving direction can be calculated by correcting the rotation angle in the X axis direction. In the figure, $Z$ is Z axis acceleration, $Y$ is Y axis acceleration, and $G$ is Gravitational acceleration. Thus, $Y'$ (corrected acceleration with respect to the bicycle moving direction) can be expressed by the following equation.

$$Y' = (Z - G \sin \theta) \cos \theta + (Y - G \cos \theta) \sin \theta \qquad (2)$$

After correcting the moving direction and redrawing the graphs to input the CNN, four out of five actual sudden braking events in Group B were classified as sudden braking by our system (80% accuracy). Thus, we found that the CNN classifier can successfully detect sudden braking with high accuracy. However, unfortunately, there are still few sudden braking events obtained on the street both in Groups A and B, although it would be natural that such events is rare in normal bicycle driving on the public streets. We would like to collect more sudden braking events to verify the classification accuracy by continuing the experiment.

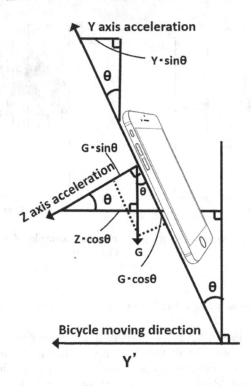

**Fig. 9.** Correction of moving direction

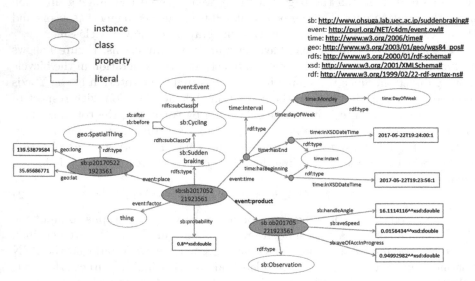

**Fig. 10.** Part of the LOD for OHD

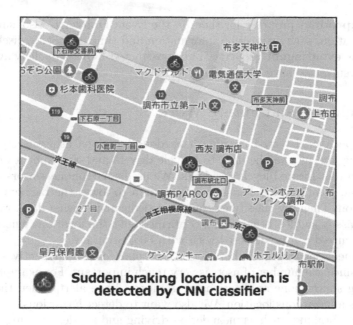

Fig. 11. Illustration of the OHD

## 5   RDF Construction

Finally, we defined an RDF schema for OHD including the detected hazardous location information and released the data as LOD to promote the secondary use by third parties. Figure 10 shows part of OHD LOD, which represent a hazard map illustrated in Fig. 11.

RDF schemata are often designed based on existing ontologies. The Semantic Sensor Network (SSN) ontology [12] proposed by the W3C Semantic Sensor Network Incubator Group is an ontology for describing sensors and their data. However, the SSN is not suitable when the sensor locations are temporally changing, as those attached to bicycles. Thus, we designed the RDF scheme based on Event Ontology[7], since the Event Ontolgy includes location, time, active agent, and factor, and then the information related to the sudden braking event can be structured as an Event class with semantic consistency. The Cycling class is a subclass of the Event class. The Cycling class means a group of bicycles driving and has a Sudden braking class as a subclass. Since the Sudden braking class means a set of sudden braking operations, the Cycling class includes driving before and after the sudden braking operation. In particular, we defined that the bicycle driving behavior included in the Cycling class has velocity, acceleration, and steering angle at specific intervals. Thus, we created

---

[7] http://motools.sourceforge.net/event/event.n3.

the **event:product** property for them in the Event Ontology. Also, since sensitive information, such as position information and time cannot be disclosed, we omitted the **event:agent** property. The constracted LOD is puclicly available.[8] This LOD currently have 770 triples.

## 6   Conclusion and Future Work

In this paper, we detected possible hazardous locations using user participatory sensing and offered the data to bicycle drivers and government officials as OHD to prevent future bicycle accidents. We used smartphone sensors to obtain acceleration, location, and handle rotation data to lower the barriers to users' participation. Then, by classifying those not-dedicated sensing data, we applied CNN to classify sudden braking and confirmed that the locations where sudden braking occurred can be detected with with 80% accuracy.

As we mentioned at the end of Sect. 2.3, we plan to design incentives for users in future, since the number of users is very much important. For example, we are considering smartphone applications or games to issue alerts when the bicycle approaches a hazardous location. We also want to detect hazardous events other than sudden braking, such as meandering driving and sudden steering wheel. In addition, we will personalize the classifier by considering users' normal driving patterns, since the riding styles for the users are different in general. Finally, by linking the constructed RDF with POI data etc., we further consider the factors of bicycle accidents. For example, we could estimate sudden braking locations in the area, where the sensor data are not collected yet by making use of the analysis result.

**Acknowledgments.** This work was supported by JSPS KAKENHI Grant Numbers 16K12411, 17H04705. I would like to thank the students cooperated with the experiment.

## References

1. Herbert, W.H.: Industrial Accident Prevention: A Scientific Approach. McGraw-Hill, New York (1931)
2. Administrative evaluation office of Ministry of Internal Affairs and Communications Administrative evaluation and monitoring of bicycle traffic safety measures. http://www.soumu.go.jp/menu_news/s-news/94984.html (in Japanese)
3. Eisenman, S., Miluzzo, E., Lane, N.: BikeNet: a mobile sensing system for cyclist experience mapping. ACM Trans. Sens. Networks **6**(1), 1–39 (2009)
4. Saito, H., Yamada, K.: Collecting, sharing and inferring user behavior for mobile recommendation with location history and its context. In: Proceedings of European Conference on Data Mining 2013 Prague Czech Republic, July 2013
5. Takeuchi, S., Tamura, S.: Human action recognition using acceleration information based on hidden markov model. In: Asia-Pacific Signal and Information Processing Association, No. 1, pp. 829-832, Sapporo, October 2009

---

[8] http://www.ohsuga.lab.uec.ac.jp/openhazarddata.

6. Kiyohara, T., Orihara, R., Sei, Y., Tahara, Y., Ohsuga, A.: Activity recognition for dogs based on time-series data analysis. In: Duval, B., Herik, J., Loiseau, S., Filipe, J. (eds.) ICAART 2015. LNCS (LNAI), vol. 9494, pp. 163–184. Springer, Cham (2015). https://doi.org/10.1007/978-3-319-27947-3_9
7. Kawamura, T., Umezu, K., Ohsuga, A.: Mobile navigation system for the elderly - preliminary experiment and evaluation. In: Proceedings of the 5th International Conference on Ubiquitous Intelligence and Computing, Oslo University College, pp. 578–590. Oslo, Norway (2008)
8. Egami, S., Kawamura, T., Ohsuga, A.: Building urban LOD for solving illegally parked bicycles in Tokyo. In: Groth, P., Simperl, E., Gray, A., Sabou, M., Krötzsch, M., Lecue, F., Flöck, F., Gil, Y. (eds.) ISWC 2016. LNCS, vol. 9982, pp. 291–307. Springer, Cham (2016). https://doi.org/10.1007/978-3-319-46547-0_28
9. Arakawa, Y., Matsuda, Y.: Gamification mechanism for enhancing a participatory urban sensing: survey and practical results. J. Inf. Process. **57**(1), 1–15 (2016)
10. Matsuda, Y., Arai, I.: An experiment of a streetlamp classifying and a vertical illuminance assessing utilizing smartphones light sensors. In: 3rd International Symposium on Technology for Sustainability (ISTS2013), No.264, Hong Kong, China, November 2013
11. Krizhevsky, A., Sutskever, I., Hinton, G.: ImageNet Classification with Deep Convolutional Neural Networks. In: Neural Information Processing Systems (NIPS 2012), p. 25 (2012)
12. Compton, M., Barnaghi, P., Bermudez, L.: The SSN ontology of the W3C semantic sensor network incubator group. Web Semant. Sci. Serv. Agents World Wide Web **17**, 25–32 (2012)

# Semantic IoT: Intelligent Water Management for Efficient Urban Outdoor Water Conservation

Trina Myers[✉] [iD], Karl Mohring [iD], and Trevor Andersen [iD]

James Cook University, Townsville, Queensland, Australia
{trina.myers,karl.mohring,trevor.andersen}@jcu.edu.au

**Abstract.** Water depletion is critical in the dry tropics due to drought, increased development and demographic or economic shifts. Although educational initiatives have improved urban indoor water-use, excessive outdoor wastage still occurs because in most urban areas residential users only have a biannual reading of quantity available to make informed or educated decisions on necessary or unnecessary consumption. For example, the average consumer will water lawns during a designated non-restricted time. The amount of water they use is determined arbitrarily (i.e., either by sight or by blocks of time). In many cases, water is wasted due to over saturation, automated sprinklers that cannot sense precipitation, poor placement of sprinkler direction, etc. Outdoor water use efficiency could be maximized if water flow was shut off when an area of lawn has had sufficient water based on a more intelligent monitoring system. This paper describes the development of an intelligent water management and information system that integrates real-time sensed data (soil moisture, etc.) and Web-available information to make dynamic decisions on water release for lawns and fruit trees. The initial pilot-prototype combines Semantic Technologies with Internet of Things to decrease urban outdoor water-use and educate residents on best water usage strategies.

**Keywords:** Semantic technologies · Internet of Things · Water conservation

## 1 Introduction

The key drivers to develop sustainable urban water management are external factors such as climate change, drought, population growth and consolidation in urban centers [1, 2]. As the era of cheap water fades, these drivers have increased the need for water industry providers to implement more sustainable strategies in urban water management and conservation. Consumer education on household water use is a strategy used to decrease excessive water consumption [3].

The current focus has been on improving water use inside the home but a large part of the problem exists in outdoor use of water and unintelligent watering systems. The methods to motivate the public to change bad water use habits are driven primarily by mandated water restrictions and initiatives to install water efficient devices (e.g., shower heads). However, to change behaviour, awareness and deeper understanding of the underlying variables, such as soil saturation, soil type, timing and quantity, must be part

© Springer International Publishing AG 2017
Z. Wang et al. (Eds.): JIST 2017, LNCS 10675, pp. 304–317, 2017.
https://doi.org/10.1007/978-3-319-70682-5_21

of the education process [1, 3]. However, to make informed decisions or to automate water consumption processes in smarter ways, one source of data to gauge home use - the water metre - is not adequate. To be successful, a conservation program must get the data to the consumer and make the change financially beneficial to them [3]. People must be given the "geo-temporal" and fiscal context of their consumption:

- How much water do I use or should I use, how much money can I save?
- How do I fare compared to my street, my neighbourhood, my city?
- Based on weather data and evapotranspiration calculations – how much should I have used outside? [3]

*Intelligent water metreing* (IWM) can transform urban water management and determine, in real-time or near real-time, water consumption to provide local or remote data on water consumption [4]. There are municipal initiatives to install smart water metres across wider communities (e.g., Townsville, Mackay and Gold Coast in Queensland) that logs a resident's water usage hourly and streams the data via wireless technologies to a main server, which can be accessed by the home owner via a Web browser to visualize daily water-use. These initiatives are building awareness of water consumption at the user level and alerts to leaks and wastage. However, the data only shows the quantity of water consumed and not whether the water was unnecessarily used in the first place.

The promotion of smarter urban water use will require more extensive data than that currently available to household residents (i.e., total quantity in a 6-month period). For example, the average consumer will water lawns during the designated non-restricted times. The amount of water they use is determined arbitrarily (i.e., either by sight or by blocks of time). In many cases, water is wasted due to over saturation, automated sprinklers that cannot sense precipitation, poor placement of sprinkler direction, etc. If that consumer were alerted or the water flow stopped when an area of lawn has had sufficient water based on a more intelligent monitoring system, outdoor water use efficiency could be maximized. There has been much work in creating smarter homes via internal Internet of Things (IoT) sensor networks for efficient power consumption [5–7]. Semantic technologies (i.e., linked data) combined with IoT could also be applied to better manage water usage in the garden.

The Lawnbot pilot study aimed to advance efficient autonomous irrigation by developing an intelligent system of aggregated data to make decisions on necessary versus unnecessary water use in outdoor watering systems (i.e., water is only used when it is required). The Lawnbot project entails a pilot irrigation management system that makes intelligent decisions on water release based on data from various in situ sensors integrated with external Web available data and information (Fig. 1). Specifically, the research objectives of this project are to (1) infer alerts and trigger autonomous decisions in residential outdoor irrigation systems to minimize waste, (2) maximize plant and fruit growth and (3) build consumer awareness for better water use habits.

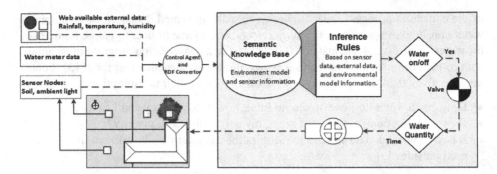

**Fig. 1.** Architecture for pilot semantically enabled urban irrigation

## 2   Background

### 2.1   Current Watering Paradigms

North Queensland has been under water restrictions since 1987 following a prolonged dry season and recently in heavy water restrictions since 2015. In Townsville, these restrictions limit watering lawns and gardens up to a maximum of four hours per week per household and in accordance to a strict schedule. In response, local municipal authorities have encouraged residences to adopt plant species with lower water requirements and less wasteful watering behaviours [4].

The Townsville Municipal Council introduced a recommended weekly lawn watering volume of 25 mm. By this recommendation, a small lawn in North Queensland of 150 m$^2$ should receive approximately 3750 L per week to promote healthy growth. Common sprinklers use up to 2100 L/h and low-flow sprinklers use under 600 L/h [8]. The water pressure would determine how long a sprinkler would take to reach this desired litre capacity. Notably, in a majority of this city, an amount of 3750 L would be reached in approximately two hours using a common sprinkler. However, residents have been observed to take advantage of the four hours of weekly watering time by running sprinklers for the entire duration. With a single, typical sprinkler, this undesired behaviour can result in a weekly water consumption of 4800–8640 L, which is in excess of what is actually needed by most lawns.

### 2.2   Factors that Influence Required Water Volume

The watering recommendation given by the Townsville Municipal Council represent a general estimate of lawn watering requirements. However, the actual amount of water required for grass depends on many factors, some include: species, sunshine, humidity, evapotranspiration, ground soil moisture, rainfall, etc. [9]. Information on these factors can come from three various sources: the sensed environment, inferred from external sources, or from user input.

Real-time information about the surrounding environment, collected by sensors or regular surveillance, is useful for finding the current conditions of the plants and surrounding soil. The current conditions can be employed and tracked mostly to determine if it is an appropriate time to water, as well as the actual amount of water that has been supplied, and how much is needed. For example, the best times to water plants are during cool and humid periods to minimize the amount of water lost to evaporation [10]. Therefore, ambient temperature and relative humidity sensors would be used to determine the best watering times. Further, soil moisture sensors can determine the saturation level of the soil to ensure the soil is not over-watered, which can lead to nutrient depletion in soil and root death from oxygen starvation [11]. Ambient light levels can assist in tracking shade and cloud cover and predicting weather events.

Plants in loose or granular soils tend to drain quickly, which means that plants must be watered for longer, as they only have a short amount of time to take in water. Conversely, cohesive soils such as clay have poor drainage, which gives more time to take in water, but put roots at higher risk of waterlogging if water is supplied too quickly. Vertical soil sensors can monitor and track how water moves through the soil to determine its drainage rate.

Environmental conditions beyond the immediate watering area/s can be inferred using external information. One of the most impactful factors that affects the required watering volume is the weather, especially rainfall, which can eliminate the need for watering entirely. A purely-sensed control system would be able to detect rainfall to halt watering, but would unable to anticipate rainfall. This lack of awareness could lead to wasted water by not taking advantage of natural resources and may put the soil at risk of waterlogging. However, this scenario can be avoided by aggregating weather forecasts, localized sensing equipment, and nearby monitoring stations to track rainfall and predict where and when rain will occur, then adjusting the watering schedule accordingly to leverage natural watering. Similarly, the physical and chemical makeup of the soil can be inferred from real-time sensor information and geographical surveys, given the approximate location of the residence [12].

Another factor that affects the water requirements of plants is evapotranspiration, which is the combined water loss through evaporation and transpiration. Evapotranspiration is specific to plant species, the surrounding environment and represents the optimum amount of water that the plant should receive for healthy growth. Calculating evapotranspiration is a complex procedure that must take multiple factors into account such as ambient temperature, relative humidity, and solar radiation [9]. However, this value, along with drainage rate of the soil and rainfall volume, informs how much water must be supplied through irrigation to meet the needs of the plants in the watering quadrant [13].

## 2.3 Resident Specific Information

Some information that affects watering volume that a garden or lawn requires cannot be easily inferred or detected and must be supplied by the user. The three user-defined factors in this study were the species of plants in the watering area, the size and location of the quadrant and the sprinkler type used for watering.

Different species of grass have different water requirements for healthy growth and can enter dormant stages during frigid or drought conditions and can enter dormant stages where they are more susceptible to over-watering. The exact location, size, and shape of the watering area can be used to infer the amount of shade cast on the watering quadrant at different points during the day, which can affect the times when watering is appropriate.

The sprinkler type, such as common, low-flow sprinklers or misters, also has an impact on selecting the best time to water plants and lawns. Airborne watering systems, such as sprayers and misters, deliver water to the entire plants including its stem and leaves. These sprinkler types are better suited to watering in the morning as leaves are susceptible to fungal infection if they are watered at a time when they are not able to dry [8].

## 2.4   Related Work

Recently developed automated irrigation systems emphasize "do-it-yourself", low-cost, and web accessibility enabled by platforms such as Arduino and Raspberry Pi. For example, Vinduino [14] uses multiple moisture sensors at different depths to determine when to water, and prevent overwatering, in vineyards. The developers of the Vinduino project claim 25% water savings across their vineyards [14]. OpenSprinkler provides smart watering control based on historic, current, and forecast weather data [15]. Neither of these projects incorporate Semantic Technologies to introduce a range of data that could enrich the outcomes of the knowledge base.

There are related work that does incorporate semantic technologies such as AGROVOC [16], Agri-IoT [17], CSIRO's Kirby Farm project [18] and the ThinkHome smart home system [6]. AGROVOC is a formal vocabulary in RDF form that allows for the linking of agricultural data. AGROVOC has evolved into a SKOS-XL linked dataset that includes hierarchies of agricultural concepts such as *organisms*, *methods*, *events*, and *processes* and links to other vocabularies about fisheries, environment, and biotechnology [16]. As such, the Agrontology is a potential resource to integrate within the Lawnbot ontology. The ThinkHome project is "smart-home" initiative that incorporates semantic technologies with IoT for improved resource management. However, the focus is predominantly on energy consumption and power management as opposed to water conservation. The Agri-IoT project and CSIRO's Kirby Farm project are semantic web and IoT-based frameworks that are capable of processing multiple data streams for more effective agricultural management [17, 18]. These projects incorporates linked data from multiple data points, including sensed, government and environmental web-based data, for informed and accurate event detection and decision making by farmers. The Agri-IoT and the Kirby Farm projects differ to this study because the focus is in the wider agricultural field rather than the smaller domain of urban lawn management.

## 3   Semantic Knowledge Base and Control Agent

Semantic technology data models aim to capture the meaning of data to represent real world situations for data integration and manipulation [19, 20]. Formal logical paradigms are applied to automate classifications of concepts and the inference of new information. The computer can make intelligent decisions based on conclusions derived through predicate and propositional logic systems embedded in explicit ontological definitions [19, 20].

The Lawnbot ontology (Appendix A) is built on top of the *Semantic Sensor Network* (SSN) ontology [21, 22]. The SSN ontology includes concepts for sensing the environment and making changes through logic-controlled actuators. That is, *Sensors* make *Observations* of *ObservedProperties* belonging to *FeaturesOfInterest* and *Actuators* cause *Actuations* that modify *ActuableProperties* of *FeaturesOfInterest*. For example, the *Observations* of specific areas would infer the *WaterValveActuator* would open the valve to release water.

The *FeaturesOfInterest* relevant to intelligent water management are *Yards*, *Quadrants*, and *WeatherAreas*. That is, each *Yard* consists of several *Quadrants* (Fig. 2) and would fall within a wider *WeatherArea*. Each *Yard* may have distinct watering requirements depending on its properties, for example: different *SoilComposition*, different *MicroClimateFactors* based on the timing and amount of shading, etc.

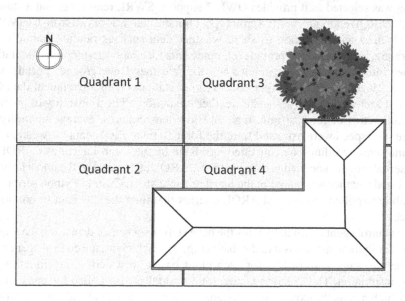

**Fig. 2.** Lawnbot test layout showing individual watering quadrants

*Quadrants* contain *Plants*, each of which has a *PlantSpecies*. The dimensions and life cycle status of plants are data-type properties for use in the inference rules to model size and possible impact on shade. *PlantSpecies* determines the crop coefficient, which combined with the dimensions and life cycle status, can together help infer evapotranspiration and determine the watering requirements for the *Quadrant*.

Local sensors gather data at the *Yard* and *Quadrant* levels. At the *Yard* level, sensors measure the *ObservedProperties* that include temperature (ambient and soil), humidity (ambient and soil) and illuminance. At the *Quadrant* level, soil moisture (both superficial and deep) is observed. The sensed data is collected via the control agent and converted to RDF form and ingested to the knowledge base.

Each *Quadrant* contains a *Plant* of a *PlantSpecies,* which determines *WateringRequirements.* *Quadrant* has *SoilComposition*, with properties that can affect watering or fertilization. We further model *Quadrant* size and shading information. *PlantSpecies* has a crop coefficient for determining evapotranspiration. *Quadrant* has a *MicroClimateFactor* affected by shading to determine evapotranspiration.

A *Sprinkler* in each *Quadrant* (or across quadrants) is supplied water by opening a water valve, which is represented as a *WaterValveOnState.* The *SprinklerType* determines the data property *WaterVolumePerMinute,* which is applied in the inference rules to toggle the *WaterValveOnState* for each *Quadrant/s.* Forecast weather data for a *WeatherArea* is modelled by *PredictedProperties* including probability of precipitation, quantity of precipitation, high and low temperature, average windspeed, and average humidity. The concept of *WateringRestrictions* is applied to *Yards* to avoid illegal watering.

The Stardog graph triplestore[1] was used to develop the semantic knowledge base. Stardog was selected as it provides OWL 2 support, SWRL reasoning, and a standard HTTP SPARQL endpoint. For the prototype, climatic data was drawn from CLIMWAT [23], which is an application to share weather data such as rainfall, humidity and temperature, and was used to provide reference data for evapotranspiration calculations. Weather forecasting data is extracted via the Weather Underground[2] portal, which combines citizen science data (personal weather stations) with government data (e.g., Bureau of Meteorology) to automate weather predictions. The probability of precipitation, millimetres of forecast rain, high and low temperatures, average humidity and average windspeed were extracted from the forecast data. JSON data for weather forecasts and sensor readings are converted into RDF by the control agent using RDFLib[3] and inserted into the knowledge base via the SPARQL endpoint. For the pilot study, raw sensory and weather was stored in the Stardog triple-store. Custom Python scripts with RDFLib are applied to create SPARQL queries that map the raw data to ontological instances.

The control agent continually polls the base station for sensor data using the requests module for Python and sends it to database (Fig. 3). Each night, the control agent calculates the net water gain or loss for each plant based on watering, precipitation, and evapotranspiration. The Weather Underground portal is also polled for updated forecasts, which means the expected evapotranspiration can be calculated, and so expected gains or losses in water in the coming days can be determined.

---

[1] http://www.stardog.com/.

[2] https://www.wunderground.com/.

[3] https://github.com/RDFLib.

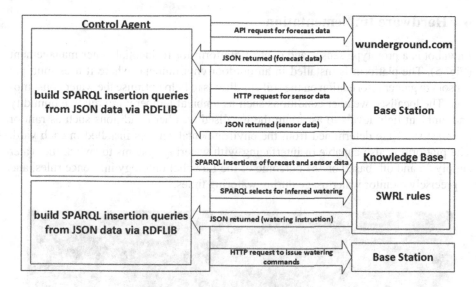

**Fig. 3.** The control agent architecture.

Based on the needs of the plants and lawn, the forecast evapotranspiration and precipitation, and watering restrictions, the system can infer whether to turn the water on and for how long (i.e., how many litres of water is required for each quadrant) (Fig. 4). For example, grass on a given quadrant may require 25 mm of water per week under typical conditions in summer due to proximity of a shading object such as a house or tree. If six dry days have passed, but a 90% chance of 40 mm of precipitation is predicted in the next three days, the system will determine that it should not water the quadrant, but instead wait for the expected rain. The soil moisture sensors will determine if the expected rain has occurred to ground truth the inference outcome.

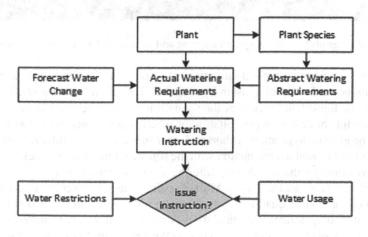

**Fig. 4.** The inference rule schema.

# 4    Hardware Implementation

Lawnbot is a prototype sensor and control platform for residential water management (Fig. 5). The platform is installed in an outdoor environment, where it uses multiple sensors to gather information on the soil conditions to help optimise the water consumption. The localised weather conditions such as ambient temperature, relative humidity and ambient light levels to ascertain the localised weather conditions such as rain or overcast skies are determined from the environmental sensors installed in each yard. The platform is also capable of interfacing with watering systems to switch the water supply on and off, based on the outcome of the Lawnbot ontology inference rules, and to precisely monitor water usage during watering times.

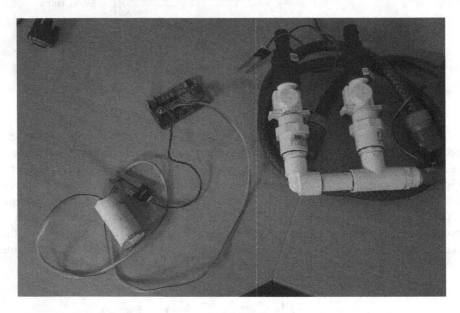

**Fig. 5.** Lawnbot hardware system showing soil sensors and water control system.

Two sensor types measure soil saturation at two different depths: a surface-level soil-moisture probe, and a buried gypsum hygrometre. The soil-moisture probe is a device that sits in the topsoil and measures the saturation of the superficial layer of the soil, which is useful for detecting precipitation or when water is otherwise pooling on the ground. The gypsum hygrometre is buried deeper in the soil to monitor moisture levels at root level and is used in conjunction with the soil-moisture probe to track the rate that water moves through the soil during differing environmental conditions, surrounding different plant species, and soil types. Multiple pairs of these sensors can effectively split up a lawn or garden into quadrants, which can be monitored and watered individually. This separation of quadrants is particularly advantageous if they have differing circumstances, such as shade, changing soil types and/or proximity to external water sources such as rivers or dams. Similarly, the system uses multiple valves and flow metres to track how much water is supplied to each watering quadrant (Fig. 2).

The platform can run in a standalone configuration, but its limited awareness of the surrounding area reduces its potential effectiveness. For example, the system may waste water by watering before a rainstorm. By incorporating linked data, the control agent transmits the sensor data to the semantic knowledge base for combination with external data sources such as local weather information to take advantage of natural rainfall and conditions for better water efficiency. In this configuration, no standalone switching occurs and all water management is handled by commands received from the semantic knowledge base.

## 5  Implementation and Discussion

Lawnbot was trialed on a residential property using four watering quadrants (Fig. 2). All four quadrants were spatially separated by the reach of the sprinkler type to avoid water spilling in from other quadrants, but were subject to the same weather conditions. Soil and grass types were consistent for all quadrants, but two quadrants received shade for most of the afternoon, while the other two were in full sun for most of the day.

For direct comparison between Lawnbot and conventional watering schemes, one shaded and one non-shaded watering quadrant were managed by the Lawnbot system, while the remaining areas were watered by typical water usage under locally-imposed timed water restrictions. These restrictions limited watering to only three days per week and for limited times during the morning or afternoon. Water metre readings before and after each hand watering period were used to calculate the total volume used during each session.

Lawnbot watering was enabled throughout the week drawing data from local and external sources to infer water use. Both the soil probe and gypsum hygrometre were installed at the center of each watering quadrant managed by Lawnbot, with hygrometres buried at a depth of 0.5 m. After a testing period of 30 days, the water usage of each of the quadrants were compared, as well as a visual check of the grass in each quadrant to observe if the grass appeared healthy. On a daily average, the Lawnbot system used less water than the manual system because it stopped the water flow after an inferred period while the manual watering occurred for the full four-hour council allotment.

The outcome is an anticipated decrease in the quantity of water used in outdoor irrigation at the residential level. Table 1 shows a six month simulation over the 2016 January to June period in Townsville and Cairns, which contrasts a dry tropical zone to a wet tropical zone. The control yards are watered 25 mm every 7 days on schedule regardless of actual rain. The lawnbot yards are watered so as to maintain 25 mm over 7 days while calculating past rain and predicted rain up to three days out. Cairns shows 32% water savings and Townsville shows 21%. Notably, a real-time long term trial is not possible at present due to drought level watering restrictions.

The immediate benefits of the system are to the council's water management program, residents who pay for water and/or users who are concerned with water depletion. The proposed output will be a pilot system that will be demonstrated by automatically managing residential outdoor irrigation for lawns and fruit trees based on various

**Table 1.** Simulated inference in Townsville and Cairns, North Queensland

| Month | Predicted average | Cairns rain | Cairns control quadrant | Cairns lawnbot quadrant | Townsville rain | Townsville control quadrant | Townsville lawnbot qaudrant |
|---|---|---|---|---|---|---|---|
| Jan | 10.35 | 4.66 | 3.57 | 2.46 | 1.83 | 3.57 | 2.25 |
| Feb | 9.11 | 3.20 | 3.57 | 2.29 | 2.23 | 3.57 | 1.96 |
| Mar | 6.29 | 5.50 | 3.23 | 2.29 | 8.84 | 3.23 | 2.29 |
| April | 8.58 | 3.11 | 4.17 | 1.57 | 0.22 | 4.17 | 3.23 |
| May | 7.53 | 2.73 | 3.23 | 2.81 | 0.02 | 3.23 | 3.42 |
| June | 6.26 | 1.21 | 3.33 | 2.97 | 0.52 | 3.33 | 3.40 |
| Total | | 20.41 | 21.09 | 14.38 | 13.66 | 21.09 | 16.56 |

disparate data input sources and a semantic system that "understands" how the variables interact.

Automating the release of water (the system manipulates the valve) will further benefit the resident and promote use of the system. The residents will visually see when water should or should not be used and money saved based on the aggregate of available data and inferred output, which are relevant to changing water consumption behaviour [3].

# 6    Conclusions

This paper presented the prototype Lawnbot water management platform, which is an automated watering system for residential lawns and gardens that applies Semantic and IoT technologies. The resulting system incorporates real-time sensor data, weather forecasts, geological and environmental information to infer the precise amount of water needed to minimize water wastage without compromising the health and wellbeing of the lawn or garden. The prototype combines a sensor-actuator system that automatically manages the water flow in yards based on semantic inference. The combination of data from multiple sources with a sensor-actuator system has the potential to make better watering decisions than other systems of its kind. A method to evaluate the system was discussed that compared the watering performance of the semantic-controlled platform to manual watering under council water restriction guidelines.

Future work of the Lawnbot semantic knowledge base includes the refinement of the Lawnbot ontology, a user dashboard and extending controls to fertilisers. The spatial accuracy of weather predictions and rainfall tracking can also be augmented by gathering information from nearby urban sensor installations, and from other IoT platforms. A visualisation tool such as a user dashboard would better inform users of their water usage habits and compare with nearby properties. The residents will visually see when water should or should not be used and money saved based on the aggregate of available data and inferred output. Users will also be able to define their own watering quadrants with specific shade areas and plants to input into the knowledge base. Further, there are plans to expand the system to manage controls of liquid fertilisers and pH balancing for improved plant health.

# Appendix A

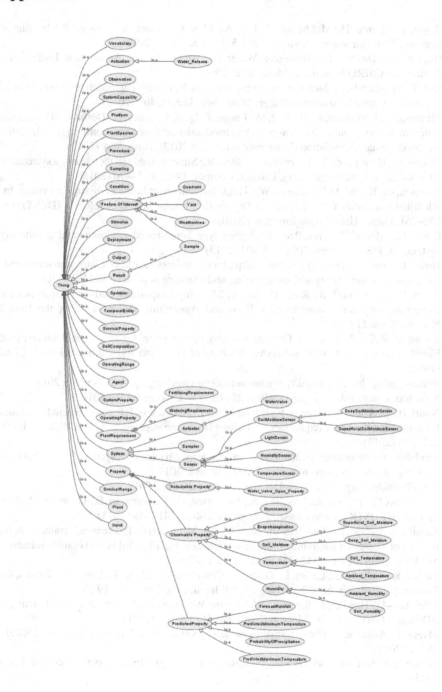

# References

1. Boyle, T., Giurco, D., Mukheibir, P., Liu, A., Moy, C., White, S., Stewart, R.: Intelligent metering for urban water: a review. Water **5**(3), 1052–1081 (2013)
2. Hussey, K., Dovers, S.: Managing Water for Australia: The Social and Institutional Challenges. CSIRO Publishing, Melbourne (2007)
3. Hill, T., Symmonds, G.: Sustained water conservation by combining incentives, data and rates to effect consumer behavioural change. Water Soc. **153**(1), 409 (2012)
4. Manning, C., Hargroves, K., Walker, M., Lange, J., Igloi, S., Bruce, G., Desha, C.: Dry tropics water smart: a community based approach to residential outdoor water consumption. In: 2013 Australian Water Association Conference (OzWater 2013), Perth, WA (2013)
5. Wilson, C., Hargreaves, T., Hauxwell-Baldwin, R.: Smart homes and their users: a systematic analysis and key challenges. Pers. Ubiquit. Comput. **19**(2), 463–476 (2015)
6. Reinisch, C., Kofler, M.J., Kastner, W.: ThinkHome: a smart home as digital ecosystem. In: 4th IEEE International Conference on Digital Ecosystems and Technologies (DEST), pp. 256–261, Dubai, United Arab Emirates (2010)
7. Cook, M., Myers, T., Trevathan, J.: A prototype home-based environmental monitoring system. Int. J. Smart Home **7**(6), 389–404 (2013)
8. How to save water at home. https://www.townsville.qld.gov.au/water-waste-and-environment/water-supply-and-dams/saving-and-consumption. Accessed Feb 2017
9. Allen, R.G., Pereira, L.S., Raes, D., Smith, M.: Crop evapotranspiration - Guidelines for computing crop water requirements. Food and Agriculture Organization of the United Nations, Rome (1998)
10. Smajstrla, A.G., Zazueta, F.: Evaporation loss during sprinkler irrigation. University of Florida Cooperative Extension Service, Institute of Food and Agriculture Sciences, EDIS (1994)
11. Water-logging. http://soilquality.org.au/factsheets/waterlogging. Accessed Jan 2017
12. Soils series. https://data.qld.gov.au/dataset/soils-series. Accessed Jan 2017
13. Nouri, H., Beecham, S., Kazemi, F., Hassanli, A.M.: A review of ET measurement techniques for estimating the water requirements of urban landscape vegetation. Urban Water J. **10**(4), 247–259 (2013)
14. Vinduino, a wine grower's water saving project. https://hackaday.io/project/6444-vinduino-a-wine-growers-water-saving-project. Accessed June 2017
15. OpenSprinkler. https://opensprinkler.com. Accessed 7 June 2017
16. Caracciolo, C., Stellato, A., Morshed, A., Johannsen, G., Rajbhandari, S., Jaques, Y., Keizer, J.: The AGROVOC linked dataset. Semant. Web **4**(3), 341–348 (2013)
17. Kamilaris, A., Gao, F., Prenafeta-Boldu, F.X., Ali, M.I.: Agri-IoT: a semantic framework for Internet of Things-enabled smart farming applications. In: IEEE 3rd World Forum on Internet of Things (WF-IoT), pp. 442–447, Reston, USA (2016)
18. Taylor, K., Griffith, C., Lefort, L., Gaire, R., Compton, M., Wark, T., Lamb, D., Falzon, G., Trotter, M.: Farming the web of things. IEEE Intell. Syst. **28**(6), 12–19 (2013)
19. Allemang, D., Hendler, J.: Semantic Web for the Working Ontologist: Effective Modeling in RDFS and OWL, 2nd edn. Morgan Kaufmann, Burlington (2011)
20. Myers, T., Atkinson, I.: Eco-informatics modelling via semantic inference. Inf. Syst. **38**(001), 16–32 (2013)
21. Semantic Sensor Network Ontology. https://www.w3.org/TR/vocab-ssn/. Accessed 4 July 2017

22. Compton, M., Barnaghi, P., Bermudez, L., García-Castro, R., Corcho, O., Cox, S., Graybeal, J., Hauswirth, M., Henson, C., Herzog, A., Huang, V., Janowicz, K., Kelsey, W.D., Le Phuoc, D., Lefort, L., Leggieri, M., Neuhaus, H., Nikolov, A., Page, K., Passant, A., Sheth, A., Taylor, K.: The SSN ontology of the W3C semantic sensor network incubator group. Web Semant. Sci. Serv. Agents World Wide Web **17**, 25–32 (2012)

23. CLIMWAT.    http://www.fao.org/land-water/databases-and-software/climwat-for-cropwat/en/. Accessed Aug 2017

# Semantically Enhanced Case Adaptation for Dietary Menu Recommendation of Diabetic Patients

Norlia Mohd Yusof and Shahrul Azman Mohd Noah[✉]

Center for Artificial Intelligence Technology,
Faculty of Information Science and Technology,
Universiti Kebangsaan Malaysia, UKM,
43600 Bangi, Selangor, Malaysia
norlia@siswa.ukm.edu.my, shahrul@ukm.edu.my

**Abstract.** Dietary menu planning for diabetic patients is a complicated tasks involving specific and common-sense knowledge. Case-based approach has been used to provide recommendation in the case where ratings were not easily available for domains such as menu planning. Among the important but yet difficult tasks in the case-based approach is case adaptation. To successfully support case adaptation, the constraint-based approach and food composition ontology were employed. Constraints knowledge were represented as production rules and exploits the food ontology to support adaptation. An ontological approach is also proposed to perform the inference process to satisfy the multiple design constraints.

**Keywords:** Dietary menu planning · Ontology · Case adaptation · Knowledge-Based recommendation

## 1 Introduction

Designing nutritious menu plan for patients, however, is not the same as designing menus for healthy individuals [1]. Patients require advice and directions from dietitians in designing their menus. Dietitians on the other hand, during the course of consultation with a patient, may need to consider specific dietary constraints and even simple common sense constraints. Among these constraints are: first, the physical constraints where the design task needs to fulfill the food groups and its serving sizes allotted in the meal exchange table (MET). Second is the personal preferences of specific patients that might be due to cultural, ethnic and religious beliefs [2]. Thus, there must be a mechanism to consider these aspect when recommending dietary menu planning. Third, is the common sense involved. There is a common-sense that some foods can go together (e.g., roast turkey with stuffing) while others do not (e.g., roast turkey with ketchup and pickles) [3].

Dietitians require to consider both physical and personal needs of the patient. This requirement is essential to help patient willingly and able to change his or her diet behavior that will benefit to their health. Thus, individualization is imperative when designing a menu for therapeutic diet.

© Springer International Publishing AG 2017
Z. Wang et al. (Eds.): JIST 2017, LNCS 10675, pp. 318–333, 2017.
https://doi.org/10.1007/978-3-319-70682-5_22

Recommender system technology is a technology that aim to generate meaningful recommendations to a specific user based on his or her profile, interests or past preferences. While it uses is mainly popular in the area of e-commerce and search applications, its applications in more specialized and personalized domain such as menu recommendation are gaining attention among researchers [4–6]. While user's ratings of items and item's content have been popularly used to support prediction in collaborative-based and content-based recommender systems respectively, there are cases where such choices of data are not best choice [7]. Therapeutic menu recommendation is one of the domain that have little to no ratings. Thus, we opted to knowledge-based recommender methods.

Case-Based Reasoning (CBR) is one of the popular approaches used for menu planning and generations as exhibited by the work of [3, 8, 9]. However, rarely the generated menus can be directly recommended to patients due to certain factors individual preferences. Thus, case adaptation is necessary. Case adaptation is considered as the most difficult task of CBR's life cycle particularly when involving constraints. Efforts in automating case adaptation have been mainly focused at the processing level rather at the data level. Thus, changes in the adaptation process require extensive changes at the implementation level. Semantic technology, however, able to leverage the burden on the processing level by incorporating intelligent at the data level.

This research, therefore, proposed an ontological approach to dietary menu planning for diabetic patients. An ontological approach is proposed to satisfy both the physical and aesthetic constraints using structural transformation and substitution adaptations.

## 2 Related Work

The CBR approach to dietary menu planning is exhibited by the work of Khairudin et al. [9]; Khan et al. [8]; Marling et al. [3] and Hinrichs [10] which involved retrieving menus for new case based on previous past cases. The limitation of the CBR approach is that the retrieved menus heavily rely on stored cases and require extensive adaptations for those cases which are only partially similar. As CBR still dealing with conventional database such as relational and Memory Organization Packets (MOP), the inferencing elements are very much limited rely heavily on the logical processing of the CBR engine. Another work worth mentioning is [11]. This work concern with developing an intelligent agent for diabetic food recommendation. The agent capable of to create a meal plan according to a person's lifestyle and particular health needs. However, the capability of the agent is still at the level of food recommendation and not menu recommendation. As such, the literature reported the use of Taiwanese food ontology and personal food ontology. It is also not that clear to what extent the semantic technology is being used in this work.

In this paper, we proposed to use the semantic technology framework namely the ontology to support case adaptation of dietary menu planning and generation for diabetic patients. The case adaptation approach proposed mainly to handle constraints specified in the requirements of the given cases. Instead of dealing constraints at the processing level, we proposed to deal constraints at the data level i.e. by modeling the cases; the domain knowledge by means of ontological models and the adaptation knowledge by means of ontological reasoning.

## 3 The Proposed Approach

Figure 1 shows the activities involved in the proposed approach. The first activity taken in this study was the modeling of the domain knowledge using ontology. The second activity involved the development of a CBR module. It consists of case base modeling, CBR engine development and ontology inferencing. The final step is the evaluation of the proposed approach by comparing them with the conventional approach. In this paper, we discuss the first and second activity in the following subsections whereas the last activity is our future works.

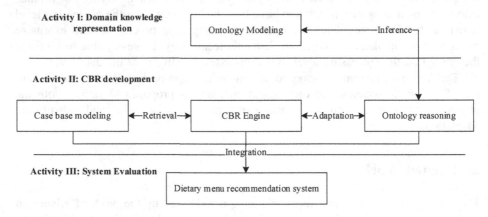

**Fig. 1.** Implementation process of the proposed approach.

### 3.1 Ontology Modeling

Ontology is divided into two main components i.e. knowledge representation and reasoning mechanisms. Ontology modeling is considered as a knowledge representation activities. The domain knowledge was modeled based on the Malaysian food composition database [12]. This database is essential tool for dietitians during designing healthy dietary menus [13]. It comprises the macronutrients, vitamins and minerals for each food that normally eaten by Malaysian. The food composition ontology was developed by following the Ontology Development 101 methods proposed by [14] and was discussed in [15]. The Protégé 4.3[1] was used to develop the ontology based on the Web Ontology Language (OWL) framework. This language is the de facto standard for ontology-based application system.

Three components of knowledge representation in OWL are class, property and individual. Two main classes of food composition ontology are raw and processed

---

[1] https://protege.stanford.edu/.

foods; and cooked foods. Figures 2 and 3 illustrate both of these classes. As shown in both figures, the subclasses in shade indicate that the class has an equivalent relationship with another class. The equivalence class is needed because the some of the food groups consist of raw; processed and prepared foods. For example, the plant-based protein food group consist of chickpeas (raw food); baked beans (processed food) and dhal gravy (prepared food). Three remaining food groups do not have equivalence class since they do not have any cooked food.

Properties are very important components in ontology modeling because it allows the relationships between classes and among individuals. The property component represent a binary relation between individuals which known as object-type property, or between individual and data type which known as data-type property. The both property of main classes have been set up differently. Due to the limited space, We discuss the design decision taken for object- and data-property of raw and processed foods only.

The personalization aspects of foods are assigns using object-type property and data-type for the significant nutrient belong to them. For example, if the food is known as local food such as rice for staple food or guava for local fruits, it has object property hasRace as Malay, Indian and Chinese. If it is belongs to specific race such as chapatti which is normally consumed by Indian, the object property of this food assign with hasRace to Indian only. Other personalization aspect considered are: the suitable time for taking the foods. For example, oatmeal is more suitable to be taken during breakfast, whereas rice is more suitable for lunch and noodles for dinner. Apart from that, superfood for diabetes (fibre value > 3 grams per serving) or foods that has been identified by the American Diabetes Association; allergy foods that it associated with; and food that is prohibited by religious laws were also included in the personalization aspects. Figure 4 illustrates such an example of personalization aspect for whole meal chapatti.

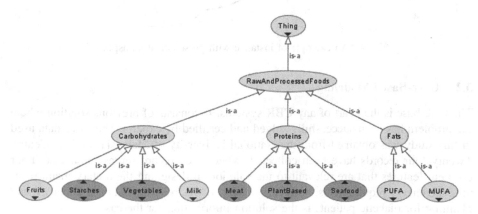

**Fig. 2.** The class hierarchy of raw and processed foods

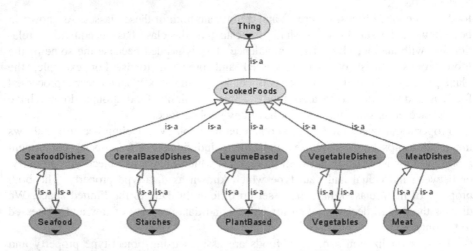

**Fig. 3.** The class hierarchy of cooked foods

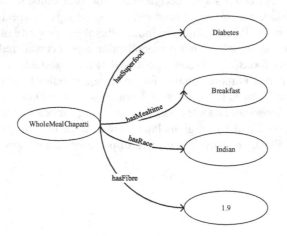

**Fig. 4.** An example of instance with personalization aspects

## 3.2    Case-Based Modeling

The case base is the heart of any CBR systems. It consists of previous solution where the problem has been successfully solved and certified by experts. The cases data used in this study was obtained from the National University of Malaysia Medical Centre. Twenty four records have been collected, where each record represents a case. Four problem features that are relevant to the solution in designing the dietary menu planning are energy requirements, body mass index (BMI), religion and race. Dietary menu planning for diabetic patients is the solution information for the case.

Since all cases can be represented uniformly, i.e. all cases have the same attribute, it can be represented by the attribute-value pair. The memory organization for this representation technique is a flat memory. It stores cases in a table with rows composed of

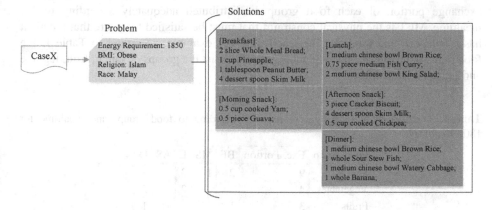

**Fig. 5.** Example of case with problem and proposed solution

cases and columns consisting of attributes. Figure 5 shows an example of a case with a problem and the corresponding solution.

### 3.3 CBR Engine

The CBR engine consists of case retrieval and case adaptation modules. The CBR shell of jCOLIBRI2[2] was chosen as the CBR framework in this work. All case-based reasoning methods have in common the following process [16, 17]:

- retrieve the most similar case (or cases) comparing the case to the library of past cases;
- reuse the retrieved case to try to solve the current problem;
- revise and adapt the proposed solution if necessary; and
- retain the final solution as part of a new case.

The case retrieval technique applied in this study is the nearest neighbor (NN). For case adaptation with ontology, the approach will retrieve the highest similarity of existing cases with the new case, which are known as the best case. This case will be adapted to meet all the patient's constraints that have been provided by a dietitian. Figure 4 illustrates the process model of the proposed recommender system.

A dietitian needs to input the patient's details and their MET information. The patient's details are divided into two parts i.e. patient's background and food restrictions. MET is a tool designed by a dietitian to guide the dietary menu planning according to the patient's energy needs, which are measured in calories (kcal). It consists of eight food groups and mealtimes. The number of mealtime are determined by the patient needs. They must eat at least three times a day which includes breakfast, lunch and dinner; whereas snacks will very much depend on the patient needs and preferences. Table 1 shows a sample of MET with 1800 calories. In this example, patient wants to have two snack time i.e. morning and afternoon snacks. The total

---

[2] http://gaia.fdi.ucm.es/research/colibri/jcolibri.

exchange portion of each food group is distributed adequately according to the mealtime. MET is the physical constraint that must be satisfied to ensure that a patient has a balanced and moderate diet each day. For example, according to Table 1, the food items in the dietary menu planning must consist the groups of starch, fruit, legume and fat during breakfast.

**Table 1.** Distribution of exchanges of portion according to food groups and mealtime for 1800 Kcal/Day.

| Food group | Exc. Portion | BF | MS | L | AS | D |
|---|---|---|---|---|---|---|
| Starch | 9 | 2 | 1 | 3 | | 3 |
| Vegetables | 4 | | | 2 | | 2 |
| Fruits | 3 | 1 | | | 1 | 1 |
| Milk | 2 | | 1 | | 1 | |
| Seafood | 4 | | | 4 | | |
| Meat | 3 | | | | | 3 |
| Legumes | 1 | 1 | | | | |
| Fat | 8 | 1 | | 3 | 1 | 3 |

[Exc. = exchange, BF = breakfast, MS = morning snack, L = Lunch, AS = afternoon snack, D = dinner]

### 3.4 Ontology Reasoning

Case adaptation represents the adaptation knowledge as task knowledge. In our case the adaptation knowledge is in the form of constraints and represented as production rules. Apart from the adaptation knowledge, case adaptation also exploits the domain knowledge to support its operation effectively.

The adaptation of best case involves ontological reasoning. The implementation of this task was made possible with Jena[3], which is a framework for the Semantic Web application. Jena provides ARQ as their SPARQL query language for Resource Description Framework (RDF) data. Jena supports several reasoners i.e. RDFS, OWL, transitive and generic rule-based reasoner. In this study, we opted for the use of the OWL reasoner to support the instance-based reasoning [18]. This type of reasoning is based on description logic (DL). Among the service include in this reasoning process are instance checking, retrieval and instance consistency checking [19]. We exploits all the instance-based reasoning in this study.

The adaptation framework consists of four constraints to be checked by the dietary recommendation system i.e. the MET, forbidden food, food accompaniment and serving size as illustrated in Fig. 6. However, due to the limited space, we focus on the discussion of the first constraints to be fulfilled during the adaptation process.

---

[3] https://jena.apache.org/.

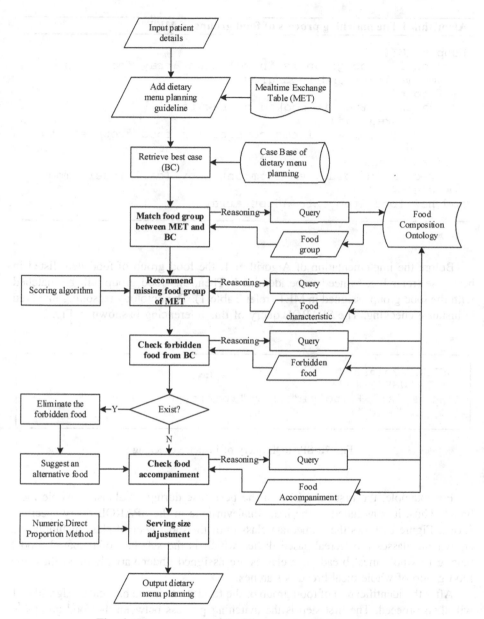

**Fig. 6.** The process model of the recommendation system

### 3.4.1   MET Fulfillment

Algorithm 1 shows the first constraints to be satisfied by the recommendation system i.e. fulfill the MET requirements. Basically, the process involved is matching of food groups between the new case and the best case. If the food group of the new case is missing from the best case, the system make a recommendation using insert operation. This operation can be considered as a transformation adaptation.

---

**Algorithm 1 The matching process of food groups in MET**

---

```
Loop by MET
    check if food group is in solution array (based on
    same mealtime and food group)
    if found
        check if item is cooked or raw food
        if cooked food
            check for all items combination food group exists
            in MET
    if not found
        recommend food item from ontology (call insert mod-
        ule)
add new food item into adapt array
```

---

Before the implementation of Algorithm 1, the food group of food item listed in best case (refer Fig. 5) need to be identified. Then, this food group will be matched with the food group assigned in MET (refer Table 1). The ontology reasoning involved is instance checking. The SPARQL query of this inferencing is shown in Fig. 7:

```
"SELECT ?class " +
      "WHERE { " +
      "x:" + individual + " rdf:type ?class . " +
      "} \n";
```

**Fig. 7.** SPARQL query of instance checking

For example, the first food item in the best case during breakfast is whole meal bread. Thus, it is assigned as the individual variable in the SPARQL query specified above. Figure 8 shows the immediate class until the most general class including the equivalent classes (i.e. cereal-based dishes which is the subclass of prepared foods) belong to whole meal bread. The classes are assigned under variable of x. thus, the food group of whole meal bread is starches.

After the identification of food group of the food items in the best case, Algorithm 1 will then proceed. The first step is the matching process between the food groups in MET with the food groups in the best case for the same mealtime. For example, the first food group in MET during breakfast is starches. This food group exist in the best case which belongs to the whole meal bread. Thus, the requirement of this food group in the MET is fulfilled by the best case. This process is done iteratively until all the food groups in MET are fulfilled by the system.

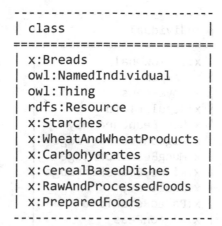

```
------------------------------
| class                      |
==============================
| x:Breads                   |
| owl:NamedIndividual        |
| owl:Thing                  |
| rdfs:Resource              |
| x:Starches                 |
| x:WheatAndWheatProducts    |
| x:Carbohydrates            |
| x:CerealBasedDishes        |
| x:RawAndProcessedFoods     |
| x:PreparedFoods            |
------------------------------
```

**Fig. 8.** Output of instance checking for whole meal bread

### 3.4.2 Inserting New Component

If the system could not match the food group required in the MET with the best case, it will consider the food group as 'missing'. The insert component process will be executed to recommend new food item under the missing food group. The first process is to retrieve the individuals belongs to the missing food group. These individuals are the candidate to be selected for the new food item. For example, the missing food group in MET during breakfast is plant-based protein. The ontology reasoner will infer the domain ontology to retrieve all the individual of this class. The inference mechanism involved is instance retrieval, which is a kind of ABox reasoning [19]. The SPARQL query is represented in Fig. 9 and the output of this inferencing is illustrated in Fig. 10 respectively.

```
"SELECT DISTINCT ?individual " +
        "WHERE { " +
        "?individual rdf:type x:" + foodgroups + " . " +
        "} \n";
```

**Fig. 9.** SPARQL query of instance retrieval

The second process is the individual filtration from the ineligible ones. For cooked foods, four criteria are considered i.e. the forbidden food, the existence of food accompaniment's food group if any, availability of food group for the main ingredients

```
-------------------------
| individual             |
=========================
| x:YellowDhal           |
| x:Tempeh               |
| x:SoyaBeans            |
| x:RegularTofu          |
| x:SoyaBeanCurdInSyrup  |
| x:PigeonPeasPorridge   |
| x:MungBeansPorridge    |
| x:KidneyBeansPorridge  |
| x:DhalGravy            |
| x:BakedRojakTofu       |
| x:PigeonPeas           |
| x:MungBeans            |
| x:KidneyBeans          |
| x:Chickpeas            |
| x:BakedBeans           |
-------------------------
```

**Fig. 10.** Output of instance retrieval of plant-based class

and the serving size for main ingredients. For raw and processed foods, filtration only performed on allergy food and the availability of food accompaniment's food group. The algorithm of this sub process is as follows.

---

**Algorithm 2 Ineligible food item filtration**

---

```
if cooked food
   check if each item combination is allergic or prohib-
   ited
     remove from item list
   check if food accompaniment is allergic or prohibited
     remove from item list
   check if there is one or more item combination does
   not exist in MET
     remove from item list
   check if there is one or more item combination ex-
   ceeds the serving size
     remove from item list
if raw food
   check if food item is allergic or prohibited
     remove from item list
   check if food accompaniment is allergic or prohibited
     remove from item list
```

---

The third process is the selection of the best food item for the missing food group. It involves two main activities. The first one is the filtration of food item in several food group. This activity has classify the food groups into three classes. The first class is for cooked foods, the second class is food group where it can be eaten directly i.e. milks and fruits. The third class is for food group that have some of the food item that can be eaten directly. Either it need to be cooked or it cannot be stand alone. For example, in plant-based protein, mung beans can be eaten directly, it must be cooked food. Another example for food that cannot be stand alone is cooking oil. For any food item that cannot be eaten alone or raw, it is assigned with the object property isEdible equal to false. Starches, vegetables, legume and fats food groups fall under this class.

The filtration only involve the third class because the inedible one needs to be filtered out from being the final candidate for the best food item. No inferencing involves because the edible food items are retrieve directly using SPARQL query as shown in Fig. 11.

```
"SELECT ?property ?value " +
       "WHERE { " +
       "x:" + individual + " ?property ?value . " +
       "FILTER( ?p = x:isEdible) . " +
       "} \n";
```

**Fig. 11.** SPARQL query for filtration of inedible food item

According to the example of plant-based protein, from the 15 retrieved food items as shown in Fig. 10, only tempe, soya bean curd in syrup, pigeon peas porridge, mung beans porridge, chick peas, kidney beans porridge, dhal gravy, baked rojak tofu, dan baked beans to be the final candidates. Apart from the inedible, another criterion adhered is the food group availability for the cooked food in MET.

The foods item filtered at this stage are the final candidate to be recommended as the new food item for the missing food group. The food recommendation criteria are applied to select the best candidate. These include the diabetes super food, mealtime suitability, race cuisine, and nutrients that are significant to the food group. Among the nutrients are fiber for starches, vegetables, fruits and the legume food group; saturated fat and cholesterol for seafood, meats and fats food group; and monounsaturated (MUFA) and polyunsaturated fat (PUFA) for the fats food group. These recommendation criteria are retrieved using the SPARQL query with no inferencing process involved (Fig. 12).

Each of the recommendation criteria has been assigned with its own weightage to calculate the score obtained by the food item candidates. The criteria of non-nutrient are calculate based on similarity concept while the nutrients criteria apply the normalization concept. The total score of food item candidates is obtained by the sum-

```
Query Description:
---------------------------------------------------------------
(null;1800;Obese;Hindu;Indian;)

Food Preferences Description:
---------------------------------------------------------------
Prohibit: Beef;
Allergic:
Like: ChickenCurry; LambCurry; FishCurry;
Dislike:

MET:
---------------------------------------------------------------
([BF]: 2 sz Starches; 1 sz Fruits; 1 sz PlantBased ; 1 sz Fats; //
[MS]: 1 sz Starches; 1 sz Milk ;  // [L]: 3 sz Starches; 2 sz Vege-
tables;  4 sz Seafood; 3 sz Fats; // [AS]:  1 sz Fruits; 1 sz Milk;1
sz Fats; // [D]: 3 sz Starches; 2 sz Vegetables; 1 sz Fruits; 3 sz
Meat; 3 sz Fats;)
```

```
Retrieved best case:
[Description: (Case09;1850;Obese;Islam;Malay;)][Solution:
(Case09;[BF]:1.5 cup Oatmeal; 0.75 Cup Chunks Pineapple;1 cup Skim-
Milk; 8 halves Walnuts;// [MS]: 0.75 cup SweetPotato; // [L]:1 me-
dium_bowl BrownRice; 1 whole_medium FishSoup; 1 medium_bowl King-
Salad;4 tbsp Avocado;// [AS]: 1 slice Papaya; 1 cup MungbeansPorridge;
//[D]:1 medium_bowl BrownRice; 1 whole_medium BeefCurry; 1 1.00
CupDiced Cucumber; //;)][Sol.Just.: null][Result: null] ->
0.7722222222222223
[BF]:
Oatmeal  : 1.5 cup
Kiwifruit  : 1.0 whole_large
SkimMilk  : 1.0 cup
Walnuts  : 8.0 halves
[MS]:
SweetPotato  : 0.75 cup
[L]:
BrownRice  : 1.0 medium_bowl
FishSoup  : 1.0 whole_medium
KingSalad  : 1.0 medium_bowl
Avocado  : 4.0 tbsp
[AS]:
Papaya  : 1.0 slice
MungbeansPorridge  : 1.0 cup
[D]:
BrownRice  : 1.0 medium_bowl
BeefCurry  : 1.0 whole_medium
Cucumber  : 1.00 CupDiced
```

```
Adaptation Result
---------------------------------------------------------------
[BF]:
WholeMealChapatti 2.00 SmallSixInchAcross  [2.0 Starches]
Kiwifruit  1.00 WholeLarge  [1.0 Fruits]
DhalGravy
    YellowDhal 0.50 Cup Cooked;  [1.00 PlantBased]
    CornOil 1.00 Teaspoon  [1.00 Fats]
```

**Fig. 12.** Example of adaptation results

```
         Carrot 0.165 CupStrips Cooked;  [0.33 Vegetables]
         Tomato 0.33 CupSlices [0.33 Vegetables]
         Brinjal 0.165 CupCubesOneInch Cooked;  [0.33 Vegetables]
[MS]:
SweetPotato  0.50 CupCubes  Cooked; [1.0 Starches]
SkimMilk 4.00 HeapedTablespoon [1.0 Milk]
[L]:
BrownRice  0.99 Cup  Cooked; [3.0 Starches]
KingSalad  2.00 Cup  [2.0 Vegetables]
FishSoup
    RedSnapper 1.00 Piece  [4.00 Seafood]
    CornOil 2.00 Teaspoon  [2.00 Fats]
Avocado 2.00 Tablespoon  [1.0 Fats]
[AS]:
Orange 1.00 Whole  [1.0 Fruits]
NonFatPlainYogurt 0.67 Cup  [1.0 Milk]
Almonds 6.00 Nuts  [1.0 Fats]
[D]:
BrownRice  0.99 Cup  Cooked; [3.0 Starches]
Cucumber  1.00 CupDiced; [2.0 Vegetables]
Guava 0.50 LargeWithoutSeeds  [1.0 Fruits]
ChickenCurry
    ChickenBreast 3.00 Matchbox Cooked;  [3.00 Meat]
    CornOil 3.00 Teaspoon  [3.00 Fats]
       Tomato 0.5 CupSlices  [0.5 Vegetables]
       Potato 0.125 Cup Cooked;  [0.25 Starches]
       Carrot 0.25 CupStrips Cooked;  [0.5 Vegetables]
--------------------------------------------------
Nutrient Data
--------------------------------------------------
Energy : 1732.25
CHO : 221.64
Protein : 93.38
Fat : 47.04
Fibre : 40.55
--------------------------------------------------
CHO: 51.18%, Pro: 21.56%, Fat: 24.44% = 97.18%
Cycle finished. Type exit to idem or enter to repeat the cycle
```

**Fig. 12.** (*continued*)

mation of similarity and normalization that multiplied by its weightage respectively. The scoring algorithm is formulated using the following equation:

$$S = S_{sim} + S_{norm} \qquad (1)$$

$$S_{sim} = \sum_{i=1}^{n} w_i \times sim_i \qquad (2)$$

$$S_{norm} = \sum_{i=1}^{n} w_i \times norm_i \qquad (3)$$

where

$$norm_i = {}^{x_i}/_{\max(x)}$$

$$sim_i \begin{cases} 1, input = food\ selection\ criteria \\ 0, input \neq food\ selection\ criteria \end{cases}$$

$$w_i = weightage\ to\ the\ i^{th}\ food\ selection\ criteria$$

$$x_i = i^{th} nutrient\ vaue$$

$$\max(x) = maximum\ value\ of\ nutrient$$

## 4  Discussion and Conclusion

In this paper, an ontological approach was developed to perform design case adaptation for the task of dietary menu planner for diabetic patients. The ontology supported both the transformation and substitution adaptation tasks. It was able to satisfy the physical, preference and common sense constraints using the adaptation tasks above. DL inference mechanisms specifically to instance-based reasoning i.e. instance checking and retrieval; and equivalence class were used to infer the new (additional) triples. SPARQL retrieved the inferred and asserted triples that matched a given query. We have performed initial evaluation of the proposed approach among dietitians and promising feedbacks have been achieved. Near further works include enhancing the case bases and provide a better interface and visual representation of the adapted results.

**Acknowledgement.**  This research is partially supported by the Malaysia Ministry of Education Grant FRGS/1/2014/ICT02/UKM/01/1 awarded to the Center for Artificial Intelligence Technology at the Universiti Kebangsaan Malaysia.

## References

1. Noah, S.A., Abdullah, S.N., Shahar, S., Abdul-hamid, H., Khairudin, N.: DietPal: a web-based dietary menu-generating and management system. J. Med. Internet Res. **6**(1), 13 (2004)
2. Mahan, L.K., Raymond, J.L.: Krause's Food & Nutrition Therapy, 14th edn. Elsevier - Health Sciences Division, Philadelphia (2017)
3. Marling, C.R., Petot, G.J., Sterling, L.S.: Integrating case-based and rule-based reasoning to meet multiple design constraints. Comput. Intell. **15**(3), 308–332 (1999)
4. Jung, H., Chung, K.: Knowledge-based dietary nutrition recommendation for obese management. Inform. Technol. Manag. **17**(1), 29–42 (2016)
5. Yang, L., Hsieh, C.-K., Yang, H.: Yum-me: a personalized nutrient-based meal recommender system. ACM Trans. Inform. Syst. **9**(4), 1–31 (2017)
6. Trang Tran, T.N., Atas, M., Felfernig, A., Stettinger, M.: An overview of recommender systems in the healthy food domain. J. Intell. Inform. Syst. 1–26 (2017)

7. Jannach, D., Zanker, M., Felfernig, A., Friedrich, G.: Recommender Systems: An Introduction. Cambridge University Press, New York (2011)
8. Khan, A.S.: Incremental Knowledge Acquisition for Case-Based Reasoning. Doctoral Dissertation, University of New South Wales (2003)
9. Khairudin, N., Noah, S.A., Azizan, A., Jelani, A.B.: Case-based diabetic dietary plan using memory organization packets. In: Proceedings of the International Conference on Information Retrieval & Knowledge Management (CAMP), pp. 91–94 (2012)
10. Hinrichs, T.R.: Problem Solving in Open Worlds: A Case Study in Design. Lawrence Erlbaum Associates, New Jersey (1992)
11. Lee, C., Wang, M., Hagras, H.A.: Type-2 fuzzy ontology and its application to personal diabetic-diet recommendation. IEEE Trans. Fuzzy Syst. **18**(2), 374–395 (2010)
12. Tee, E. S., Noor, I., Azudin, N., Idris, K.: Nutrient Composition of Malaysian Foods, 4th edn. Institute for Medical Research, Kuala Lumpur (1997)
13. Pennington, J.A.T., Stumbo, P.J., Murphy, S.P., McNutt, S.W., Eldridge, A.L., Chenard, C.A.: Dietetic practice and research. J. Am. Diet. Assoc. **2**, 2105–2113 (2007)
14. Noy, N.F., McGuinness, D.L.: Ontology Development 101 : A Guide to Creating Your First Ontology. Technical report SMI-2001-0880, Stanford Medical Informatics (2001)
15. Yusof, N., Noah, S.A., Wahid, S.T.: Ontology modeling of malaysian food composition. In: Proceedings of the 3rd International Conference on Information Retrieval and Knowledge Management, Melaka, pp. 149–154 (2016)
16. Aamodt, A., Plaza, E.: Case-based reasoning: foundational issues. Methodological Var. Syst. Approaches Artif. Intell. Commun. **7**(1), 39–52 (1994)
17. Goel, A.K., Diaz-Agudo, B.: What's hot in case-based reasoning. In: Proceedings of the Thirty-First AAAI Conference on Artificial Intelligence, pp. 5067–5069 (2017)
18. Apache Jena. Apache Jena Reasoners and rule engines: Jena inference support. https://jena. apache.org/documentation/inference/index.html (2017)
19. Baader, F., Calvanese, D., McGuinness, D.L., Nardi, D.: Patel-schneider, P.F. (eds.): The Description Logic Handbook. Cambridge University Press, Cambridge (2010)

# Linked Urban Open Data Including Social Problems' Causality and Their Costs

Shusaku Egami[1]([✉]), Takahiro Kawamura[1,2], Kouji Kozaki[3], and Akihiko Ohsuga[1]

[1] University of Electro-Communications, Tokyo, Japan
egami.shusaku@ohsuga.lab.uec.ac.jp
[2] Japan Science and Technology Agency, Tokyo, Japan
[3] Osaka University, Osaka, Japan

**Abstract.** There are various urban problems, such as suburban crime, dead shopping street, and littering. However, various factors are socially intertwined; thus, structural management of the related data is required for visualizing and solving such problems. Moreover, in order to implement the action plans, local governments first need to grasp the cost-effectiveness. Therefore, this paper aims to construct Linked Open Data (LOD) that include causal relations of urban problems and the related cost information in the budget. We first designed a data schema that represents the urban problems' causality and extended the schema to include budget information based on QB4OLAP. Next, we semi-automatically enriched instances according to the schema using natural language processing and crowdsourcing. Finally, as use cases of the resulting LOD, we provided example queries to extract the relationships between several problems and the particular cost information. We found several causes that lead to the vicious circle of urban problems and for the solutions of those problems, we suggest to a local government which actions should be addressed.

**Keywords:** Linked Open Data · Knowledge graph · Urban problem · Causality · Crowdsourcing

## 1 Introduction

Local governments have a number of urban problems, such as suburban crimes, dead shopping street, and littering. Thus, local government representatives have been discussing solutions to these problems. However, because various factors are socially intertwined, the problems are difficult to solve without understanding the causal relations of the problems. Thus, structural management of the related urban problem data and their causality are required for visualizing and solving such problems. Causal relations and "causality" are relationship between factor and result between two things. Then, to implement the action plans, local governments need to grasp the cost-effectiveness. Since most local governments are very cost sensitive, new businesses for solving urban problems will not be

© Springer International Publishing AG 2017
Z. Wang et al. (Eds.): JIST 2017, LNCS 10675, pp. 334–349, 2017.
https://doi.org/10.1007/978-3-319-70682-5_23

implemented without clearly estimating their effects, such as cost reduction. In fact, this was a comment we received from an official in the Yokohama Policy Bureau, Kanagawa Prefecture, Japan.

Therefore, this paper aims to construct linked open data (LOD) that includes causal relations of urban problems and the related cost information in their budget sheet. This LOD can help predict the impact of urban problems by tracing the causality and hierarchical links in background ontologies. The LOD can also help local governments to consider solutions of the urban problems with their cost effectiveness. In this study, we first designed a LOD schema that represents the causality of the urban problems and extended the schema to include budget information based on QB4OLAP [1], which is an extension of RDF Data Cube Vocabulary[1]. Next, we semi-automatically constructed the LOD based on the schema. Specifically, we extracted causality words from web pages using Japanese dependency structure analysis, and then we selected the causality words using crowdsourcing. Finally, as use cases of the resulting LOD, we provided example queries to extract the relationships between several problems and the particular cost information. As a result, our contributions are as follow:

1. Designing a schema of urban problem causality and local governments' budgets;
2. Proposing a method for semi-automatically constructing Linked Data with causality;
3. Publishing the data on the web as LOD; and
4. Presenting example queries to consider solutions of urban problems and to reduce the budgets.

The remaining sections of this paper are organized as follows. In Sect. 2, an overview of LOD relating to urban problems, city data and crowdsourcing is described. In Sect. 3, the schema design is described. In Sect. 4, the method for constructing LOD related to urban problems with causality and its evaluation are described. In Sect. 5, example queries are presented, and we discuss the solutions of urban problems. Finally, Sect. 6 concludes this paper with some feasible future extensions.

## 2    Related Work

### 2.1    Knowledge Graph for Solving Social Issues

Some studies have proposed use of linked data for solving social issues. Szekely et al. [2] built linked data from crawled sites for combating human trafficking and developed the lost children search system. The system has been deployed by six law enforcement agencies and several NGOs. Szekely et al. built linked data related to a specific domain of social problems, whereas this paper aims to build LOD related to multiple urban problems with causality.

---

[1] https://www.w3.org/TR/vocab-data-cube/.

In our previous work [3], we built and visualized LOD for solving illegally parked bicycles, which is one of the urban problems that needs to be resolved in Japan. In addition, we proposed a methodology for designing LOD schema of the urban problems that are occurring on a daily basis, such as illegally parked bicycles. In this methodology, all the steps were accomplished manually. Since the number of web documents collected was limited and the task relied on the knowledge of workers, the previous approach suffered from low coverage with respect to the extraction of the causality of urban problems. Thus, we here propose a method of semi-automatically extracting the causality of urban problems [4]. Moreover, since the constructed LOD was based on the schema extended from Event Ontology[2], and thus it was difficult to search urban problem causality using OWL inference rules. In addition, the LOD did not contain the budget information. Thus, in this paper we defines a new LOD schema representing urban problem causality and the budget information.

Shiramatsu et al. [5] proposed LOD to share goals to solve social issues. A goal-matching method using LOD has been proposed for facilitating civic technology (Civic Tech). The Civic Tech is aimed at solving social issues using information technology through collaboration between citizens and local governments. Furthermore, it has been reported that a web application called Goal-Share was developed and applied in domestic Civic Tech events. However, Shiramatsu's LOD mainly describes the public goals for solving social issues and does not describe causalities. Associating them with the LOD proposed in our study will facilitate social problem solving in the future.

## 2.2 Knowledge Graph for Analyzing City Indicators

Santos et al. [6] defined city knowledge graph as owl in order to analyze various city indicators. They proposed the quality of experience (QoE) Indicators Ontology representing calculated numerical values and supporting convenient visualization. The dashboard application, which can generate widgets visualizing knowledge graph data was also developed. Pileggi et al. [7] defined the ontological framework and implemented it as OWL-DL ontology to represent dynamic fine-grained urban indicators. In order to simplify the understanding of the data structure as well as the facilitation of its usability, the ontology was partitioned in five sub-ontologies; Indicator, Data, Profiling, Computations and Geographic Context based on the function of the scope within the model.

LinkedSpending [8] are linked data based on OpenSpending[3], which is an open platform for public financial information, including budgets, spending, balance sheets, procurement, etc. As of May 2017, 1,104 data sets from 75 countries have been registered. The country with the largest number of registered data sets is Japan of 415 data sets. The data is modeled based on RDF Data Cube vocabulary[4], which is designed for modeling multidimensional data such

---

[2] http://motools.sourceforge.net/event/event.html.
[3] https://openspending.org.
[4] https://www.w3.org/TR/vocab-data-cube/.

as statistical data. However, these linked data do not describe urban problems and cannot be directly used to solve urban problems.

## 2.3 Crowdsourcing and NLP for Linked Data

Demartini et al. [9] proposed an entity linking method using crowdsourcing. Crowdsourcing was used to improve the quality of the links, and they developed the probabilistic framework to integrate inconsistent results. Celino et al. [10] developed a mobile application to link Point of interests (POIs) data to pictures using crowdsourcing. They introduced the method of game with a purpose (GWAP) [11] to give users incentives. However, there is no study to construct LOD related to urban problem causality using crowdsourcing.

Nguyen et al. [14] proposed a method for constructing Linked Data concerning users' activities. Conditional Random Field (CRF) was used to extract the users' activities from Japanese weblogs, and then triples related to *action*, *object*, *time* and *location* were constructed. This Linked Data would be applied to analize users' activities at the time of an earthquake. LODifier [15] also extracted entities from unstructured text using a Named Entity Recognition (NER) system Wikifier [16], and combined the entities to DBpedia and WordNet. There are many other studies using NLP techniques to construct Linked Data sets. However, the accuracy of the methods to extract the necessary words from unstructured documents in those studies is still unsatisfiable. Thus, in this paper, we combined an NLP technique and crowdsourcing in order to extract urban problem causality.

## 3 Designing a Schema of Problem Causality and Costs

Our LOD is mainly for investigation of solutions to urban problems. Specifically, querying the LOD enables local governments to consider effects of the solutions and their necessary budgets. Thus, we designed the LOD schema shown in Fig. 1 to represent urban problem causality and local governments' budgets.

The upper half of the figure defines vocabularies representing urban problem causality. In this part, all resources are classified as UrbanProblem or NotUrbanProblem. There are two main causality properties, upv:factor and upv:affect. Both properties are owl:TransitiveProperty and are subproperties of the upv:related property. Since all urban problems are not events that have time and spatial thing, we did not reuse the event:factor property in the Event Ontology. Also, there are sub-properties of upv:factor and upv:affect to represent agreements of crowdsourcing. By dividing the causality properties into upv:factor and upv:affect, it is possible to reason forward or backward chainings with agreement levels restricting domain or range. For example, when users extract the strong causality, the upv:factor_level4 and the upv:affect_level4 properties can be used, and when users extract the causality regardless of the agreement, the upv:factor and the upv:affect properties can be used in SPARQL queries.

The lower half of the figure defines vocabularies representing budget information. Since most local governments' budget information is published as tabular data, such as Microsoft Excel and PDF, we described it using QB4OLAP, which is an extension of RDF Data Cube Vocabulary[5]. The QB4OLAP has been used in the data model of business intelligence tools and introduces qb4o:LevelProperty and qb4o:AggregateFunction in order to support aggregation operations. Hence, users can query the total budget of each department and an urban problem, which has the highest budget. The range of the business property is the Business class, and the Business class has at least one dcterms:subject property.

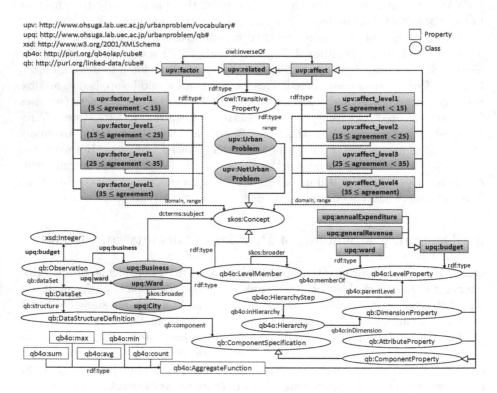

**Fig. 1.** Linked Data schema of urban problem causality and its costs

# 4  Building LOD Based on Designed Schema

## 4.1  Extraction of Urban Problem Causality

In this section we propose a method of semi-automatically extracting the causality of urban problems as follows.

---

[5] https://www.w3.org/TR/vocab-data-cube/.

(1)  Collect web documents using a search engine.
(2)  Extract causality words from the collected documents using NLP.
(3)  Generate word clouds based on the extracted words.
(4)  Filter the extracted words using crowdsourcing.

We collected documents by search engine using names of urban problems and synonyms of "factor" as keywords. For example, the first keyword is "Suburban crime," and the second keyword is "factor" and its synonyms, such as "element", "origin", and "cause". We obtained the synonyms of the second keyword from Japanese WordNet[6]. Then, we obtained the document lists using Google Custom Search API[7] and Bing Web Search API[8]. We separately collected 50 HTML files and 50 PDF files for each keyword sets (different combinations of the first and second keywords). We collected HTMLs and PDFs separately to collect reports published by governments and citizens' voices. However, many unrelated documents were collected in this step; thus, the documents containing few words related to urban problem name were excluded.

Next, we extracted noun words using Japanese morphological analysis. In order to facilitate the subsequent crowdsourcing process, we concatenated the verbal nouns and constructed noun phrases. For example, the phrase "preventing delinquency" was split into "preventing" and "delinquency" using morphological analysis, but these words were concatenated to a noun phrase in this study.

Then, we extracted noun phrases that have causal relationships to the synonyms of the "factor" based on dependency relations in each sentence using Japanese dependency analysis [12].

Likewise, words related to the influence of urban problems were extracted, using the synonyms of "influence" as the second keywords, such as "affect", "effect", and "evoke".

## 4.2   Filtering Causality Words Using Crowdsourcing

We generated word clouds based on extracted possible causality words and filtered the words using crowdsourcing. We here assumed that the word clouds raise the impression of words and make it easier to extract important words.

First, similar words obtained in Sect. 4.1 were integrated by edit distances. We used Jaro-Winkler distance [13] to calculate the similarity of words, and the threshold was empirically set to 0.8. When similar words were found, the number of occurrences of words was integrated to the longest word. Figure 2 shows a word cloud of suburban crime factors. When the frequency of a specific word was high, the word size became large and was placed close to the center of the cloud. The color of words was set randomly. Then, we ordered two tasks for crowdsourcing: "Select 10 words that are considered factors of suburban crime" and "Select 10 words that are considered influence affected by suburban crime."

---

[6]  http://compling.hss.ntu.edu.sg/wnja/index.en.html.

[7]  https://developers.google.com/custom-search/?hl=en.

[8]  https://azure.microsoft.com/en-us/services/cognitive-services/
bing-web-search-api/.

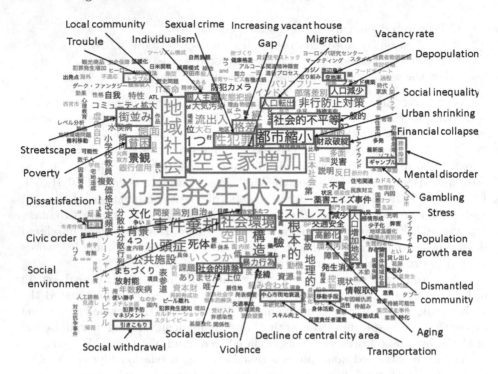

**Fig. 2.** A word cloud of suburban crime factors

In this paper, we used Lancers[9] as the crowdsourcing service. We set the reward for the two tasks at 50 JPY, and asked up to 50 people to work on each problem. Then, we took words that were selected by more than 10% of the workers. The selected words were those translated in Fig. 2.

Furthermore, in order to enrich the causality of urban problems, we repeated our method using the extracted causality words. The repetition of this method increased the intermediate nodes in our knowledge graph. However, all the words do not have causal relations. Thus, we also extracted co-occurring words from top 50 web documents related to the causality words. If there are more than 20 co-occurring words, such as "cause", "factor", "influence", and "effect", we apply our extraction method to the causality words.

### 4.3   Building LOD Based on the Extracted Causality Words

We built LOD based on designed schema using the extracted words. Since the Lancers can export the results of crowdsourcing in a CSV format, we converted the CSV to an RDF file based on the designed schema using Apache Jena[10]. Specifically, urban problem resources were created as sub-classes of the `upv:UrbanProblem` class, and others were created as sub-classes

---

[9] http://www.lancers.jp.
[10] https://jena.apache.org/.

of the `upv:NotUrbanProblem` class. The causality links were created using the `upv:factor` and the `upv:affect` properties corresponding to the number of the agreement. In addition, we extracted noun words from the name of each class and created hyper classes based on them.

## 4.4   Building LOD Based on Budget Data of Local Government

Osaka is a city designated by a government ordinance of Japan, and it is the capital city of Osaka Prefecture. Osaka City has published various open data in Osaka City Open Data Portal Site[11]. Most open data containing budget information published on the portal site are CC-BY 4.0 licensed. First, we converted their tabular data and PDF files to RDF files based on the designed schema using Apache Jena and Apache POI[12].

Next, we linked the local government business resources to the causality resources. Since there were not the detailed descriptions of businesses in the source budget sheet, it was necessary to link the business resources to the causality resources using the names of the business resources. However, it was difficult to acquire the relations, such as the one between the "Business for solving the street smoking" and the "cigarette". Thus, we linked the business resources to the causality resources using the Algorithm 1.

In the algorithm, we first extracted noun words except for the stop words from the name of the local government businesses. Then, we obtained synonym words corresponding to extracted noun words from the Japanese WordNet as the candidates of linking. In addition, we also obtained the *glosses* of the noun words. The *gloss* consists of more than one short sentences describing the word sense and the use of the word. Thus, we extracted the noun words as the candidates of linking from the short sentences describing the word sense. If the candidate words matched to the causality resources, we linked the business resources to the causality resources using the `dct:subject` property.

Figure 3 shows part of the LOD finally constructed in this study. The resulting LOD is accessible in our website[13]. There are 49,386 triples in the LOD. We validated our LOD using RDFUnit [17], which is a test driven data-debugging framework. The 68 test cases were automatically generated, and then all test cases passed. There were no timeout, no error, and no violation instances. Therefore, the result showed that we correctly reused existing vocabularies without violation of domains and ranges restrictions. All resources are linked, and there is no independent resources. In addition, we defined the a Semantic Web Rule Language (SWRL) rule, which represents that an influence caused by a thing should propagate to its super concept as follows.

$$(?x\ upv{:}affect\ ?y),\ (?y\ rdfs{:}subClassOf\ ?z) \text{-} > (?x\ upv{:}affect\ ?z) \qquad (1)$$

?x and a ?y are causality resources including urban problems. ?z is a super class of the ?y.

---

[11] https://data.city.osaka.lg.jp/.
[12] https://poi.apache.org/.
[13] http://www.ohsuga.lab.uec.ac.jp/urbanproblem/.

---

**Algorithm 1.** Linking the business resources to the causality resources

---

**Require:** $businessList, causalityList, stopwordList$
**Ensure:** $RDF$
 1: **for** each $business \in businessList$ **do**
 2:     $morphemeList_{business} \Leftarrow morphologicalAnalysis(business.label)$
 3:     **for** each $morpheme_{business} \in morphemeList_{business}$ **do**
 4:         **if** $morpheme_{business}.partOfSpeech == Noun$
            $\&\& \ !stopwordList.contains(morpheme_{business}.text)$ **then**
 5:             **if** $causalityList.contains(morpheme_{business}.text)$ **then**
 6:                 $causality \Leftarrow causalityList.get(morpheme_{business}.text)$
 7:                 $RDF.addStatement(business, dct : subject, causality)$
 8:             **end if**
 9:             //Obtain synonym words of the noun words in the business name
10:             $synonymList \Leftarrow getSynonymList(morpheme_{business}.text)$
11:             **for** each $synonym \in synonymList$ **do**
12:                 **if** $causalityList.contains(synonym)$ **then**
13:                     $causality \Leftarrow casaulityList.get(synonym)$
14:                     $RDF.addStatement(business, dct : subject, causality)$
15:                 **end if**
16:             **end for**
17:             //Obtain short sentences describing the word sense
18:             $gloss \Leftarrow getGloss(morpheme.text)$
19:             $morphemeList_{gloss} \Leftarrow morphologicalAnalysis(gloss)$
20:             **for** each $morpheme_{gloss} \in morphemeList_{gloss}$ **do**
21:                 **if** $morpheme_{gloss}.partOfSpeech == Noun$
                    $\&\& \ !stopwordList.contains(morpheme_{gloss}.text)$ **then**
22:                     **if** $causalityList.contains(morpheme_{gloss}.text)$ **then**
23:                         $causality \Leftarrow casaulityList.get(morpheme_{gloss}.text)$
24:                         $RDF.addStatement(business, dct : subject, causality)$
25:                     **end if**
26:                 **end if**
27:             **end for**
28:         **end if**
29:     **end for**
30: **end for**

---

## 4.5  Evaluation and Discussion

**Evaluation of NLP.** Table 1 shows the statistics of causality words extraction. Since there were many synonyms of the "affect", the number of the documents related to influences was 2,465, and larger than the number of the documents related to factors (1,438). We also excluded some synonyms of "factor", since some synonyms of "factor" such as "procatarxis" are rarely used. Hence, the search results contained many unrelated documents. As a result, the number of influence words became large, and their agreements of selection became lower than the factor.

The omission of words related to urban problems was a cause of lowering the agreement in a process of causality word extraction process using morphological analysis and Japanese dependency structure analysis. We found that there are cases, in which the chunk extracted by our method does not match the chunk describing causality. Since there are many complex sentences in the documents published by governments, we could not extract the chunk containing causality words in many cases. To solve this problem, we will define extraction rules. There were also cases, where causality words were extracted from descriptions not related to the urban problem. To exclude these errors, we will extract words

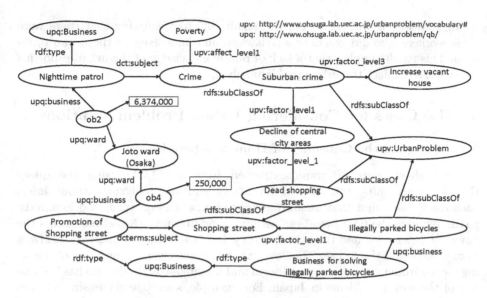

**Fig. 3.** Part of the constructed LOD

**Table 1.** Statistics of causality words extraction

|  | # of documents including urban problem words | # of sentences including synonyms of "factor" and "influence" | # of extracted words |
|---|---|---|---|
| Factor | 1,438 | 4,481 | 3,110 |
| Influence | 2,465 | 9,082 | 4,661 |

that appear in multiple documents, instead of words that appear many times in a single document.

**Evaluation of Crowdsourcing.** To calculate the agreement of causality word selection by crowdsourcing, we used Fleiss's kappa [18]. The number of users was 50. The average number of extracted words related to urban problem factors was 0.291, and the number of words related to influence was 0.212. The total average of agreement was 0.256, and it was fair agreement according to the benchmark [19]. Therefore, the extraction method of the causality word and the display of the word cloud can be considered to be appropriate to some extent.

The agreement of factor of traffic accidents was high (0.443), since the various instances of traffic accidents are reported by the Metropolitan Police Department, educational institutions, and news organizations, which resulted in the workers having extensive background knowledge. On the other hand, the high agreement of noise factors (0.468) would be because the workers have had their own experience affected by noise; thus, the word displayed in the word cloud was easy for them to select.

In addition, it was seen from the results of crowdsourcing that there were some workers who did not have any background knowledge of the urban problems. This results implies that a lack of public knowledge of the urban problems can adversely affect the urban problem solving.

# 5   Use Cases for Considering Urban Problem Solutions

## 5.1   Querying the Causality of Homeless people

Figure 4 shows part of graph extracted from our LOD using the query (Fig. 5) of searching causality of Homeless and the budgets without inference rules. We found that Gambling dependence, Unemployed, and Economic pains increase the poor Homeless people. In addition, the Homeless causes the Poverty business, and then the Poverty business leads to increase Welfare recipients and the Crime. The Poverty business means a business of screwing money from poor people using illegal and inhumane methods, and has become one of the social problems in Japan. For example, such poverty business forces homeless people to apply for the livelihood protection and mulcts their money. The livelihood protection in Japan is a kind of safety nets that guarantees the minimum standard of living. The eligible citizens can receive a certain amount of living expenses until the economic independence is achieved. Recently, however, the total cost of the livelihood protection becomes more than 2.88 trillion JPY and accounts for an enormous financial burden.

In fact, the total cost of the livelihood protection in Osaka city is more than 290 billion JPY according to the statistics data in 2015, and thus Osaka city is the municipality with the largest livelihood protection expenses in Japan. In particular, Nishinari ward in Osaka city has 62.4 billion JPY for the welfare

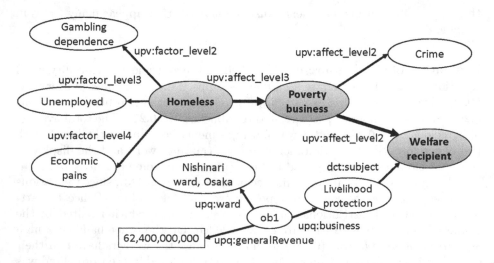

**Fig. 4.** Causality of homeless

```
PREFIX upr: <http://www.ohsuga.lab.uec.ac.jp/urbanproblem/resource/>
PREFIX urv: <http://www.ohsuga.lab.uec.ac.jp/urbanproblem/vocaburary#>
PREFIX upq: <http://www.ohsuga.lab.uec.ac.jp/urbanproblem/qb/>
PREFIX dct: <http://purl.org/dc/terms/>
SELECT DISTINCT * FROM <http://www.ohsuga.lab.uec.ac.jp/urbanproblem>
WHERE {
{SELECT ?factor WHERE {
    upr:Homeless upv:factor_level2|upv:factor_level3|upv:factor_level4 ?factor .
}} UNION {
SELECT ?influence ?business ?budget ?ward  WHERE {
    ?ob1 upq:business ?business; ?upq:generalRevenue ?budget; upq:ward ?ward .
    { SELECT ?influence ?business (MAX(?gr) AS ?budget) WHERE {
        upr:Homeless (upv:affect_level2|upv:affect_level3|upv:affect_level4)+ ?influence .
        ?business dct:subject ?influence .
        ?ob1 upq:business ?business; upq:ward ?ward; upq:generalRevenue ?gr .
    } group by ?influence ?business }
}}}
```

**Fig. 5.** SPARQL query for searching Homeless causality and their budgets

budget, and it is 2.5 times more than the second largest ward. In Nishinari ward, there are also many poverty businesses and crimes targeting to homeless people. For example, poor people are forced to rent a room with dishonest dealers, and then they takes the margin of the welfare money. In fact, some dishonest dealers and crime groups have been arrested. To solve these negative chains starting from the Homeless, local governments must first solve the insufficient employment and the gambling dependence, which are the major causes of the Homeless. Unfortunately, we could not confirm the corresponding activities in Osaka city, but such activities have a possibility that reduces welfare recipients and the budgets for them.

### 5.2  Querying the Causality of Littering

Figure 6 shows the part of graph extracted from our LOD using a SPARQL query of searching causality of Littering and their budgets (Fig. 7). We found that insufficient Installation of ashtrays affects the Littering, and the Littering affects the Groundwater pollution. Thus, the direct relation with the upv:affect property from the Littering to the Groundwater is inferred by our predefined rule (Eq. 1) that an influence caused by a thing propagates to a super concept. Furthermore, the causality extraction for the Noise indicated that the Groundwater also affects the Noise. As the result, we found that as a chain reaction the insufficient Installation of ashtrays may affect Littering, the Groundwater pollution, and also the Noise according to the transitive relations of the upv:affect property. In fact, there is a case that cigarettes littered by street smokers flow into the groundwater; thus, the groundwater is polluted, and then the water flow becomes noisy, and finally, the plumbing to improve the water quality causes the further noise.

Regarding the relations between urban problems and the budgets, we found that Suminoe ward in Osaka city has 336,000 JPY as the business for improving water and soil qualities. In contrast, the business for solving the street smoking

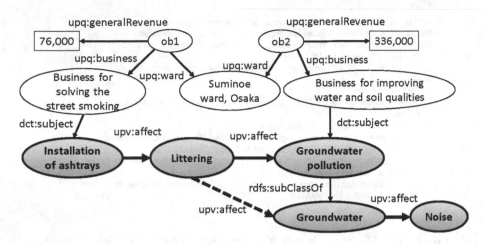

**Fig. 6.** Causality of littering

```
PREFIX upr: <http://www.ohsuga.lab.uec.ac.jp/urbanproblem/resource/>
PREFIX urv: <http://www.ohsuga.lab.uec.ac.jp/urbanproblem/vocaburary#>
PREFIX upq: <http://www.ohsuga.lab.uec.ac.jp/urbanproblem/qb/>
PREFIX dct: <http://purl.org/dc/terms/>
SELECT DISTINCT * FROM <http://www.ohsuga.lab.uec.ac.jp/urbanproblem>
WHERE {
upr:Littering urv:affect+ ?influence ;
     urv:factor+ ?factor .
OPTIONAL {
?b1 dct:subject ?influence .
?ob1 upq:business ?b1 ;
     upq:ward upq:Suminoe-ku_Osaka_City ;
     upq:generalRevenue ?gr1 .
} OPTIONAL {
?b2 dct:subject ?factor .
?ob2 upq:business ?b2 ;
     upq:ward upq:Suminoe-ku_Osaka_City ;
     upq:generalRevenue ?gr2 .
}}
```

**Fig. 7.** SPARQL query for searching Littering causality and their budgets

has just 76,000 JPY. Thus, we can suggest that increasing the budget for solving street smoking may lead to more effective solutions for these serial problems since Littering is located at upstream side of the problems. The reduction of the street smoking may reduce the Littering, the Groundwater pollution, and also the Noise. Consequently, they can reduce the budget for improving water and soil qualities.

## 5.3  Querying the Causality of Suburbanization

Figure 8 shows part of graph extracted from our LOD using a query of searching causality of Suburbanization and their budgets. We found in this graph the vicious Suburbanization $\xrightarrow{\text{upv:affect}}$ Deterioration of security $\xrightarrow{\text{upv:affect}}$ Depopulation circle. Since suburbanization (the hollowing out of city centers) causes a decrease of pedestrian traffic and increase empty houses, and the city security becomes worse. The deterioration of security causes crimes more and makes residents move out of that area. Consequently, the suburbanization is further accelerated. In addition, as this vicious circle is repeated, the other urban problems, such as Traffic accident, Dead shopping street, and Heinous crime are also caused, and thus the risk of such secondary damages may expand. Therefore, the suburbanization should be addressed as soon as possible. On the other hand, we found that Truancy and the Graffiti as factors to cause the vicious circle of suburbanization. The truancy increases bad juveniles and leads to the deterioration of the security. Abandoned graffitis give the public sense that the local government is not functioning, and lead to the deterioration of the security. This phenomenon is well-known as the broken window theory [20].

However, at Abeno ward in Osaka city, the budget for solving the truancy is 15,000 JPY, and the budget for promoting the beautification of the town is only 12,000 JPY. Even in the other wards, the budget for solving the truancy is 15,083 JPY on average, and the budget for promoting the beautification of the town is 14,000 JPY on average. Thus, we found that those businesses are not

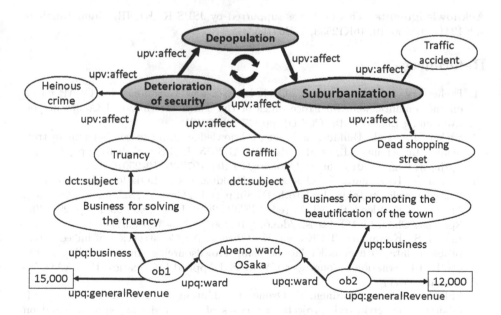

**Fig. 8.** Causality of suburbanization

emphasized in Osaka, however, increasing the budgets of those businesses has a possibility that the risk of entering the vicious circle can be reduced.

# 6 Conclusion and Future Work

In this paper, a schema design for urban problem causality as well as budgets and the building of LOD based on that schema were described. The causality part of the designed schema enabled search for factors and influences of urban problems. The budget part based on QB4OLAP also enabled search for budgets to solve such urban problems. Furthermore, SPARQL query examples that contribute to local government were presented. In future, we will add more rules in order to provide advanced search function.

In this study, the urban problem causality was constructed based on government analysis, research results of the sociology domain, human knowledge, facts, and association of ideas. Therefore, we can explain the causality by confirming the source data. We will increase intermediate nodes in the case that it is difficult to understand the causality. We are now cooperating with the Osaka City Citizen Bureau through Osaka City Citizen Activity Promotion Council to consider several urban problem solutions based on our LOD. In fact, one of the authors of this paper is a member of this council. We will also negotiate with the Osaka City ICT Strategy Office. This Office has provided several governmental data to domestic Open Data challenges; thus, we will convert and add those data to our LOD in future.

**Acknowledgments.** This work was supported by JSPS KAKENHI Grant Numbers 16K12411, 16K00419, 16K12533, 17H04705.

# References

1. Etcheverry, L., Vaisman, A.A.: QB4OLAP: a new vocabulary for OLAP cubes on the semantic web. In: Proceedings of the Third International Conference on Consuming Linked Data (COLD), pp. 27–38 (2012)
2. Szekely, P., et al.: Building and using a knowledge graph to combat human trafficking. In: Arenas, M., et al. (eds.) ISWC 2015. LNCS, vol. 9367, pp. 205–221. Springer, Cham (2015). https://doi.org/10.1007/978-3-319-25010-6_12
3. Egami, S., Kawamura, T., Ohsuga, A.: Building urban LOD for solving illegally parked bicycles in Tokyo. In: Groth, P., Simperl, E., Gray, A., Sabou, M., Krötzsch, M., Lecue, F., Flöck, F., Gil, Y. (eds.) ISWC 2016. LNCS, vol. 9982, pp. 291–307. Springer, Cham (2016). https://doi.org/10.1007/978-3-319-46547-0_28
4. Egami, S., Kawamura, T., Kozaki, K., Ohsuga, A.: Construction of linked urban problem data with causal relations using crowdsourcing. In: Proceeding of the 6th IIAI International Congress on Advanced Applied Informatics (IIAI-AAI), (to appear) (2017)
5. Shiramatsu, S., Tossavainen, T., Ozono, T., Shintani, T.: Towards continuous collaboration on civic tech projects: use cases of a goal sharing system based on linked open data. In: Tambouris, E., Panagiotopoulos, P., Sæbø, Ø., Tarabanis, K., Wimmer, M.A., Milano, M., Pardo, T.A. (eds.) ePart 2015. LNCS, vol. 9249, pp. 81–92. Springer, Cham (2015). https://doi.org/10.1007/978-3-319-22500-5_7

6. Santos, H., Dantas, V., Furtado, V., Pinheiro, P., McGuinness, D.L.: From data to city indicators: a knowledge graph for supporting automatic generation of dashboards. In: Blomqvist, E., Maynard, D., Gangemi, A., Hoekstra, R., Hitzler, P., Hartig, O. (eds.) ESWC 2017. LNCS, vol. 10250, pp. 94–108. Springer, Cham (2017). https://doi.org/10.1007/978-3-319-58451-5_7
7. Pileggi, S.F., Hunter, J.: An ontological approach to dynamic fine-grained Urban Indicators. Procedia Comput. Sci. **108**, 2059–2068 (2017)
8. Höffner, K., Martin, M., Lehmann, J.: LinkedSpending: openspending becomes linked open data. Semant. Web J. **7**(1), 95–104 (2016)
9. Demartini, G., Difallah, D.E., Cudré-Mauroux, P.: Large-scale linked data integration using probabilistic reasoning and crowdsourcing. Int. J. Very Large Data Bases **22**(5), 665–687 (2013)
10. Celino, I., Contessa, S., Corubolo, M., Dell Aglio, D., Valle, E. D., Fumeo, S., Krüger, T.: Linking smart cities datasets with human computation - the case of UrbanMatch. In: Proceedings of the 11th International Semantic Web Conference (ISWC), pp. 34–49 (2011)
11. Ahn, L.V.: Games with a purpose. IEEE Comput. **39**(6), 92–94 (2006)
12. Kudo, T., Matsumoto, Y.: Japanese dependency analyisis using cascaded chunking. In: Proceedings of the 6th Conference on Natural Language Learning, vol. 20, pp. 1–7 (2002)
13. Winkler, W.: The state record linkage and current research problems. Technical report, Statistics of Income Division, Internal Revenue Service Publication (1999)
14. Nguyen, T.M., Kawamura, T., Tahara, Y., Ohsuga, A.: Self-supervised capturing of users' activities from weblogs. Int. J. Intell. Inf. Database Syst. **6**(1), 61–76 (2012)
15. Augenstein, I., Padó, S., Rudolph, S.: LODifier: generating linked data from unstructured text. In: Simperl, E., Cimiano, P., Polleres, A., Corcho, O., Presutti, V. (eds.) ESWC 2012. LNCS, vol. 7295, pp. 210–224. Springer, Heidelberg (2012). https://doi.org/10.1007/978-3-642-30284-8_21
16. Milne, D., Witten, I.H.: Learning to Link with Wikipedia. In: Proceedings of the 17th ACM Conference on Information and Knowledge Management (CIKM), pp. 509–518 (2008)
17. Kontokostas, D., Westphal, P., Auer, S., Hellmann, S., Lehmann, J., Cornelissen, R., Zaveri, A.: Test-driven evaluation of linked data quality. In: Proceedings of the 23rd International Conference on World Wide Web (WWW), pp. 747–758 (2014)
18. Fleiss, J.L., Cohen, J.: The equivalence of weighted kappa and the intraclass correlation coefficient as measures of reliability. Educ. Psychol. Measur. **33**(3), 613–619 (1973)
19. Viera, A.J., Garrett, J.M.: Understanding interobserver agreement: the kappa statistic. Fam. Med. **37**(5), 360–363 (2005)
20. Wilson, J.Q., George, L.K.: Broken windows. Critical issues in policing: contemporary readings, pp. 395–407 (1982)

# Erratum to: Refinement-Based OWL Class Induction with Convex Measures

David Ratcliffe[1,2(✉)] and Kerry Taylor[1]

[1] College of Engineering and Computer Science, Australian National University,
Canberra, ACT 2601, Australia
{david.ratcliffe, kerry.taylor}@anu.edu.au
[2] CSIRO Data61, GPO Box 1700, Canberra, ACT 2601, Australia

**Erratum to:**
**Chapter "Refinement-Based OWL Class Induction with Convex Measures" in: Z. Wang et al. (Eds.): Semantic Technology, LNCS 10675, https://doi.org/10.1007/978-3-319-70682-5_4**

In the original version of the paper the results reported for OWL-Miner and DL-Learner in Section 5 were incorrect, because the authors had made errors in preparing the data for the mutagenesis experiment. In the updated version of the paper, corrections were made (mainly in Section 5) to report results based on the corrected mutagenesis data.

The updated online version of this chapter can be found at
https://doi.org/10.1007/978-3-319-70682-5_4

© Springer International Publishing AG 2017
Z. Wang et al. (Eds.): JIST 2017, LNCS 10675, p. E1, 2017.
https://doi.org/10.1007/978-3-319-70682-5_24

# Author Index

Printed in the United States
By Bookmasters